New Trends in Mathematical Programming

Applied Optimization

Volume 13

The titles published in this series are listed at the end of this volume.

New Trends in Mathematical Programming

Homage to Steven Vajda

Edited by

Franco Giannessi
Department of Mathematics,
University of Pisa

Sándor Komlósi
Faculty of Business and Economics,
Janus Pannonius University

and

Tamás Rapcsák
Computer and Automation Institute,
Hungarian Academy of Sciences

SPRINGER-SCIENCE+BUSINESS MEDIA, B.V.

A C.I.P. Catalogue record for this book is available from the Library of Congress.

ISBN 978-1-4419-4793-2 ISBN 978-1-4757-2878-1 (eBook)
DOI 10.1007/978-1-4757-2878-1

Printed on acid-free paper

Originally published by Kluwer Academic Publishers in 1998

Steven Vajda
1901-1995

CONTENTS

PREFACE

Though the volume covers 22 papers by 36 authors from 12 countries, the history in the background is bound to Hungary where, in 1973 András Prékopa started to lay the foundation of a scientific forum, which can be a regular meeting spot for experts of the world in the field. Since then, there has been a constant interest in that forum. Headed at present by Tamás Rapcsák, the Laboratory of Operations Research and Decisions Systems of the Computer and Automation Institute, Hungarian Academy of Sciences followed the tradition in every respect, namely conferences were organized almost in every second year and in the same stimulating area, in the Mátra mountains. The basic fields were kept, providing opportunities for the leading personalities to give voice to their latest results. The floor has been widened recently for the young generation, ensuring this way both a real location for the past, present and future experts to meet and also the possibility for them to make the multicoloured rainbow of the fields unbroken and continuous.

The volume is devoted to the memory of Steven Vajda, one of the pioneers on mathematical programming, born is Hungary. In 1992 he took part in the XIth International Conference on Mathematical Programming at Mátrafüred where, with his bright personality, he greatly contributed to the good spirituality of the event. We thank Jakob Krarup for his reminiscence on the life and scientific activities of late Steven Vajda.

The Editors hope that this collection will provide the reader with a glimpse of the recent advances of mathematical programming and its features.

We thank John Martindale, Kluwer Academic Publishers, for his encouraging assistance in the production of the volume.

Finally, special thanks are due to Péter Dombi for his excellent work on typesetting and editing the manuscripts.

Steven Vajda, 1901 – 1995

Jakob Krarup
DIKU (Dept. of Computer Science,
University of Copenhagen)
Universitetsparken 1, DK-2100 Copenhagen, Denmark
E-mail: krarup@diku.dk

It is a gift of grace to enjoy a long life without suffering the physical horrors of old age and even more so to preserve both a warm heart and a brilliant mind to the end. Those gifts were granted to Professor Steven Vajda who passed away after a short illness on the 10th December 1995.

His fiftieenth book, *A Mathematical Kaleidoscope*, co-authored with Emeritus Professor Brian Conolly, came out just a few weeks before his death. The following biographical notes are based on the section "About our Authors" found therein, on conversations with Professor Conolly, on an interview in *OR Newsletter* (Bather, 1995), and on the citation prepared by Professor Maurice Shutler for the Award of the Companionship of the Operational Research Society to Steven Vajda (Shutler, 1995).

Steven Vajda was born in Budapest in 1901. He studied mathematics primarily in Vienna with shorter visits paid to Berlin and Göttingen, obtaining degrees in actuarial science and mathematics. After qualifying, he worked as an actuary in Hungaria, Romania and Vienna. In 1939, just before the outbreak of World War II, he moved to England. Like many others arriving from continental Europe at that time, Steven Vajda was interned for six months on the Isle of Man where he taught mathematics and participated in establishing a "do-it-yourself" university. During most of World War II he worked for an insurance company at Epsom but in 1944 was invited to join the British Admiralty as a statistician, soon rising to Assistant Director of Physical Research, and later of Operational Research. In 1952 he was promoted Head of Mathematics Group at the Admiralty Research Laboratory.

Pat Rivett was the first Professor of OR in the UK (Lancaster University, 1963). Steven Vajda became the second one when he joined Birmingham University in 1965, a position he held until his retirement in 1968 when he became a Fellow. In 1967 he was invited by Sussex University to become a Fellow, and in 1973 became Visiting Professor of Mathematics, in which role he has continued actively, teaching and writing research papers, for about 22 years, a record which is unsurpassed in the UK and probably anywhere outside the UK as well.

F. Giannessi et al. (eds.), New Trends in Mathematical Programming, 1–4.

Steven Vajda was awarded an honorary degree (D.Tech. h.c.) by Brunel University. His eminence was also recognized by the Operational Research Society (ORS) in the award of its Silver Medal, followed in 1995 by a Companionship.

The point of embarkation as regards my personal reminiscences must be the year 1954. To those who graduated from a Danish upper secondary school at that time, only three options seemed realistic. The bloodthirsty ones joined the medical school or became dentists, others enrolled at the Technical University, and the remaining few including myself enjoyed two years of compulsory military service before deciding what to do next.

A strong interest in zeros and ones - to the exclusion of almost everything between these two extremes - appeared to be fundamental to my later career. This interest was aroused neither by air defense, nor by subsequent studies of electrical engineering, but dates back to the early years in which I was involved (1958-64) at *Regnecentralen*, The Danish Institute of Computing Machinery, where Danish edp was born. As a graduate student with a limited flair for electrical gadgets and with no professional niche in sight, I found the climate around the first digital computer in Denmark and its users appealing in every conceivable respect. Even an *OR division* materialized. From conceptually simple scenario analyses we moved on via *linear programming* and were soon spellbound by the magic of the computationally "intractable" integers.

The prime influence of Steven Vajda's textbook *Mathematical Programming* (Vajda, 1961) cannot be overrated in this context. It was instrumental when the above-mentioned OR division was created and, *sans comparaison*, our key reference during these first years of development. Not only did it introduce the various families of optimisation problems and the algorithmic tools employed for their solution but also the scope and limitations of mathematical programming as a means for investigating managerial and other decision problems using normative models were accounted for.

These directions were futher pursued in my Ph.D. work at The Technical University. I am grateful to my thesis supervisor at that time, not primarily for his supervision as such since our professional interests hardly intersected, but for his emphasis on the significance of building a personal network and his creativity in finding support for conference participation and the like. The most important event of that period was a two-week *NATO Summer School on Contemporary Methods of Discrete Mathematics* (Varenna, Italy, 1966) directed by two leading scholars in the field, Professors F. Harary and B. Roy. An invited speaker was awaited with particular anticipation, a certain Prof. S. Vajda whose visions and thought-provoking viewpoints left a strong impression on all those present.

We met again many years later at the *11th International Conference on Mathematical Programming* (Mátrafüred, Hungary, 1992) where Steven delivered the opening lecture. Under the title *"Farkas Alternatives and Applications"*, the significance of *duality* as a central concept within mathematical programming was highlighted in a very suggestive way. Afterwards, we discussed the first occurrence of duality in general nonlinear programming. I was able to recollect most of Harold Kuhn's findings as presented at the Mathematical Programming Symposium (Budapest, 1976) and, to my great surprise, thus to tell Steven just a little bit about duality which was new to him. This fortunate coincidence gave rise to other meetings and to the exchange of letters resulting in two forthcoming papers (Krarup and Vajda 1996a, 1996b). Even more valuable, however, was the friendship that evolved between us through these endeavours.

Similar debts to Steven are in one way or another owed by many others. Since Steven joined the Admiralty, he has spent about 50 years consciously or unconsciously motivating the careers of numerous OR workers. He has exerted this influence directly by teaching and conference presentations, and indirectly by his writings and by the example of his life.

A group of friends and colleagues joined forces and suggested to the Mathematical Programming Study Group, ORS, that a special meeting should be organised to celebrate the work of Steven Vajda, the true founding father of mathematical programming in the UK. Focussing on duality, the meeting was eventually held in London on 10 February 1995. Among the highlights were the award of the Companionship of ORS and the warm speeches delivered afterwards at the dinner. Once again, the influence of *Mathematical Programming* was acknowledged and an even earlier volume (Vajda, 1956) was evoked, the very first book in Europe on linear programming, also translated into French, German, Japanese, and Russian. It is indeed Steven Vajda who can rightly claim to have introduced the subject to both Europe and Asia.

A report on the festive 10 February appeared shortly after (Simons, 1995). Another visible outcome is the forthcoming *Special Edition of Journal of Mathematics in Business and Industry* edited by S. Powell and H.P. Williams (Powell and Williams, 1996).

With a warm heart, a brilliant mind, and an amazing mobility, Steven remained a frequent traveller to conferences and symposia. Insight and an implacable curiosity marked his active contributions as a lecturer or discussant. Likewise, no effort was spared outside the conference halls. Scenic places, historical landmarks: no steps were too high to be climbed. Not even the extremely hot weather in Israel in July 1995 discouraged Steven from taking part in the physically quite demanding 2-days' excursion organised in conjunction with EURO XIV. Not a single detail was missed, let alone the celebrations of EURO's 20th Anniversary. IFORS 96 in Vancouver was supposed to be "next time".

A gift of grace to "shuffle off this mortal coil" in the way it happened. May the same be granted to all of us — when the time comes.

P.S. Being curious about words, I asked Professor L. B. Kovács, my nativeborn Hungarian colleague at DIKU, whether the name *Vajda* actually means something in Hungarian. It appears that the word can be found in Hungarian texts dating back to about 950 A.D. A *Vajda* is sort of a *regional leader* or a *reigning prince* and close variants of the word are known in ancient Greek, Polish, ancient Russian, Serbo-Croatian, and Slovenian. *Vajdaság* or *the region led by Vajda* is the Hungarian designation of a region in the northern part of Yugoslavia.

Acknowledgements

This obituary has been written at the request of Karen Aardal, Features Editor of *OPTIMA*, The Newsletter of *The Mathematical Programming Society.* I thank Karen Aardal for this opportunity to commemorate a distinguished colleague and friend and for her permission also to publish the obituary elsewhere. Likewise, the kind assistance provided by Professors B. Conolly and L.B. Kovács is gratefully acknowledged.

References

1. Bather, J., "An interview with Steven Vajda", *OR Newsletter*, January 1995.
2. Conolly, B., and S. Vajda, *A Mathematical Kaleidoscope*, Albion Publishing, Chichester, 1995.
3. Krarup, J., and S. Vajda (1996a), *"On Torricelli's geometrical solution to a problem of Fermat"* to appear in (Powell and Williams, 1996).
4. Krarup, J., and S. Vajda (1996b), *"Visualizing duality"*, submitted for publication.
5. Powell, S., and H. P. Williams, Eds. (1996), *Special Edition of Journal of Mathematics in Business and Industry: Duality: In celebration of work of S. Vajda.*
6. Shutler, M. F. (1995), "Steven Vajda", citation for the Companionship of ORS, to appear in (Powell and Williams, 1996).
7. Simon, R., "Founding father of British LP is honoured. Celebration of Steven Vajda's work", *OR Newsletter*, April 1995.
8. Vajda, S., *Theory of Games and Linear Programming*, Methuen, 1956.
9. Vajda, S., *Mathematical Programming*, Addison-Wesley, 1961.

The Fermat–Weber problem and the Lagrangian duality theory

Carla Antoni
Lecturer, Naval Accademy, Mathematics Group, Viale Italia, Livorno and Department of Mathematics, Università di Pisa
Via Buonarroti 2, 56100 Pisa

Angela Dalena
Associate Professor, Naval Accademy, Mathematics Group, Viale Italia, Livorno, Italy

Abstract. Witgzgall and Rockafellar have discovered a dual of the Fermat–Weber problem. Our question is:

"What is the origin of this formulation?"

The aim of this paper is to consider this formulation from the viewpoint of Lagrangian duality. More precisely, through the use of a relaxation technique proper to Lagrangian theory, it demonstrates how the above-mentioned dual formulation is reached. We desire to clarify that the purpose is not to establish connection with a theory of duality, bat rather to propose an "origin". The section 4 gives same remarks on Fasbender duality.

1. Introduction

The Fermat–Weber Problem [2], [8], also known as The General Fermat Problem [4], [5], is the following: to locate a single facility in the plane such that the sum of the weighted Euclidean distances to n given points is minimized.

In order to consider in the context of Lagrangian duality the problem found by Witzgall and Rockafellar, and considered by Kuhn [4], we will make a few initial observations.

Consider the mathematical programming

$$\inf_{x\in R} f(x)\,, \tag{$\hat{P}$}$$

where $f : A \longrightarrow \mathbb{R}$ and R is a subset of A. If the constraint R is "relaxed", that is the set R is replaced by any superset $\widetilde{R} \supseteq R$, the problem

$$\inf_{x\in \widetilde{R}} f(x)\,, \tag{1.1}$$

gives a lower bound for $(\hat{P})$. Then, if a class $\mathcal{R}$ of supersets of R is fixed, we look for the "best relaxation" in $\mathcal{R}$; namely we look for the

$$\sup_{\widetilde{R}\in\mathcal{R}} \left(\inf_{x\in\widetilde{R}} f(x)\right). \tag{1.2}$$

F. Giannessi et al. (eds.), New Trends in Mathematical Programming, 5–12.

We call any problem such as in (1.1) a *Relaxation* of $(\hat{P})$ and such as in (1.2) a *Relaxed Dual* of $(\hat{P})$.

Technical comment

The Lagrangian Duality can be seen in terms of the "best relaxation" (for the effective utility in the applications, see for instance [7]).
In fact, consider a mathematical programming $(\hat{P})$ where R denotes the set $\{x \in X : g(x) = 0\}$, with $g : X \to: \mathbb{R}^m$. Let us observe the following facts:
i) for any vector $\lambda \in \mathbb{R}^m$, the set $\widetilde{R}(\lambda) = \{x \in X : \langle \lambda, g(x) \rangle = 0\}$ (for every $v = (v_1, \ldots, v_m)$, $w = (w_1, \ldots, w_m)$ in $\mathbb{R}^m$, $\langle v, w \rangle$ denotes the quantity $\sum_{i=1}^m v_i w_i$) contains R and hence gives a Relaxation of $(\hat{P})$;
ii) the problem

$$\sup_{\lambda \in \mathbb{R}^m} \Big(\inf_{x \in \widetilde{R}(\lambda)} f(x) \Big) \tag{1.3}$$

is a Relaxed Dual of $(\hat{P})$;
iii) the definition of $\widetilde{R}(\lambda)$ enables to replace $f(x)$ with $f(x) - \langle \lambda, g(x) \rangle$ in (1.3) and to produce the problem known as Lagrangian Relaxation of $(\hat{P})$:

$$\sup_{\lambda \in \mathbb{R}^m} \inf_{x \in \widetilde{R}} (f(x) - \langle \lambda, g(x) \rangle) . \tag{1.4}$$

iv) finally, the set X instead of $\widetilde{R}(\lambda)$ in (1.4) yields the problem known as the *Lagrangian Dual* of $(\hat{P})$, that is

$$\sup_{\lambda \in \mathbb{R}^m} \inf_{x \in X} (f(x) - \langle \lambda, g(x) \rangle) .$$

2. Relaxed Dual of the Fermat–Weber problem

To obtain a Lagrangian Dual Problem of the Fermat–Weber Problem, we consider it as a particular case of the following one.

Let S_0, $S_1, \ldots S_n$ be $n+1$ subset of $\mathbb{R}^m$. We look for a point of S_0 which minimizes the sum of the weighted distances to the sets S_i; that is, we consider the problem

$$\min_{X \in S_0} \sum_{i \in I} w_i \, d(X, S_i) , \tag{P}$$

where I denotes the set $\{1, \ldots, n\}$ and where, for a set A, $d(\,\cdot\,, A)$ is the Euclidean distance to A. This problem is a particular case of that considered in [3], where a dual problem is developped.

When $S_0 = \mathbb{R}^2$ and $S_i = \{X_i\}$, $i \in I$, the problem (P) coincides with the Fermat-Weber Problem. We will denote this particular case by (P_W).
In this way, the given points X_i are the constraints of (P_W); hence, a Relaxation can be obtained replacing each set $S_i = \{X_i\}$ with a straight line R_i through X_i. Therefore, $R_1, \ldots, R_n$ fixed, the Relaxation is the following:

$$\min_{X \in \mathbb{R}^2} \sum_{i \in I} w_i \, d(X, R_i) \, .$$

Finally, we look for the "best" n-uple $R_1, \ldots, R_n$, where every R_i belongs to the sheaf $\mathcal{F}_i$ of straight lines through X_i; namely, we consider the problem

$$\max_{(R_1, \ldots, R_n) \in \mathcal{F}} \left(\min_{X \in \mathbb{R}^2} \sum_{i \in I} w_i \, d(X, R_i) \right) ,$$

where $\mathcal{F}$ denotes the product $\mathcal{F}_1 \times \cdots \times \mathcal{F}_n$. If, for each $i \in I$, U_i denotes an orthogonal versor to R_i, the last problem is

$$\max_{(U_1, \ldots, U_n) \in \mathcal{B}} \left(\min_{X \in \mathbb{R}^2} \sum_{i \in I} w_i |\langle U_i, X - X_i \rangle| \right) , \qquad (P_W^*)$$

where $\mathcal{B}$ denotes the set $\{(U_1, \ldots, U_n) : U_i \in \mathbb{R}^2, \ \|U_i\| = 1, \ i \in I\}$ (the function $\|\cdot\|$ is the euclidean norm and $|\cdot|$ is the absolute value). The problem (P_W^*) is a Relaxed Dual of (P_W).
Now, for each X in $\mathbb{R}^2$, and for each n-uple $(U_1, \ldots, U_n)$ in $\mathcal{B}$, put

$$F(X) := \sum_{i \in I} w_i \, d(X, X_i) = \sum_{i \in I} w_i \parallel X - X_i \parallel ,$$

$$m(U_1, \ldots, U_n) := \min_{X \in \mathbb{R}^2} \sum_{i \in I} w_i \, |\langle U_i, X - X_i \rangle| \, . \qquad (2.1)$$

The following result holds.

THEOREM 1. *For each X in $\mathbb{R}^2$ and for each n-uple $(U_1, \ldots, U_n)$ in $\mathcal{B}$, it holds*

$$F(X) \geq m(U_1, \ldots U_n) \, . \qquad (2.2)$$

Moreover, (P_W^*) has maximum and

$$\min_{X \in \mathbb{R}^2} F(X) = \max_{(U_1, \ldots, U_n) \in \mathcal{B}} m(U_1, \ldots, U_n) \, . \qquad (2.3)$$

Proof. The relation (2.2) follows from

$$|\langle U_i, X - X_i \rangle| \leq \|X_i - X\| , \qquad \forall i \in I.$$

To prove the equality (2.3), first suppose that the solution $\bar{X}$ of (P_W) does not coincide with any of the given points X_i. In this case, set

$$\bar{U}_i = \frac{X_i - \bar{X}}{\| X_i - \bar{X} \|} \quad \forall i \in I,$$

and check that $m(\bar{U}_1, \ldots, \bar{U}_n) = F(\bar{X})$ because $\sum_{i \in I} w_i \bar{U}_i = \nabla F(\bar{X}) = 0$. In fact, it holds

$$m(\bar{U}_1, \ldots \bar{U}_n) \geq \min_{X \in \mathbf{R}^2} \sum_{i \in I} w_i \langle \bar{U}_i, X_i - X \rangle = \sum_{i \in I} w_i \langle \bar{U}_i, X_i - \bar{X} \rangle = F(\bar{X}).$$

Thanks to (2.2), (2.3) holds. When the soluton $\bar{X}$ coincides with one of the given points, for example if $\bar{X} = X_j$, it is enough to modify the j–th versor $\bar{U}_j$. To this aim, put

$$\widetilde{W} = \frac{-\sum_{i \neq j} w_i \bar{U}_i}{w_j}, \quad \bar{U}_j = \frac{\widetilde{W}}{\| \widetilde{W} \|}.$$

Because $\| \widetilde{W} \| \leq 1$ (see [4]), it holds that

$$\begin{aligned} m(\bar{U}_1, \ldots, \bar{U}_n) \geq\ & \min_{X \in \mathbf{R}^2} \left(\sum_{i \neq j} w_i |\langle \bar{U}_i, X_i - X \rangle| + w_j |\langle \widetilde{W}, X_j - X \rangle| \right) \geq \\ & \min_{X \in \mathbf{R}^2} \left(\sum_{i \neq j} w_i \langle \bar{U}_i, X_i - X \rangle + w_j \langle \widetilde{W}, X_j - X \rangle \right) = \\ & \sum_{i \neq j} w_i \langle \bar{U}_i, X_i - \bar{X} \rangle + w_j \langle \widetilde{W}, X_j - \bar{X} \rangle = F(X_j) , \end{aligned}$$

The thesis follows from (2.2).

3. From the Relaxed Dual to the formulation proposed by Witzgall and Rockafellar

The dual problem discovered by Witzgall and Rockafellar is the following

$$\max_{(V_1, \ldots, V_n) \in \mathcal{V}} \sum_{i \in I} \langle V_i, X_i \rangle , \tag{D}$$

where $\mathcal{V}$ denotes the set

$$\{(V_1, \ldots, V_n) : V_i \in \mathbb{R}^2, \ \|V_i\| \leq w_i, i \in I, \ \sum_{i \in I} V_i = 0\}.$$

Starting from the Relaxed Dual (P_W^*), the problem (D) is found. First of all, the elimination of the absolute value in (P_W^*) gives the following inequality

$$\max_{(U_1, \ldots, U_n) \in \mathcal{B}} m(U_1, \ldots, U_n) \geq \max_{(U_1, \ldots, U_n) \in \mathcal{B}} \min_{X \in \mathbf{R}^2} \sum_{i \in I} w_i \langle U_i, X_i - X \rangle . \tag{3.1}$$

Moreover, since $\inf_X \sum_{i\in I} w_i\langle U_i, X_i - X\rangle = -\infty$, except if $\sum_{i\in I} w_iU_i = 0$, the second member of (3.1) gives the problem

$$\max_{(U_1,\dots,U_n)\in\mathcal{U}} \sum_{i\in I} w_i \langle U_i, X_i\rangle \,, \tag{3.2}$$

where $\mathcal{U} = \{(U_1,\dots,U_n) : U_i \in \mathbb{R}^2,\ \|U_i\| = 1,\ i \in I,\ \sum_{i\in I} w_iU_i = 0\}$. We will see that the extremum value of this problem does not coincide with the extremum of (P_W). To achieve this fact it is sufficient to extend the set $\mathcal{U}$ to the following one

$$\{(U_1,\dots,U_n) : U_i \in \mathbb{R}^2,\ \| U_i \|\leq 1\,,\ i\in I,\ \sum_{i\in I} w_iU_i = 0\}\,.$$

In this way, we found the dual (D).

The following result holds.

PROPOSITION 2. *The problem* (D) *has an unique maximum point.*

Proof. Let $l : (\mathbb{R}^2)^n \to \mathbb{R}$ be the linear function defined by

$$l(V_1,\dots,V_n) = \sum_{i\in I} \langle V_i, X_i\rangle.$$

The continuity of l and the compactness of $\mathcal{V}$ assure that the solution exists. We prove that the solution is unique. Ab absurdo, let

$$(\widetilde{V}_1,\dots,\widetilde{V}_n),\quad (\widehat{V}_1,\dots,\widehat{V}_n)$$

be two distinct solutions. Then, there are at least two indexes -the first two, without any loss of generality- such that $\widetilde{V}_1 \neq \widehat{V}_1$ and $\widetilde{V}_2 \neq \widehat{V}_2$. Set

$$W_i = \lambda\widetilde{V}_i + (1-\lambda)\widehat{V}_i\,,\ i = 1,\dots,n\,,$$

where λ is an element of $]0,1[$, and observe that $(W_1,\dots,W_n)$ is a maximum point for (D) with $\| W_1 \|< w_1$, and $\| W_2 \|< w_2$.
Let η be the normalized vector of $X_1 - X_2$ and let t be a real number in $(0, \min\{w_1-\| W_1 \|, w_2-\| W_2 \|\})$. Then, the point $(W_1 + t\eta, W_2 - t\eta, W_3,\dots,W_n)$ belongs to $\mathcal{V}$ and

$$l(W_1 + t\eta, W_2 - t\eta, W_3,\dots,W_n) = \mathcal{M} + t \| X_1 - X_2 \|> \mathcal{M}\,,$$

where $\mathcal{M} = \max_{(V_1,\dots,V_n)\in\mathcal{V}} l(V_1,\dots,V_n)$. This contradicts the optimality of $(W_1,\dots,W_n)$. The following example shows that the problem in (3.2)

does not have the same extremum of (D).

EXAMPLE 3. Consider the problem (P_W) with $w_1 = w_2 = w_3 = 1$, and the angle $X_1\widehat{X_2}X_3$ exceeding 120°. In this case, the vectors

$$U_i = \frac{X_i - X_j}{\| X_i - X_j \|}, \quad i = 1,2 \qquad U_3 = -\sum_{i=1,2} U_i ,$$

give the maximum in (D). Because $\| U_3 \| < 1$, the extremum of the problem in (3.2) is less than that of (P_W).

4. Remarks on Fasbender duality

A dual formulation of the Fermat Problem

$$\min_{X \in \mathbf{R}^2} \sum_{i=1}^{3} \| X_i - X \| ,$$

was proposed by Fasbender [1]: he asks an equilateral triangle circumscribing the triangle with vertices X_1, X_2, X_3 which has the maximum altitude.

Kuhn [4] maintains the dual (D) "generalizes in a precise fashion the geometric duality discovered by Fasbender". In fact, Kuhn claims a one-to-one correspondence between feasible 3-uple $(V_i)_i$ for the dual problem (D) and equilateral triangles circumscribing the given one. It seems that the triangle associated with the 3-uple $(V_i)_i$ is that with i–th side containing X_i and orthogonal to V_i.

We remark the triangle does not circumscribe the given one necessarily. In fact, consider, for example, the three vertices in the plain: $X_1 = (1,0)$, $X_2 = (1/4, \sqrt{3}/4)$, $X_3 = (-1, -\sqrt{3}/2)$. The tern $V_1 = (0,-1)$, $V_2 = (\sqrt{3}/2, 1/2)$, $V_3 = (-\sqrt{3}/2, 1/2)$ determines a non–circumscribing equilateral triangle.

Nevertheless, when $w_i = 1$, $i = 1,2,3$, we observe the optimal 3-uple of (D) determines a circumscribing triangle. On the contrary, for any weights, the optimal 3-uple in (D) does not even determine a circumscribing triangle. This claim follows from the following

PROPOSITION 4. *Let X_1, X_2, X_3 be three points in the plain such that the angle $X_1\widehat{X_2}X_3$ is greater than 90°. Let $\bar{X}$ be a point in the interior of the triangle $X_1X_2X_3$ such that the angle $\bar{X}\widehat{X_2}X_1$ is greater than 90°.*

Then, positive weights $\bar{w}_1, \bar{w}_2, \bar{w}_3$ exist such that $\bar{X}$ is the solution of the Weber Problem (P_W).

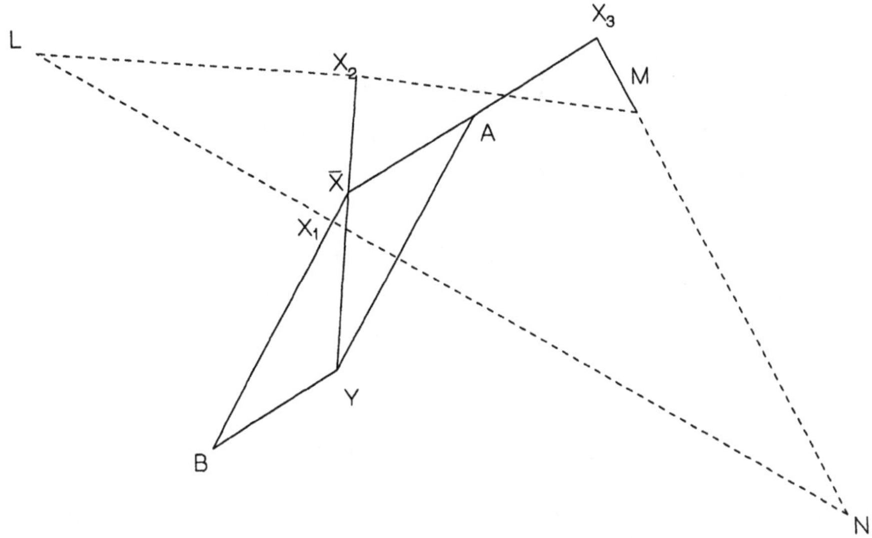

Figure 1.

Proof. Consider the parallelogram $\bar{X}AYB$ (see fig. 1), where the point Y belongs to the straight line through $\bar{X}$ and X_2. Then, since

$$Y-\bar{X}=(A-\bar{X})+(B-\bar{X}), \tag{4.1}$$

it is enough to set

$$\bar{w}_1=\| B-\bar{X} \|, \quad \bar{w}_2=\| Y-\bar{X} \|, \quad \bar{w}_3=\| A-\bar{X} \| . \tag{4.2}$$

We will prove that $\bar{X}$ is the solution of

$$\min_{X\in\mathbf{R}^2} \sum_{i\in I} \bar{w}_i d(X,X_i).$$

To show this fact, put

$$\bar{V}_i=\bar{w}_i\frac{X_i-\bar{X}}{\| X_i-\bar{X} \|}, \quad i=1,2,3,$$

and observe that $\sum_{i\in I}\bar{V}_i=0$. Then

$$\begin{aligned} F(\bar{X}) &= \textstyle\sum_{i=1}^3 \bar{w}_i \| X_i-\bar{X} \| = \sum_{i=1}^3 \bar{w}_i\langle X_i-\bar{X}, \frac{X_i-\bar{X}}{\|X_i-\bar{X}\|}\rangle \\ &= \textstyle\sum_{i=1}^3 \langle X_i-\bar{X},\bar{V}_i\rangle = \sum_{i=1}^3 \langle X_i,\bar{V}_i\rangle = l(\bar{V}_1,\bar{V}_2,\bar{V}_3). \end{aligned}$$

Since for each X in $\mathbf{R}^2$ and for each (V_1,V_2,V_3) in $\mathcal{V}$,

$$F(X)\geq l(V_1,V_2,V_3),$$

$\bar{X}$ is the solution of (P_W) (and $(\bar{V}_1,\bar{V}_2,\bar{V}_3)$ is the solution of (D)).

Remark 5. The weights in (4.2) satisfy the triangular inequalities [6].

Remark 6. The 3-uple $(\bar{V}_1, \bar{V}_2, \bar{V}_3)$ determines a non–circumscribing triangle T. In fact, if T is the triangle with i–th side containing X_i and orthogonal to $\bar{V}_i$, we observe that the angle $\widehat{\bar{X}X_2M}$ is equal to 90° and $\widehat{\bar{X}X_2X_3}$ exceeds 90°. Hence (see fig 1) X_3 is exterior to the triangle T.

Remark 7. The above mentioned triangle T is similar to the triangle with sides w_1, w_2, w_3. This happens when the solution $\bar{X}$ of (P_W) does not coicide with any X_i, $i = 1, 2, 3$, and hence the solution of the dual (P_W^*) is given by $V_i = w_i \frac{X_i - \bar{X}_i}{\|X_i - \bar{X}_i\|}$, $i = 1, 2, 3$.

5. Concluding remarks

5.1. The preceding remarks make the geometric interpretation of the dual problem (P_W^*) an open question. More precisely, given an n-uple $R_1, \ldots, R_n$ of straight lines trough $X_1, \ldots, X_n$ respectively, that is an n-uple $(U_1, \ldots, U_n)$ in $\mathcal{B}$, we look for the meaning of the quantity $m(U_1, \ldots, U_n)$ (defined in section 2) for any weights w_i.

5.2. The same happens for the Relaxed Dual (P_W^*).

5.3. Given a problem (P), a duality theory τ define a dual problem (P_τ^*); if it possible to formulate the problem $(P_\tau^*)_\tau^*$ we detect the relations between (P) and $(P_\tau^*)_\tau^*$. The same analisys can be made for the Fermat–Weber Problem.

References

1. E. Fasbender, "Über die gleichseitigen Dreiecke, welche um ein gegebenes Dreieck gelegt werden können", J. f. Math., 30 (1846), pp. 230–231.
2. O. L. Franksen and I. Grottan–Guinness, "The earliest contribution to location theory? Spatio–economic equilibrium with Lamé and Clapeyron", Mathematics and computers in simulation, 31 (1989), pp. 195–220.
3. H. Idrissi, O. Lefebre and C. Michelot, "Duality for constrained multifacility location problems with mixed norm and applications", Annals of Operations Research, 18 (1989), pp. 71–92.
4. H. W. Kuhn, "Nonlinear Programming", J. Abadie, North–Holland, 1967.
5. H. W. Kuhn, "A note on Fermat's problem", Mathematical Programming, 4 (1973), pp. 98–107.
6. M. Martelli, "Geometrical solution of weighted Fermat problem about triangles", this volume.
7. J. F. Shapiro, "A survey of Lagrangean technique for discrete optimization", Annals of Discrete Mathematics, North–Holland , 5 (1979), pp. 113–138.
8. G. O. Wesolowsky, "The Weber problem: history and perspective", Location Science, 1 (1993), pp. 5–23.

Longest Fragment First Algorithms for Data Compression*

József Békési and Gábor Galambos
Juhász Gyula Teacher Training College,
Department of Computer Science, P.O. Box 396,
H-6701 Szeged, Hungary

Timo Raita
University of Turku,
Department of Computer Science,
Lemminkaisenkatu 14A, SF-20520, Turku, Finland

Abstract. On–line text–compression algorithms are considered, where compression is done by substituting substrings of the text according to some fixed dictionary (code book). Due to the long running time of optimal compression algorithms, several on–line heuristics have been introduced in the literature. In this paper we analyse two modified version of an old algorithm introduced by Shuegraf and Heaps [9]. We will investigate the worst case behaviour of this Longest Fragment First (LFF) algorithm for several types of special dictionaries.

1. Introduction

Networking, file systems and communication are among the most popular words in the computer science. Each of these subjects has a very important feature: Indirectly or directly they are in connection with *data transfers* among or within computers. Obviously, the longer the length of the transmitted bit–string, the longer the transmission time and the larger the probability of making a mistake during the transfer. Since the data transfer is being influenced by the software and the hardware in the same way, essentially, there are two possibilities to increase its performance: Either to use better (and more expensive) hardware, or to *compress* the data before doing a transfer. We will investigate the latter case. There are different methods to execute a compression: using some classical results from the information and probability theory already twenty years ago a great progress on the field of character coding has been made. Such well known results were born like the Shannon-Fano code [3], the Huffmann-code [5] and the Arithmetic coding [8]. These methods are based on the relative frequency of the characters in a given string and so they are usefull if one can analyze the whole string before encoding it, or we know the stochastical behav-

* This research was supported by a grant from the National Scientific Research Foundation (OTKA Nr. 16349).

F. Giannessi et al. (eds.), New Trends in Mathematical Programming, 13–28.

ior of the source. The other possibility to perform a compression on a given source–string is to substitute pieces of the string with the help of a *dictionary*. A dictionary consists of pairs of strings over two finite alphabets (*source–words, code–words*), which are used to replace substrings in the source–string. The name of dictionary encoding reflects its origin. The elements of two purposefully constructed word-sets are put into correspondance, not necessarily ont-to-one way. The second big class of applications, namely in cryptography, dates back to the XV-XVI century. Applications to text compression are also old. Short after the introduction of the commercial Morse-telegraphy, the lawer Smith constructed a book purposefully for enshortening commercial texts.

One class of the dictionary encoding methods uses a *static dictionary*, i.e. a fixed dictionary that cannot be changed or extended during the encoding–decoding procedure. In this case our aim is to translate (encode) the source–string with the help of dictionary strings into a code–text with minimal length; in other words, we want to find a space–optimal encoding procedure.

Schuegraf and Heaps [9] proved that the above setup is equivalent to the problem of finding a *shortest path* in a related directed edge–weighted graph: For a source–string $S = s_1 s_2 \ldots s_n$, we define a graph $N = (V, A)$ on the vertex-set $V = \{v_0, v_1, \ldots, v_n\}$. There is an edge $(v_i, v_{i+d}) \in A$ iff in the dictionary there exists a pair (source–word, code–word) such that the source-word consists of d characters that exactly match the original source–string in the positions $i+1, \ldots, i+d$. The weight of this edge is the length of the corresponding code–word. (Usually in practice the code alphabet is binary. In this case the weight is the number of bits of the corresponding code-word.) It is easy to see that a shortest path from v_0 to v_n in the graph N corresponds to an optimal compression of the source–string S.

EXAMPLE 1.1 Consider the string THIS_IS_IT and the dictionary [7]

source–word	T	H	I	S	_	TH	HIS	IS	IS_	IT
code-word	t	h	i	s	a	b	c	d	e	f
weight	x	x	x	x	x	x	x	x	x	x

The corresponding graph is shown by Figure 1.

Of course, an on–line heuristic will generate only a suboptimal compression. In order to know how much worse its performance can possibly

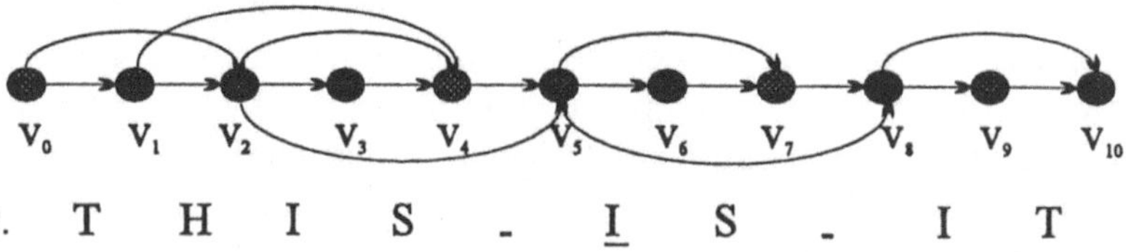

Figure 1.

be compared to the optimal compression a *worst–case analysis* has to be performed.

The *worst–case behaviour* of a heuristic is generally measured by its *asymptotic worst–case ratio* which is defined as follows: Let $D = \{(w_i, c_i) : i = 1, \ldots, k\}$ be a static dictionary and consider an arbitrary data compression algorithm A. Let $A(D, S)$ resp. $OPT(D, S)$ denote the compressed string produced by algorithm A resp. the optimal encoding for a given source–string S. The length of these codings will be denoted by $\|A(D, S)\|$ resp. $\|OPT(D, S)\|$. Then the *asymptotic worst–case ratio* of algorithm A is defined as

$$R_A(D) = \lim_{n\to\infty} \sup \left\{ \frac{\|A(D,S)\|}{\|OPT(D,S)\|} : S \in S(n) \right\},$$

where $S(n)$ is the set of all text–strings with exactly n characters.

The first worst–case analysis for an on–line data compression method was performed by Katajainen and Raita [7]. They analyzed two simple on–line heuristics, the *longest matching* and the *differential greedy* algorithm which will be defined exactly later.

In this paper, we will continue their investigations using new algorithms. We answer several problems and provide matching upper and lower bounds for special dictionary types.

The paper is organized as follows: Section 2 gives definitions of the heuristic algorithms. In Section 3 we analyze the Longest Fragment First algorithm. Some concluding remarks are presented in Section 4.

2. Preliminaries

Four parameters are used in the literature to investigate and state asymptotic worst–case ratios:

$$\begin{aligned} Bt(S) &= \text{length of each symbol of the source–string } S \text{ in bits}, \\ lmax(D) &= \max\{|w_i| \; i = 1, \ldots, k\}, \\ cmin(D) &= \min\{\|c_i\| \; i = 1, \ldots, k\}, \\ cmax(D) &= \max\{\|c_i\| \; i = 1, \ldots, k\}, \end{aligned}$$

where $|w_i|$ denotes the length of a string w_i in characters and $||c_i||$ the length of a code word c_i in bits. If the meaning is clear from the context, we will simply denote the bit length of each input character by Bt and omit the reference to the dictionary by using $lmax$ instead of $lmax(D)$. At the above definitions we use some facts, which comes from practice. For example that the code alphabet is binary, and that the characters are represented by binary coding. We will not use these assumptions during the proof of the theorems, so they are true in case of more general definitions as well. (For example if $||c_i||$ is not measured by bits.) We just want to represent with this, that the problems come from practice.

Not surprisingly, the worst–case behaviour of a heuristic strongly depends on the features of the available dictionary. The following types of dictionaries will be examined in this paper:

A dictionary is called *general* if it contains all symbols of the input alphabet as source–words (this ensures that every heuristic will in any case reach the sink of the underlying graph and thus will terminate the encoding with a feasible solution). In this paper we will only deal with general dictionaries.

A word-set is callled block set if all words are of the same length, and variable length set if they are not of the same length. Accordingly, dictionaries can be classified as block-to-block, block-to-variable, variable-to-block and variable-to-variable depending on the source words and the code words. For the variable-to-block case the best result is due to Jelinek and Schneider [6].

A general dictionary is called

1. *code uniform* dictionary, if all code–words are of equal length (i.e. $||c_i|| = ||c_j||$, $1 \le i, j \le k$),
2. *nonlengthening* dictionary, if the length of any code–word never exceeds the length of the corresponding source–word (i.e. $||c_i|| \le |w_i|Bt$, $1 \le i \le k$),
3. *suffix* dictionary, if with every source–word w also all of its proper suffices are source–words (i.e. if $w = \omega_1\omega_2\cdots\omega_q$ is a source–word $\Rightarrow \omega_h\omega_{h+1}\cdots\omega_q$ is a source–word for all $2 \le h \le q$),
4. *prefix* dictionary, if with every source–word w also all of its proper prefices are source–words (i.e. if $w = \omega_1\omega_2\cdots\omega_q$ is a source–word $\Rightarrow \omega_1\omega_2\cdots\omega_h$ is a source–word for all $1 \le h \le q-1$).

Two on–line heuristics were analyzed in [7] with respect to their worst–case behaviour:

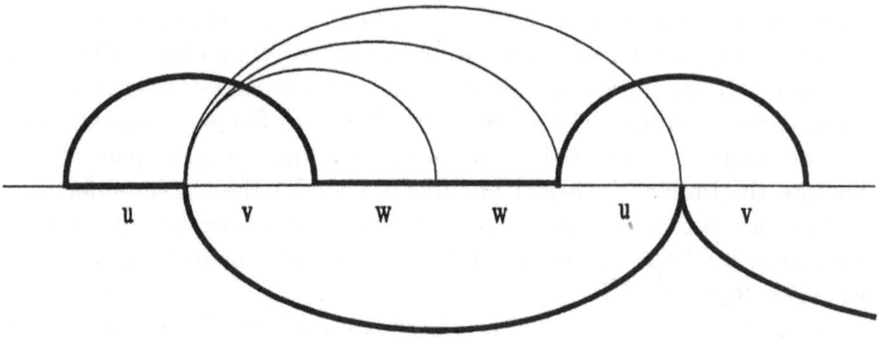

Figure 2.

- The *longest matching* heuristic *LM* chooses at each vertex of the underlying graph the longest outgoing arc, i.e. the arc corresponding to the encoding of the longest substring starting at the current position.
- The *greedy* heuristic, which we will call *differential greedy DG* (introduced by Gonzalez–Smith and Storer [4]) chooses at each position the arc implied by the dictionary entry (w_i, c_i) yielding the maximal "local compression", i.e. the arc maximizing $|w_i|Bt - \|c_i\|$. Ties are broken arbitrarily.

EXAMPLE 2.1 Consider the following dictionary, and the source string $S = u\,(vwww)^i$.

source–word	u	v	w	uv	vw	vww	$vwwu$
code–word	a	b	c	d	e	f	g
weight	$cmax$	$cmax$	$cmax$	$cmax$	$cmax$	$cmax$	$cmin$

Notice that $lmax = 4$ in this case. Figure 2 illustrates how the *LM* algorithm works. The optimal path goes under the horizontal line, and the *LM*-path goes above it.

The time complexity of this algorithm is $O(|S|)$.

More than twenty years ago Shuegraf and Heaps [9] introduced the *longest fragment first (LFF)* heuristic. They supposed that the file we want to compress is divided into records of the same lengths. The idea is the following: the longest word fragment — within the actual record — which matches a word from the dictionary is chosen and encoded. Ties are broken arbitrarily. Then the remaining part of the record is compressed with an on-line algorihm (they choosed the LM heuristic). If the file has no record structure, we can consider a buffer, which always contains the actual part of the text (so we will talk about buffers instead of records). Now, before reading the next part of the file into the buffer, the algorithm always encodes the whole buffer. We will call a whole buffer-coding procedure as a *step*. We will denote this algorithm by LFF_{LM}. To avoid double indexing instead $R_{LFF_{LM}}(D)$ we write $R_{LFF}(LM, D)$.

As you can see the LFF_{LM} algorithm gives up the strict on-line property, because it looks ahead in the buffer, getting more information about its content. One has a feeling: the more information about the text to be compressed the better worst–case behaviour of a good algorithm. This suggests us that this algorithm behaves better than the on-line ones. The experimental results showed in [9] that the LFF type algorithms give better compression ratio as the longest matching or the differential greedy heuristics. However until now there has not been exact proofs about the worst-case efficiency of the given algorithms.

The paper is organised as follows. The next section contains an exhaustive analysis of the LFF_{LM} algorithm for different type of dictionaries, and at the end we mention some open questions for the future research.

3. Results for LFF_{LM} algorithm

Let us suppose that we use a buffer of length $t \cdot lmax$, where $t \geq 2$ is a parameter. First we present theorems for general dictionaries.

THEOREM 3.1 *Let D be a general dictionary. Then for the LFF_{LM} algorithm*

$$R_{LFF}(LM, D) \leq \frac{(t-1)lmax - (t-3)}{t} \frac{cmax}{cmin} .$$

Proof. Let S be an arbitrary string over the source alphabet given by D. Let the length of the buffer $t \cdot lmax$ where $t \geq 2$ is a parameter. We are interested in asymptotic worst-case ratio, so we can assume that $|S| = s \cdot t \cdot lmax$, where $s \geq 1$. So $s = \frac{|S|}{t \cdot lmax}$ is the number of steps used by the LFF_{LM} algorithm. Since $t \geq 2$ so in each step the buffer contains at least one arc from the optimal coding of S. (Of course the optimal code usually do not follow the block pattern.) Let us denote the lengths of the longest arcs of the optimal coding resp. the arcs first chosen by LFF_{LM} by $l_1, l_2, ..., l_s$ resp. $t_1, t_2, ..., t_s$ in each step. So

$$OPT(D,S) \geq \left(s + \frac{|S| - \sum_{i=1}^{s} l_i}{lmax}\right) cmin =$$

$$\left(\frac{|S|}{t \cdot lmax} + \frac{|S| - \sum_{i=1}^{s} l_i}{lmax}\right) cmin .$$

It is clear that $t_i \geq l_i$, $i = 1, ..., s$, so we get that the content of a buffer of length $t \cdot lmax$ in the ith step can be encoded in the worst case to

$$(t \cdot lmax - t_i - (t-2) + 1)\, cmax \leq (t \cdot lmax - l_i - (t-2) + 1)\, cmax \tag{3.1}$$

bits, therefore

$$\begin{aligned} LFF_{LM}(D,S) &\leq (s + s \cdot t \cdot lmax - \textstyle\sum_{i=1}^{s} l_i - s\,(t-2))\, cmax \\ &= \left(\tfrac{|S|}{t \cdot lmax} + |S| - \textstyle\sum_{i=1}^{s} l_i - \tfrac{|S|(t-2)}{t \cdot lmax}\right) cmax. \end{aligned}$$

If we introduce the notation $L(s,t) = (\sum_{i=1}^{s} l_i)/\,|S|$ we get that

$$\begin{aligned} R_{LFF}(LM,D) &\leq \lim_{|S|\to\infty} \frac{\frac{|S|}{t \cdot lmax} + |S| - |S|\, L(s,t) - \frac{|S|(t-2)}{t \cdot lmax}}{\frac{|S|}{t \cdot lmax} + \frac{|S| - |S|\, L(s,t)}{lmax}} \frac{cmax}{cmin} \\ &= \frac{\frac{1}{t \cdot lmax} + 1 - L(s,t) - \frac{(t-2)}{t \cdot lmax}}{\frac{1}{t \cdot lmax} + \frac{1 - L(s,t)}{lmax}} \frac{cmax}{cmin}. \end{aligned}$$

It is obvious that $\frac{1}{t \cdot lmax} \leq L(s,t) \leq \frac{1}{t}$, and the previous expression takes its maximum at $L(s,t) = \frac{1}{t}$. Therefore

$$R_{LFF}(LM,D) \leq \frac{(t-1)lmax - (t-3)}{t} \frac{cmax}{cmin}.$$

□

THEOREM 3.2 *The bound given in Theorem 3.1 is tight for LFF_{LM}.*

Proof. The proof can be reached with the following construction. Let D be the following dictionary:

source–word	u	v	w	vw	wv
code-word	a	b	c	d	e
weight	$cmax$	$cmax$	$cmax$	$cmax$	$cmax$

source–word	u^{lmax}	v^{lmax}	w^{lmax}	$u^{lmax-1}v$	
code-word	f	g	h	j	
weight	$cmin$	$cmin$	$cmin$	$cmax$	

Consider the following strings:

$$S_i = \begin{cases} \left(u^{lmax}\left(v^{lmax}w^{lmax}\right)^{\frac{t-2}{2}} v^{lmax}\right)^i & \text{if } t \text{ is even,} \\ \left(u^{lmax}\left(v^{lmax}w^{lmax}\right)^{\frac{t-1}{2}}\right)^i & \text{if } t \text{ is odd.} \end{cases}$$

The optimal encoding of S_i is the following:

$$OPT(D, S_i) = \begin{cases} \left(f\,(gh)^{\frac{t-2}{2}}\,g\right)^i & \text{if } t \text{ is even,} \\ \left(f\,(gh)^{\frac{t-1}{2}}\right)^i & \text{if } t \text{ is odd.} \end{cases}$$

The LFF_{LM} algorithm encodes S_i in the following way:

$$LFF_{LM}(D, S_i) = \begin{cases} \left(ajx^{\frac{t-2}{2}}b^{lmax-1}\right)^i & \text{if } t \text{ is even,} \\ \left(ajx^{\frac{t-3}{2}}b^{lmax-2}dc^{lmax-1}\right)^i & \text{if } t \text{ is odd,} \end{cases}$$

where $x = b^{lmax-2}dc^{lmax-2}e$.

We used that in case of tie the LFF algorithm can choose arbitrary source word, and now we have first choosen the word $u^{lmax-1}v$.

$$\begin{aligned} R_{LFF}(LM, D) &\geq \lim_{i\to\infty} \frac{\|LFF_{LM}(D, S_i)\|}{\|OPT(D, S_i)\|} \\ &= \lim_{i\to\infty} \frac{i\left((t-1)\,lmax - (t-3)\right)cmax}{i \cdot t \cdot cmin} \\ &= \frac{(t-1)lmax - (t-3)}{t}\frac{cmax}{cmin}. \end{aligned}$$

□

COROLLARY 3.3 *Let D be a code-uniform general dictionary. Then*

$$R_{LFF}(LM, D) = \frac{(t-1)lmax - (t-3)}{t}.$$

THEOREM 3.4 *Let D be a non-lengthening general dictionary. Then*

$$R_{LFF}(LM, D) = \begin{cases} \dfrac{(t-1)lmax - (t-3)}{t} \dfrac{cmax}{cmin} & \textit{if } cmax \leq Bt \\ T & \textit{if } Bt < cmax < 2Bt \\ \dfrac{(t-1)lmax\ Bt + cmax}{t \cdot cmin} & \textit{if } 2Bt \leq cmax \end{cases}$$

where

$$T = \frac{(t-1)lmax\ Bt - t(2Bt - cmax) + 4Bt - cmax}{t \cdot cmin}.$$

Proof. We will distinguish different subcases according to the relation between $cmax$ and Bt.

Case 1: Let us suppose that $cmax \leq Bt$. Under this condition every dictionary is nonlengthening and the results of Theorem 3.1 can be applied without changes (including the worst-case example).

Case 2: Suppose that $Bt < cmax \leq 2Bt$. We can use the same idea as in the proof of Theorem 3.1, but instead of (3.1) we can write the following:

The total weight of the arcs on the LFF path in the ith step is at most

$$(t \cdot lmax - t_i - 2(t-2))\, Bt + (t-2)\, cmax + cmax.$$

Using the same technique as in the proof of Theorem 3.1 we get that

$$R_{LFF}(LM, D) \leq \frac{(t-1)lmax \cdot Bt - t(2Bt - cmax) + 4Bt - cmax}{t \cdot cmin}.$$

Case 3: Let $2Bt < cmax \leq lmax \cdot Bt$. In this case the total weight of the arcs on the LFF path in the ith step is at most $(t \cdot lmax - t_i)\, Bt + cmax$. From this we get that

$$R_{LFF}(LM, D) \leq \frac{(t-1)lmaxBt + cmax}{t \cdot cmin}.$$

To prove that the bound is tight we need to modify the weights in the dictionary given in Theorem 3.2, so that if $cmax > l \cdot Bt$ then instead of $cmax$ we write $l \cdot Bt$, where l is the length of the correspondig source-word. This can be done easily and it is left to the reader. □

COROLLARY 3.5 *Let D be a code uniform and non-lenghtening general dictionary. Then*

$$R_{LFF}(LM, D) = \frac{(t-1)lmax - (t-3)}{t}.$$

Proof. We use that if the dictionary is non-lengthening and code uniform, then the length of each code-word is Bt. □

The next theorems will show us that in case of prefix dictionaries the worst-case behaviour is the same as for general ones.

THEOREM 3.6 *Let D be a prefix general dictionary. Then*

$$R_{LFF}(LM, D) = \frac{(t-1)lmax - (t-3)}{t} \frac{cmax}{cmin}.$$

Proof. Since the upper bound given in Theorem 3.1 is valid for any general dictionaries, so we need to give a prefix dictionary, which reaches the given bound.

source–word	u	v_1	v_2	w_1	w_2	v_2w_1
code-word	a	b	c	d	e	f
weight	$cmax$	$cmax$	$cmax$	$cmax$	$cmax$	$cmax$

source–word	w_2v_1	u^i	$v_1v_2^j$	$w_1w_2^j$	$u^{lmax-1}v_1$	
code-word	g	h_i	l_j	m_j	n	
weight	$cmax$	$cmin$	$cmin$	$cmin$	$cmax$	

where $i = 1, ..., lmax, j = 1, ..., lmax - 1$.
Consider the following strings:

$$S_i = \begin{cases} \left(u^{lmax}\left(v_1v_2^{lmax-1}w_1w_2^{lmax}\right)^{\frac{t-2}{2}} v_1v_2^{lmax-1}\right)^i & \text{if } t \text{ is even,} \\ \left(u^{lmax}\left(v_1v_2^{lmax-1}w_1w_2^{lmax}\right)^{\frac{t-1}{2}}\right)^i & \text{if } t \text{ is odd.} \end{cases}$$

The optimal encoding of S_i is the following:

$$OPT(D,S_i) = \begin{cases} \left(h_{lmax}\left(l_{lmax-1}m_{lmax-1}\right)^{\frac{t-2}{2}} l_{lmax-1}\right)^i & \text{if } t \text{ is even,} \\ \left(h_{lmax}\left(l_{lmax-1}m_{lmax-1}\right)^{\frac{t-1}{2}}\right)^i & \text{if } t \text{ is odd.} \end{cases}$$

The LFF_{LM} algorithm encodes S_i in the following way:

$$LFF_{LM}(D,S_i) = \begin{cases} \left(anx^{\frac{t-2}{2}} c^{lmax-1}\right)^i & \text{if } t \text{ is even,} \\ \left(anx^{\frac{t-3}{2}} c^{lmax-2} fe^{lmax-1}\right)^i & \text{if } t \text{ is odd,} \end{cases}$$

where $x = c^{lmax-2} fe^{lmax-2} g$. From this we get the result in a similar way like in Theorem 3.2. □

COROLLARY 3.7 *Let D be a prefix, non-lengthening general dictionary. Then*

$$R_{LFF}(LM,D) = \begin{cases} \dfrac{(t-1)lmax - (t-3)}{t} \dfrac{cmax}{cmin} & \text{if } cmax \leq Bt \\ T & \text{if } Bt < cmax < 2Bt \\ \dfrac{(t-1)lmax\ Bt + cmax}{t \cdot cmin} & \text{if } 2Bt \leq cmax \end{cases}$$

where

$$T = \frac{(t-1)lmax\ Bt - t(2Bt - cmax) + 4Bt - cmax}{t \cdot cmin},$$

Proof. If we combine the results of Theorem 3.4 and 3.6 we get the desired result. □

COROLLARY 3.8 *Let D_1 be a prefix, code-uniform and D_2 be a prefix, code uniform and non-lengthening general dictionary. Then*

$$R_{LFF}(LM,D_i) = \frac{(t-1)lmax - (t-3)}{t} \quad \textit{for } i = 1,2.$$

Proof. We can combine again the results of Corrolary 3.3 and Theorem 3.4 and 3.6. We also use that if the dictionary is code uniform and non-lengthening, then the length of each code-word is Bt. □

In the following we present some results for suffix dictionaries.

THEOREM 3.9 *Let D be a suffix general dictionary. Then*

$$R_{LFF}(LM,D) = \begin{cases} \left(1 + \frac{2\,(lmax-1)}{t}\right)\frac{cmax}{cmin} & \textit{if } t \geq 3 \\ \left(\frac{lmax+1}{2}\right)\frac{cmax}{cmin} & \textit{if } t = 2. \end{cases}$$

Proof. We know that in case of suffix dictionaries the number of arcs on the optimal path can not be less than that of the LFF_{LF}-path. (See Katajainen and Raita [7]). Unfortunately it is possible that at the end of the buffer the LFF_{LM} algorithm can not see an arc and its suffices. This means that in the worst-case we can get $lmax - 1$ arcs with length $cmax$. It is also possible that the longest arc selected by LFF_{LM} cuts an arc with its suffices. In the worst-case this can result in further $lmax - 1$ arcs with length $cmax$. Both situations can occur when $t \geq 3$, but only one of them in case of $t = 2$. Let us suppose that the optimal algorithm uses n arcs to encode the string S. This means that LFF_{LM} uses at most $n+2\,(lmax-1)$ if $t \geq 3$, and $n+(lmax-1)$ if $t = 2$. Let us define p as follows:

$$p = \begin{cases} 1 & \text{if } t = 2 \\ 2 & \text{if } t \geq 3. \end{cases}$$

Since $n \geq \frac{|S|}{lmax}$,

$$\begin{aligned} R_{LFF}(LM,D) &\leq \lim_{s\to\infty} \frac{n + ps\,(lmax-1)}{n}\frac{cmax}{cmin} \\ &\leq \lim_{s\to\infty}\left(1 + \frac{p\frac{|S|}{t\cdot lmax}\,(lmax-1)}{\frac{|S|}{lmax}}\right)\frac{cmax}{cmin} \qquad (3.2) \\ &= \left(1 + \frac{p\,(lmax-1)}{t}\right)\frac{cmax}{cmin}\,. \end{aligned}$$

Now we have to give some examples when the given bounds are reached. We distinguish two cases.

Case 1: Let us suppose that $t = 2$. Let D be the following suffix dictionary:

source–word	u	v	$u^k v$	$w^k v$
code-word	a	b	c_k	d_k
weight	$cmax$	$cmax$	$cmax$	$cmax$

source–word	w^j	vw^{lmax-1}	$u^{lmax-1}v$	$w^{lmax-1}v$
code-word	e_j	f	g	h
weight	$cmax$	$cmax$	$cmin$	$cmin$

where $k = 1, ..., lmax - 2$ and $j = 1, ..., lmax - 1$.
Consider the following strings:

$$S_i = \left(u^{lmax-1}vw^{lmax-1}v\right)^i, i \geq 1.$$

Now $LFF_{LM}(D, S_i) = \left(a^{lmax-1}fb\right)^i$ and $OPT(D, S_i) = (gh)^i$. Hence we have that

$$\begin{aligned} R_{LFF}(LM, D) &\geq \lim_{i\to\infty} \frac{\|LFF_{LM}(D, S_i)\|}{\|OPT(D, S_i)\|} \\ &= \lim_{i\to\infty} \frac{i\,(lmax+1)\,cmax}{2\,i \cdot cmin} \\ &= \frac{(lmax+1)}{2}\frac{cmax}{cmin}. \end{aligned}$$

Case 2: Now let $t \geq 3$. Let D be the following suffix dictionary:

source–word	u	v	$v^k u$	$w^k u$
code-word	a	b	c_k	d_k
weight	$cmax$	$cmax$	$cmax$	$cmax$

source–word	w^j	uw^{lmax-1}	$v^{lmax-1}u$	$w^{lmax-1}u$
code-word	e_j	f	g	h
weight	$cmax$	$cmax$	$cmin$	$cmin$

where $k = 1, ..., lmax - 2$ and $j = 1, ..., lmax - 1$.

Consider the following strings:

$$S_i = u\left(v^{lmax-1}u\left(w^{lmax-1}u\right)^{t-2}v^{lmax-1}u\right)^i, \text{if } i \geq 1.$$

Now $LFF_{LM}(D, S_i) = \left(ab^{lmax-1}f^{t-2}ab^{lmax-1}\right)^i a$ and $OPT(D, S_i) = a\left(gh^{t-2}g\right)^i$. Hence we have that

$$\begin{aligned} R_{LFF}(LM, D) &\geq \lim_{i\to\infty} \frac{\|LFF_{LM}(D, S_i)\|}{\|OPT(D, S_i)\|} \\ &= \lim_{i\to\infty} \frac{i\,(2\ lmax + t - 2)\,cmax + cmax}{i \cdot t \cdot cmin + cmin} \\ &= \left(1 + \frac{2\,(lmax-1)}{t}\right)\frac{cmax}{cmin}. \end{aligned}$$

□

THEOREM 3.10 *Let D be a suffix, non-lengthening general dictionary. Then*

$$R_{LFF}(LM, D) = \begin{cases} \left(1 + \frac{2\,(lmax-1)}{t}\right)\frac{cmax}{cmin} & \text{if } t \geq 3 \text{ and } cmax \leq Bt \\ \left(\frac{lmax+1}{2}\right)\frac{cmax}{cmin} & \text{if } t = 2 \text{ and } cmax \leq Bt \\ \frac{cmax}{cmin} + \frac{2\,(lmax-1)}{t}\frac{Bt}{cmin} & \text{if } t \geq 3 \text{ and } cmax > Bt \\ \frac{cmax}{cmin} + \frac{lmax-1}{2}\frac{Bt}{cmin} & \text{if } t = 2 \text{ and } cmax > Bt \end{cases}$$

Proof. If $cmax \leq Bt$ then the proof of Theorem 3.9 can be applied without any changes. If $cmax > Bt$ then in the worst-case the total weight of the extra arcs on the LFF-path is $2\,(lmax - 1)\,Bt$ if $t \geq 3$, and $(lmax - 1)\,Bt$ if $t = 2$. This gives the desired results. □

COROLLARY 3.11 *Let D_1 be a code-uniform, suffix and D_2 be a code-uniform, suffix and non-lengthening general dictionary. Then*

$$R_{LFF}(LM, D_i) = \left\{ \begin{array}{ll} 1 + \frac{2(lmax-1)}{t} & \text{if } t \geq 3 \\ \frac{lmax+1}{2} & \text{if } t = 2. \end{array} \right\} \text{for } i = 1, 2.$$

4. Conclusions

We analyzed an "almost" *on-line* data compression algorithm with static dictionaries from a worst-case point of view. Four properties of dictionaries were investigated: code uniform, nonlengthening, suffix and prefix. The case of a dictionary that is prefix and suffix at the same time does neither seem practical nor interesting to us, and thus there remain twelve "reasonable" combinations of these properties. Each such cases have been analyzed. It worth mentioning that if the buffer-length tends to infinity then our results converge always to the adequate LM results. More exactly, we got the following worst-case ratios: If D is one of the considered dictionaries then

$$\lim_{t \to \infty} R_{LFF}(LM, D) = R_{LM}(D).$$

This result is evident since the longer buffer the fewer influence of the first choice. On the other hand it is also obvious that our choice (LM) was absolutaly authocratic. We could also use the differential greedy or any other on-line algorithm. Vividly, it is more or less also trivial that if A is an arbitrary on-line algorithm and D is a given dictionary then

$$\lim_{t \to \infty} R_{LFF}(A, D) = R_A(D),$$

but how can we prove it?

Secondly, we can observe that we reach the limes from below. That means that we get the best results for the smallest buffer-lenght, i.e. if $t = 2$. Every such cases result in a factor 2 improving the worst-case ratios of the LM heuristics in hand, and for any longer buffer the worst case behaviour of LFF is getting worse, depending on the buffer length. The question is obvious: can we give a variation of LFF which improves the worst-case ratio of LM with a constant factor independently of buffer-lengths? Our suggestion is the following:

Let us use first the LFF for encoding the buffer. After the first choice — instead of using the LM algorithm — we encode the remaining parts of the buffer iteratively using the LFF method. We call this algorithm as Iterated Longest Fragment First. ($ILFF$). The time complexity of this algorithm is $O(n \log n)$. Our conjecture states that if D is a dictionary with one of the considered characteristics then

$$R_{ILFF}(LM, D) = \frac{1}{2} R_{LM}(D).$$

Is it true? There are other exciting questions raising at the same time. One of them is the following: If we consider the class of those algorithms which use a limited size of buffer — and we call them as *space-bounded algorithms* — then we can ask whether there exists such a space-bounded algorithm which has a better worst-case ratio than the conjectured value of $ILFF$? The interesting reader can find further tips for the future research in [2] and [1].

Acknowledgement

The authors wish to thank the two anonymous referees for their helpful comments on an earlier version of this paper.

References

1. J. Békési, G. Galambos, U. Pferschy, G. J. Woeginger, The Fractional Greedy Algorithm for Data Compression, *Computing* **56(1)**, 1996, 29-46.
2. J. Békési, G. Galambos, U. Pferschy, G. J. Woeginger, Greedy Algorithms for On-Line Data Compression, *Operations Research Proceedings*, Ed. U. Derigs, A. Bachem, A. Drexl, Berlin, 1994, 76-80.
3. R. M. Fano, Transmission of information, Wiley, New York, 1961.
4. M. E. Gonzalez–Smith and J. A. Storer, Parallel algorithms for data compression, *Journal of the ACM* **32**, 1985, 344–373.
5. D. A. Huffman, A method for the construction od minimum-redundancy codes, *Proc Institute of Electrical and Radio Engineers*, **40(9)**, 1952,1098-1101.
6. F. Jelinek, K. Schneider, On variable length-to-block coding, *IEEE-IT* **18**,1972,765-774
7. J. Katajainen and T. Raita, An analysis of the longest matching and the greedy heuristic in text encoding, *Journal of the ACM* **39**, 1992, 281–294.
8. J. J. Risassen, G. G. Langdon, Arithmetic coding, *IBM J. Research and Development*, **23(2)**,1979,149-162.
9. E. J. Schuegraf and H. S. Heaps, A comparison of algorithms for data base compression by use of fragments as language elements, *Inf. Stor. Ret.* **10**, 1974, 309–319.
10. J. Ziv and A. Lempel, A universal algorithm for sequential data compression, *IEEE Trans. Inf. Theory* **23**, 1977, 337–343.

Linear Operators and Stochastic Dominance

Erio Castagnoli
Bocconi University - Milano (Italy)

Abstract. This paper aims to show how various types of stochastic dominance, both in the vector space of Borel signed measures and in its subset of distribution functions of random variables, may be described through linear operators. This allows to obtain in a very natural way (old and new) characterizations of linear functionals preserving stochastic dominance: such functionals are the familiar expected utilities.

1. Preliminaries

We summarize briefly the contents of Castagnoli and Li Calzi (1995) to which we refer for further details and notation. General references on this and related subjects are Aliprantis and Border (1994) and Bourbaki (1987).

1. Let (X, V) and (Y, U) be two dual pairs of real vector spaces with respect to the bilinear functions $B_1 : X \times V \to \mathbb{R}$ and $B_2 : Y \times U \to \mathbb{R}$ respectively. This means that:

$$B_1(x, v) = 0 \quad \forall x \in X \Longrightarrow v = 0$$

$$B_1(x, v) = 0 \quad \forall v \in V \Longrightarrow x = 0$$

with the same properties holding for B_2. B_1 and B_2 are called *duality relations.*

2. Let Y be endowed with a partial order $\geq$, that will be called *natural order.* To ensure consistency with the vector space structure, we assume that:

$$\begin{aligned} y^1 \geq y^2 &\Longrightarrow y^1 + y^3 \geq y^2 + y^3 \quad \forall y^3 \in Y \\ y^1 \geq y^2 &\Longrightarrow \alpha y^1 \geq \alpha y^2 \quad \forall \alpha > 0 \end{aligned} \tag{1.1}$$

As a consequence, $\geq$ results to be a cone order; in particular:

$$y^1 \geq y^2 \Longleftrightarrow y^1 - y^2 \geq 0$$

Whenever $\geq$ is a preorder, it suffices to replace Y with the quotient space $Y/\sim$.

F. Giannessi et al. (eds.), New Trends in Mathematical Programming, 29–39.

3. Let A be a linear operator from X onto Y and suppose that it is continuous with respect to the weak topologies $\sigma(X,V)$ and $\sigma(Y,U)$ induced on X and Y by V and U, respectively: they are the coarsest topologies that make continuous every linear functional on X and Y, respectively.

   We will assume that the natural order $\geq$ is $\sigma(Y,U)$-continuous:

$$y^n \geq 0,\ y^n \to y \Longrightarrow y \geq 0\,.$$

4. As a consequence of the preceding points, every continuous (in the above topologies) linear functional on Y may be written as $B_2(\cdot,u)$ with $u \in U$: furthermore, such u is unique. The same occurs for X: every continuous linear functional on X may be written as $B_1(\cdot,v)$ with a unique $v \in V$.

5. It is quite natural to define cone orders in the remaining three spaces. It suffices to define their positive cones.

   In U:

$$u \geq^* 0 \Longleftrightarrow B_2(y,u) \geq 0 \quad \forall y \geq 0$$

   Such $\geq^*$ will be called *dual natural order.*

   In X:

$$x \succeq 0 \Longleftrightarrow Ax \geq 0$$

   In general, $\succeq$ is a partial preorder and it will be termed *dominance relation*[1]

   In V:

$$v \succeq^* 0 \Longleftrightarrow B_1(x,v) \geq 0 \quad \forall x \succeq 0$$

   Also $\succeq^*$ is a partial preorder and it will be called *dual dominance.*

6. There exists a unique linear operator A^* from U onto V defined by:

$$B_2(Ax,u) = B_1(x,A^*u)$$

   where $x \in X$, $u \in U$ (and hence $Ax \in Y$, $A^*u \in V$). The operator A^* is the *adjoint* of A.

7. It is easy to prove (see Castagnoli and Li Calzi, 1995) that:

$$u \geq^* 0 \Longleftrightarrow A^*u \succeq^* 0$$

   or:

$$v \succeq^* 0 \Longleftrightarrow v = A^*u \quad with \quad u \geq^* 0$$

[1] $\succeq$ is a partial order if and only if A is one-to-one.

8. As we said above, any continuous linear functional on X is of the form $B_1(\cdot, v)$ with $v \in V$. It is said to *preserve the dominance relation* $\succeq$ if:

$$x^1 \succeq x^2 \Longrightarrow B_1\left(x^1, v\right) \geq B_1\left(x^2, v\right)$$

that is, setting $x = x^1 - x^2$:

$$x \succeq 0 \Longrightarrow B_1(x, v) \geq 0$$

What precedes contains a characterization of linear functionals preserving the dominance relation $\succeq$.

Such a class is given simply by $V^+ = \{v : v \succeq^* 0\} = \{v : v = A^*u \quad with \quad u \geq^* 0, \quad u \in U\}$:

$$x \succeq 0 \Longleftrightarrow B_1(x, v) \geq 0 \quad \forall v \in V^+$$

We may conclude that the dominance relation $\succeq$ is *represented* by the class of linear functionals $B_1(\cdot, v)$, $\quad v \in V^+$.

In many cases of interest, it is possible[2] to deal with linear functionals that "strictly" preserve the dominance relations $\succeq$.

Write $y^1 > y^2$ whenever $y^1 \geq y^2$ but not $y^2 \geq y^1$and[3] attach the same meaning to $>^*$ (in U), $\succ$ (in X) and $\succ^*$ (in V).

Moreover, define strict orders as follows:

$$y \gg 0 \Longleftrightarrow B_2(y, u) > 0 \quad \forall u >^* 0$$

$$u \gg^* 0 \Longleftrightarrow B_2(y, u) > 0 \quad \forall y > 0$$

$$x \succ\!\succ 0 \Longleftrightarrow B_1(x, v) > 0 \qquad \forall v \succ^* 0$$

$$v \succ\!\succ^* 0 \Longleftrightarrow B_1(x, v) > 0 \qquad \forall x \succ 0$$

A linear functional $B_1(., v)$ is said to *strictly preserve* $\succeq$ if it preserves $\succeq$ and, moreover:

$$x \succ 0 \Longrightarrow B_1(x, v) > 0$$

[2] See Aliprantis and Border (1994) p.221 and foll., for a discussion on this point.

[3] F.i., take $Y = L_p[a, b]$, $\quad U = L_q[a, b]$, with $p^{-1} + q^{-1} = 1$ and $1 < p, q < +\infty$, under duality $\int_a^b y(t)\, u(t)\, dt$, $y \in Y$, $u \in U$.
Assume $\geq$ to be the a.e. pointwise order in Y. In this case:
$y \geq 0$ denotes an a.e. non negative function,
$y > 0$ is an a.e. non negative function taking positive values in a non-null set,
$y \gg 0$ denotes an a.e. positive function.

The class of linear functionals strictly preserving $\succeq$ is $B_1(.,v)$ with $v \in V^{++} = \{v : v \succ\!\succ^* 0\} = \{v : v = A^*u \quad \textit{with} \quad u \gg^* 0, \quad u \in U\}$: A is required to be a strictly positive operator ($x \succ\!\succ 0 \Rightarrow Ax \gg 0$).

2. Stochastic dominance in one dimension

2.1. General setting

Take $X = NBV[a,b]$, the space of normalized (right-continuous and vanishing in a) functions of bounded variation defined in $[a,b]$. It may be interpreted as the space of Borel signed measures on $[a,b]$.

Let $U = B[a,b]$, the space of Borel measurable bounded functions defined in $[a,b]$.

(X,U) is a dual pair under both duality relations:

$$B_1(x,u) = \int_a^b u(t)\,\mathrm{d}x(t)$$

$$B_2(x,u) = \int_a^b u(t)\,x(t)\,\mathrm{d}t$$

For the moment, take $Y_0 = X$ and $V_0 = U$ and consider the dual pairs (X, V_0) with duality B_1 and (Y_0, U) with duality B_2.

Consider a linear operator on X with the following structure:

$$Ax = \int_a^b K(s,t)\,\mathrm{d}x(t) \tag{2.2}$$

with K bounded on $[a,b]^2$.

We will assume that K is such that the image set $Y = AX$ is $\sigma(Y_0, U)$-dense in Y_0 and that A is continuous with respect to the weak topologies $\sigma(X, V_0)$ and $\sigma(Y_0, U)$. As a consequence, (Y, U) is a dual pair under duality B_2.

The adjoint of A is defined by:

$$A^*u = \int_a^b K(s,t)\,u(s)\,\mathrm{d}s = \int_a^b K(s,t)\,\mathrm{d}\overline{u}(s)$$

where $u \in U$ and $\overline{u}$ is its integral function.

Its image set $V = A^*U$ is $\sigma(X, V_0)$-dense in U. Hence X and V form another dual pair under duality B_1.

Finally, define the natural order $\geq$ in Y as the a.e. pointwise order. It follows that also the dual natural order $\geq^*$ in U is the a.e. pointwise order.

The dominance relation $\succeq$ on X:

$$x^1 \succeq x^2 \iff Ax^1 \geq Ax^2$$

will be called *A-stochastic dominance.*

We may immediately conclude that:

The linear functionals $\varphi : X \to \mathbb{R}$ *preserving A-stochastic dominance are all and only:*

$$\varphi(x) = B_1(x, v) = \int_a^b v(t)\,\mathrm{d}x(t) \qquad with \quad v \succeq^* 0 \qquad (2.3)$$

that is:

$$v = A^* u = \int_a^b K(s,t)\,u(s)\,\mathrm{d}s \qquad with \quad u \geq^* 0$$

In the same way, the only linear functionals strictly preserving A-stochastic dominance are given by (2.3), but with $v \succ\!\succ^* 0$.

Briefly, the class of linear functionals preserving (strictly or not) A-stochastic dominance is given by "expected utilities" with utility functions belonging to an appropriate cone.

Under obvious restrictions on domains and image sets, given the linear operators $A_1, A_2, ..., A_n$, we may define the compound operator $A = A_n A_{n-1} ... A_1$.

In particular, we may define the iterated operators $A^k = AA^{k-1}$, $k = 2, 3, ...$, and the corresponding A^k-stochastic dominances.

Usually we are interested in giving concepts of stochastic dominance only on the convex subset D of X containing distribution functions.

The restriction of the domain to a convex set allows us to enlarge the class of order preserving linear functionals. Observe infact that a functional φ appears to be a linear on a convex subset C of a vector space S if and only if it is an affine transformation of a linear functional f on S:

$$\varphi(x) = af(x) + b \qquad x \in S; \quad a, b \in \mathbb{R}$$

Moreover, when f is order preserving, it has to be $a \geq 0$ to grant that φ is also order preserving.

This amounts to say that the class of linear functionals on D that preserve A-stochastic dominance (on D) is given by (2.3) where v is determined up to an increasing affine transformation (i.a.t.).

2.2. Classical stochastic dominance

Take:

$$K(s,t) = \begin{cases} 1 \; for & t > s \\ 0 \; for & t \leq s \end{cases} \qquad s, t \in [a, b]$$

Denoting with A_{up} the corresponding linear operator, we have:

$$A_{up}x = \int_s^b \mathrm{d}x(t) \qquad A^*_{up}u = \int_a^t u(s)\,\mathrm{d}s$$

The dominance relation becomes:

$$x^1 \succeq_{up} x^2 \Longleftrightarrow \int_s^b \mathrm{d}x^1(t) \geq \int_s^b \mathrm{d}x^2(t) \qquad \forall s \in [a,b]$$

that we may call *Upper First Order Stochastic Dominance* (UFSD).

The only linear functionals preserving it are:

$$\varphi(x) = \int_a^b v(t)\,\mathrm{d}x(t)$$

where $v = A^*_{up}u = \int_a^t u(s)\,\mathrm{d}s$ with $u \geq^* 0$, that is v is a positive and increasing function.

A similar conclusion holds for the operators:

$$A_{low}x = \int_a^s -\mathrm{d}x(t) \qquad A^*_{low}u = \int_t^b -u(s)\,\mathrm{d}s$$

The dominance relation becomes:

$$x^1 \succeq_{low} x^2 \Longleftrightarrow \int_a^s \mathrm{d}x^1(t) \leq \int_a^s \mathrm{d}x^2(t) \qquad \forall s \in [a,b]$$

that we may call *Lower First Order Stochastic Dominance* (LFSD).

The linear functionals preserving it are again:

$$\varphi(x) = \int_a^b v(t)\,\mathrm{d}x(t)$$

but now $v = A^*_{low}u = \int_t^b -u(s)\,\mathrm{d}s$ with $u \geq^* 0$, that is v is negative and increasing.

Remark - Strict preservation occurs, both for UFSD and LFSD, if and only if v is strictly increasing (positive or negative, respectively).

UFSD and LFSD are different on X but they collapse in the same concept, the classical FSD, on the subset D of distribution functions.

Example. Let:

$$x^1(t) = \alpha t \quad (\alpha > 0) \quad ; \quad x^2(t) = 1 - e^{-t} \quad t \in [0,1]$$

The functions $A_{up}x^1$ and $A_{up}x^2$ are respectively:

$$\alpha(1-s) \quad ; \quad e^{-s}-e^{-1} \qquad s\in[0,1]$$

and x^1 dominates x^2 if and only if $\alpha \geq 1-e^{-1}$.

The functions $A_{low}x^1$ and $A_{low}x^2$ are:

$$-\alpha s \quad ; \quad e^{-s}-1 \qquad s\in[0,1]$$

and x^1 dominates x^2 if and only if $\alpha \leq 1-e^{-1}$.

UFSD and LFSD agree only in the case $\alpha = 1-e^{-1}$, that is if and only if $x^1(1)=x^2(1)$.

The conclusion is well known: FSD is preserved (in D) by all and only the linear functionals:

$$\varphi(x)=\int_a^b v(t)\,\mathrm{d}x(t) \qquad x\in D$$

where v is increasing and determined up to an i.a.t.

Let us turn to Second Order Stochastic Dominance (SSD). We may define an *Upper SSD* through the operator $-A_{up}^2$ and a *Lower SSD* through $-A_{low}^2$.

Again they are different on X but coincide on D. USSD and LSSD are preserved by functionals (2.3) with v positive, increasing and concave (USSD) or v negative, increasing and concave (LSSD).

On D the linear functionals preserving the (unique) SSD are characterized by v increasing and concave.

The story goes on plainly considering $(-1)^{k+1}A_{up}^k$ and $-A_{low}^k$, $k=3,4,\ldots.$

2.3. Two examples

1. Maintain to X, U and A the above meaning and choose a subinterval $[\alpha,\beta]\subset[a,b]$.

   Define in $Y=AX$ the following natural order: $y\geq 0$ means that $y(s)\geq 0 \quad \forall s\in[\alpha,\beta]$, no matter of what happens outside $[\alpha,\beta]$. The dual natural order results to be: $u\geq^* 0$ if $u(s)\geq 0$ $\forall s\in[\alpha,\beta]$ and $u(s)=0$ outside $[\alpha,\beta]$.

   The dominance relation $x^1\succeq x^2 \Leftrightarrow Ax^1\geq Ax^2$ coincides with A-stochastic dominance but only in $[\alpha,\beta]$: what happens outside is immaterial.

The linear functionals preserving this kind of dominance are given by (2.3) with $v = A^*u$, $u \geq^* 0$.

Such v's take value zero outside $[\alpha, \beta]$. Thus, functionals like in (2.3) may be written as:

$$\varphi(x) = \int_\alpha^\beta v(t)\,\mathrm{d}x(t)$$

with $v = A^*u$, $\quad u \geq^* 0$.

2. Take A to be the functional:

$$Ax = \int_a^b I_{[a,\alpha]}(t)\,\mathrm{d}x(t)$$

where α is fixed in $[a, b]$ and $I_{[a,\alpha]}(t)$ is the indicator function of $[a, \alpha]$.

In this case the dominance relation is a complete preorder: x^1 and x^2 are ordered simply according to $x^1(\alpha)$ and $x^2(\alpha)$. In particular, $x^1 \sim x^2$ whenever $x^1(\alpha) = x^2(\alpha)$. Each class of equivalence in the quotient set $X/_\sim$ may be identified by:

$$x(t) = x(\alpha)\,\Delta_\alpha(t)$$

where Δ_α is the Heaviside function making a unit jump in α:

$$\Delta_\alpha(t) = \begin{cases} 0 \;\; for & t < \alpha \\ 1 \;\; for & t \geq \alpha \end{cases}$$

The adjoint operator is:

$$A^*u = \int_a^b I_{[a,\alpha]}(t)\,u(s)\,\mathrm{d}s = I_{[a,\alpha]}(t) \cdot c$$

where c denotes the overall integral of u.

Taking $u \geq^* 0$, c is a non negative constant and thus the linear functionals preserving such an order are:

$$\varphi(x) = c \cdot x(\alpha) \qquad with \quad c \geq 0$$

3. Stochastic dominance in several dimensions

Something changes in two or more dimensions.

Now let $[a,b] = \times_{i=1}^{n} [a_i, b_i] \subset \mathbb{R}^n$. The general setting, apart from slight and obvious modifications, still works as in sect. 2.1.

If we specialise the argument to classical stochastic dominance, as in sect. 2.2, we may define in the same way a UFSD and a LFSD that are different not only on X but also on D.

Let us consider this point.

Define again:

$$A_{up}x = \int_s^b \mathrm{d}x(t) \qquad a \le s \le b; \quad a, b \in \mathbb{R}^n$$

$$A_{low}x = \int_a^s -\mathrm{d}x(t) \qquad a \le s \le b; \quad a, b \in \mathbb{R}^n$$

As expected, we have two different FSD:

- UFSD: $x^1 \succeq_{up} x^2 \Leftrightarrow A_{up}x^1 \ge A_{up}x^2$ preserved by all and only the linear functionals of the form (2.3) with $v = A^*_{up}u = \int_a^t u(s)\,\mathrm{d}s$, $u \ge^* 0$: v is positive, increasing and supermodular (increasing as a n-dimensional distribution function).

- LFSD: $x^1 \succeq_{low} x^2 \Leftrightarrow A_{low}x^1 \ge A_{low}x^2$ preserved by all and only the same linear functionals but with $v = A^*_{low}u = \int_a^t -u(s)\,\mathrm{d}s$, $u \ge^* 0$, that is v negative, increasing and submodular (increasing as the negative of a n-dimensional survival function).

Confining our attention to D, the convex subset of n-dimensional distribution functions, UFSD and LFSD remain different.

In fact, when $[a,b] \subset \mathbb{R}$:

$$A_{up}x = 1 + A_{low}x \qquad \forall x \in D$$

and thus the same inequalities hold for $A_{up}x$ and $A_{low}x$ making UFSD and LFSD equivalent.

When $[a,b] \subset \mathbb{R}^n$, $n > 1$, the previous identity does not hold: this creates different dominances.

- In D, UFSD is preserved by the usual linear functionals with v increasing and supermodular (that is, v is proportional to a n-dimensional distribution function).

- In D, LFSD is preserved by the same linear functionals with v increasing and submodular (that is, v is proportional to the negative of a n-dimensional survival function).

In any case, v is determined up to an i.a.t.

The gap among UFSD and LFSD is pointed out, f.i., in Marshall and Olkin (1979) and in Castagnoli and Li Calzi (1994).

But the story does not end here. We may define a "mixed up and low" operator (built as A_{up} with respect to some components and as A_{low} with respect to the remaining ones), giving rise to new types of stochastic dominance.

Hoping to be more clear, we will discuss the case only in two dimensions.

Let $s = (s_1, s_2)$, $t = (t_1, t_2) \in \mathbb{R}^2$. Define:

$$A_{up,low}x = \int_{s_1}^{b_1} \mathrm{d}_{t_1} \int_{a_2}^{s_2} \mathrm{d}_{t_2} x(t_1, t_2)$$

yielding to the *Upper-Lower First Order Stochastic Dominance* (ULFSD)[4]:

$$x^1 \succeq_{up,low} x^2 \Longleftrightarrow A_{up,low}x^1 \geq A_{up,low}x^2$$

preserved by the linear functionals:

$$\varphi(x) = \int_{a_1}^{b_1} \int_{a_2}^{b_2} v(t_1, t_2)\, \mathrm{d}_{t_1} \mathrm{d}_{t_2} x(t_1, t_2)$$

with:

$$v = A^*_{up,low}u = \int_{a_1}^{t_1} \mathrm{d}s_1 \int_{t_2}^{b_2} u(s_1, s_2)\, \mathrm{d}s_2 \qquad (u \geq^* 0)$$

that is positive, increasing with respect to t_1 and decreasing with respect to t_2.

To give a meaning to such dominance, let us refer to the set D of (bidimensional) distribution functions.

Let $\xi^1 = (\xi_1^1, \xi_2^1)$ and $\xi^2 = (\xi_1^2, \xi_2^2)$ be two random variables with distribution functions x^1 and x^2 respectively.

The dominance according to ULFSD may be written as:

$$\Pr\left[\xi_1^1 > s_1 \bigwedge \xi_2^1 \leq s_2\right] \geq \Pr\left[\xi_1^2 > s_1 \bigwedge \xi_2^2 \leq s_2\right] \qquad \forall s_1, s_2 \in \mathbb{R}$$

that is ξ_1^1 dominates, according to one-dimensional FSD, ξ_1^2 while ξ_2^2 dominates, again according to one-dimensional FSD, ξ_2^1.

This suggests the following interpretation for ULFSD.

[4] In a similar way we may define LUFSD.

The problem is to substitute a random variable to a pre-existing one[5]: any pair (ξ_1^i, ξ_2^i) has to be intended as the substitution of ξ_1^i to ξ_2^i.

Given two pairs of possible substitutions (ξ_1^1, ξ_2^1) and (ξ_1^2, ξ_2^2), ULFSD says that the first is to be preferred to the second when, among the entering r.v.'s, ξ_1^1 dominates ξ_1^2 and, among the r.v.'s to drop, ξ_2^2 dominates ξ_2^1.

Remark - In particular, the assumption on v are fulfilled by:

$$v(t_1, t_2) = -v(t_2, t_1)$$

v being increasing with respect to the first variable. This choice reminds the SSB utility theory of Fishburn (1984).

4. Acknowledgements

Research partly supported by M. U. R. S. T. and Bocconi University. I thank M. Li Calzi for his helpful comments.

References

1. Aliprantis, C.D. and Border, K.C.: 1994, *Infinite Dimensional Analysis,* Springer-Verlag, Berlin.
2. Bourbaki, N.: 1987, *Topological Vector Spaces,* Springer-Verlag, Berlin.
3. Castagnoli, E. and Li Calzi, M.: 1995, " 'Direct' and 'Indirect' Duality for Dominance Relations", in E. Castagnoli and G. Giorgi (eds.), *Scalar and Vector Optimization in Economic and Financial Problems,* Egea, Milano, 65-84.
4. Castagnoli, E. and Li Calzi, M.: 1996, "Expected Utility without Utility", *Theory and Decision* **41**, 281-301.
5. Fishburn, P.C.: 1984, "Dominance in SSB Utility Theory", *Journal of Economic Theory* **14**, 130-148.
6. Marshall, A.W. and Olkin, I.: 1979, *Inequalities: Theory of Majorization and its Applications,* Academic Press, Orlando.

[5] Many convincing examples may be found in insurance or financial problems: reciprocal re-insurance treaties, choice among different insurance protections, partial hedging of portfolios etc.

Some Properties of Dini-Derivatives of Quasiconvex Functions

Jean-Pierre Crouzeix
LIMOS, Université Blaise Pascal,
63177 Aubière Cedex, France

Abstract. The most general optimality conditions in quasiconvex programming are expressed in terms of normal cones to the level sets of functions. Then Kuhn-Tucker type conditions are derived by expressing these normal cones in terms of some generalized subdifferentials which often are related to some generalized derivatives.

We study some properties of the Dini-derivatives of quasiconvex and pseudoconvex functions and we show that these derivatives are useful for our purposes.

1. Introduction

Let us consider the optimization problem:

$$\inf[f(x) : x \in C].$$

Set $\tilde{S}(a) = \{x : f(x) < f(a)\}$, then $a \in C$ is an optimal solution of the problem if and only if

$$C \cap \tilde{S}(a) = \emptyset.$$

When these sets are convex, optimality conditions are obtained via classical separation theorems. These separation theorems involve the normal cones to the sets C and $\tilde{S}(a)$ at point a. Quasiconvex functions are precisely those functions for which the sets $\tilde{S}(a)$ are convex for all a such that $f(a)$ is finite.

We recall some definitions and notation.

Given D a convex subset of $\Re^n$, a function $f : D \to \Re$ is said to be *quasiconvex* on D if for all $x, y \in D, x \neq y$ and $t \in (0,1)$ one has

$$f(tx + (1-t)y) \leq \max(f(x), f(y)).$$

Geometrical equivalent definitions of quasiconvexity are expressed in terms of the lower level sets of the function. For all $\lambda \in \Re$ define:

$$S_\lambda(f) = \{x \in D : f(x) \leq \lambda\} \text{ and } \tilde{S}_\lambda(f) = \{x \in D : f(x) < \lambda\}.$$

Then f is quasiconvex if and only if all $S_\lambda(f)$ (all $\tilde{S}_\lambda(f)$), $\lambda \in \Re$ are convex. When f is differentiable on D, a necessary and sufficient condition for f to be quasiconvex on D is that

$$\langle \nabla f(a), x - a \rangle \leq 0 \text{ for all } x, a \in D \text{ such that } f(x) < f(a).$$

F. Giannessi et al. (eds.), New Trends in Mathematical Programming, 41–57.

A slight modification of this condition leads to the definition of pseudoconvexity. Assume again that D is convex and f is differentiable on D, f is said to be *pseudoconvex* on D if

$$\langle \nabla f(a), x-a\rangle < 0 \text{ for all } x, a \in D \text{ such that } f(x) < f(a).$$

For pseudoconvex functions, the condition

$$\langle \nabla f(a), x-a\rangle \geq 0 \text{ for all } x \in D$$

is a necessary and sufficient condition for optimality of f at a on D, for quasiconvex functions it is only necessary.

From now on, and as usual in convex analysis, we consider functions f defined on the whole space $\Re^n$, setting $f(x) = +\infty$ at points x where f is not defined. This allows to avoid mentioning the effective domain of the function. Pseudoconvexity can be extended to nondifferentiable functions, using for instance the concept of Dini-derivatives, see for instance, different works by Komlosi and Giorgi [6],[7],[8],[11], [12],[13]. Assume that $f(a)$ is finite and $h \in \Re^n$, the *upper* and *lower Dini-derivatives* of f at $a \in D$ in direction h are respectively,

$$f'_+(a,h) = \limsup_{t\to 0_+} \frac{f(a+th)-f(a)}{t},$$

$$\text{and } f'_-(a,h) = \liminf_{t\to 0_+} \frac{f(a+th)-f(a)}{t}.$$

Then, following Diewert, [5] f is said to be *pseudoconvex at* a if

$$f'_-(a, x-a) < 0 \text{ for all } x \text{ such that } f(x) < f(a).$$

(A stronger form uses the upper instead of the lower Dini-derivative [2, 10]).

In all what follows, $f : \Re^n \to \bar{\Re}$ is quasiconvex, $f(a)$ is finite, we consider the sets

$$\tilde{S}(a) = \{x : f(x) < f(a)\},$$

$$K(a) = \{d : a + td \in \tilde{S}(a) \text{ for some } t > 0\},$$

$$\text{and } K^\circ(a) = \{d^\star : \langle d, d^\star\rangle \leq 0 \text{ for all } d \in K(a)\}.$$

Then, $\tilde{S}(a)$ is convex, $K(a)$ is a convex cone and $K^\circ(a)$ is a closed convex cone. We assume, of course, that $\tilde{S}(a)$ is nonempty.

Given $C \subseteq \Re^n$ nonempty and convex, we denote by $\mathrm{int}(C), \mathrm{ri}(C)$ and $\mathrm{cl}(C)$ respectively the interior, the relative interior and the closure of C. Let $a \in C$, the normal cone to C at a is the set

$$N_C(a) = \{x^\star : \langle x^\star, x-a\rangle \leq 0 \text{ for every } x \in C\}.$$

Finally, we denote by $\|x\|$ the euclidean norm of x and by $B(h,r)$ the open ball of center h and radius r.

2. First properties of the Dini-derivatives of a quasiconvex function

It is not difficult to see that functions $f'_+(a,d)$ and $f'_-(a,d)$ are positively homogeneous in d. It is also clear that both $f'_-(a,h)$ and $f'_+(a,h)$ are nonnegative when $h \notin K(a)$ and nonpositive when $h \in K(a)$. Furthermore, $f'_+(a,d)$ is quasiconvex in d, and therefore can be expressed, roughly speaking, as the maximum of two convex positively homogeneous functions [2]. This convexity structure has been exploited by Crouzeix [2] and Komlosi [11],[12],[13] who use it to define generalized subdifferential for quasiconvex functions and by Gromicho [9] who uses it for algorithmic purposes. The lower Dini-derivatives do not enjoy such nice properties, but in view of the inequality $f'_-(a,d) \leq f'_+(a,d)$ appear better candidates for optimality conditions. Compare the necessary conditions for optimality of f at a on C: $f'_-(a,x-a) \geq 0$ for all $x \in C$ and $f'_+(a,x-a) \geq 0$ for all $x \in C$. In this paper, we shall treat both lower and upper Dini-derivatives at once. The following theorem gives two fundamental inequalities. The argument of its proof is partially adapted from [2].

THEOREM 1. *Assume that* $h \in \operatorname{ri}(K(a))$. *Set*

$$\rho(a,h) = \sup[r : B(h,r) \cap (K(a) - K(a)) \subseteq K(a)].$$

i) If $f'_-(a,h) > -\infty$, *then for all* d

$$f'_-(a,d) \geq \frac{\|d\|}{\rho(a,h)} f'_-(a,h).$$

ii) If $f'_+(a,h) > -\infty$, *then for all* d

$$f'_+(a,d) \geq \frac{\|d\|}{\rho(a,h)} f'_+(a,h).$$

Proof: If $d \notin K(a)$, then both $f'_-(a,d)$ and $f'_+(a,d)$ are nonnegative the inequalities hold since $f'_-(a,h)$ and $f'_+(a,h)$ are nonpositive.

Assume that $d \in K(a)$. Let

$$r \in (0,\rho(a,h)) \text{ and } k = h + \frac{r}{\|h-d\|}(h-d).$$

Then $k \in \text{ri}(K(a))$ and

$$h = k + \lambda(d - k), \text{ with } \lambda = \frac{r}{r + \|h - d\|}.$$

Also $k \in K(a)$ and therefore $\bar{t} > 0$ exists so that

$$c = a + \bar{t}k \in \tilde{S}(a).$$

For all $t \in (0, \bar{t})$, define

$$m(t) = a + th, \text{ and } n(t) = a + \xi(t)d,$$

where $\xi(t)$ is determined in such a way that $c, m(t)$ and $n(t)$ are on the same straight line. More precisely, we have

$$m(t) = c + \mu(t)(n(t) - c),$$

$$\text{where } \mu(t) = 1 - (1 - \lambda)\frac{t}{\bar{t}},$$

$$\text{and } \xi(t) = \frac{\lambda t}{\mu(t)}.$$

Since $\mu(t) \in (0, 1)$ and f is quasiconvex, we have

$$f(m(t)) \leq \max[f(c), f(n(t))],$$

$$\frac{f(m(t)) - f(a)}{t} \leq \max[\frac{f(c) - f(a)}{t}, \frac{f(n(t)) - f(a)}{t}].$$

The quantity $f(c) - f(a)$ is negative because $c \in \tilde{S}(a)$. Hence,

$$\frac{f(c) - f(a)}{t} \to -\infty \text{ when } t \to 0_+.$$

On the other hand,

$$\frac{\xi(t)}{t} \to \lambda \text{ when } t \to 0_+.$$

Because the concluding argument applies similarly to both Dini-derivatives, we develop it below only for the lower Dini-derivative. Since $f'_-(a, h) > -\infty$, it follows that for $t > 0$ small enough

$$\frac{f(m(t) - f(a)}{t} \leq \frac{f(n(t) - f(a)}{t},$$

and finally $f'_-(a,h) \leq \lambda f'_-(a,d)$. Since for all $r \in (0, \rho(a,h))$

$$f'_-(a,d) \geq (1 + \frac{\|h-d\|}{r}) f'_-(a,h),$$

it follows that

$$f'_-(a,d) \geq (1 + \frac{\|h-d\|}{\rho(a,h)}) f'_-(a,h).$$

On the other hand for all positive θ, $\theta h \in \mathrm{ri}(K(a))$ and

$$f'_-(a,\theta h) = \theta f'_-(a,h), \quad \rho(a,\theta h) = \theta \rho(a,h).$$

Hence, replacing h by θh, we get

$$f'_-(a,d) \geq (\theta + \frac{\|\theta h-d\|}{\rho(a,h)}) f'_-(a,h).$$

Then, passing to the limit when $\theta \to 0_+$, we finally obtain

$$f'_-(a,d) \geq \frac{\|d\|}{\rho(a,h)} f'_-(a,h). \quad \diamond$$

The following corollary corresponds to proposition 1 of [2]. Here, it straightforwardly derives from the theorem.

COROLLARY 1.
i) If $f'_-(a,h) = -\infty$ ($f'_+(a,h) = -\infty$) for some h, then $f'_-(a,d) = -\infty$ ($f'_+(a,d) = -\infty$) for all $d \in \mathrm{ri}(K(a))$.
ii) If $f'_-(a,h) = 0$ ($f'_+(a,h) = 0$) for some $h \in \mathrm{ri}(K(a))$, then $f'_-(a,d) = 0$ ($f'_+(a,d) = 0$) for all $d \in K(a)$.

Another quite direct corollary of the theorem gives a condition for pseudoconvexity.

COROLLARY 2. *Assume that f is upper semi-continuous on $\tilde{S}(a)$ and there is some d so that $f'_-(a,d) < 0$, then f is pseudoconvex at a.*

Proof: Assume that $f(x) < f(a)$. Since f is upper semi-continuous at $x \in \tilde{S}(a)$, then $h = x - a \in \mathrm{int}(K(a))$ and $f'_-(a, x-a) < 0$ by part i) of Theorem 1. $\diamond$

3. Quasiconvexity and Monotonicity

The connection between quasiconvexity and monotonicity is clear for functions of one real variable: on the one hand, non-decreasing and non-increasing functions are quasiconvex and on the other hand, the domain of a quasiconvex function can be partitioned in two intervals, the function being non-increasing on the first and non-decreasing on the second. For functions of several variables, the situation is as follows. Assume that f is quasiconvex, $f(a)$ is finite and λ exists so that $a \notin \text{cl}(S_\lambda(f))$ and $\text{int}(S_\lambda(f)) \neq \emptyset$. Then (Chabrillac-Crouzeix [1]) there are a neighborhood V_λ of a and a convex cone K_λ with nonempty interior such that f is monotone on V_λ according to K_λ, i.e.

$$f(x+tl) \leq f(x) \text{ for all } l \in K_\lambda, t > 0 \text{ and } x \text{ such that } x, x+tl \in V_\lambda.$$

As a consequence of the inequality just above, we straightforwardly obtain the monotonicity of the Dini-derivatives of f at a in the following sense

$$f'_+(a, d+l) \leq f'_+(a, d) \text{ and } f'_-(a, d+l) \leq f'_-(a, d),$$

for all $d \in \Re^n$ and $l \in K_\lambda$. This set K_λ is actually contained in the cone $K(a)$. We state now a more general result.

THEOREM 2. *(On the monotonicity of the Dini-derivatives)*
i) For all $d \in \Re^n$ *and* $l \in \text{int}(K(a))$ *one has*

$$f'_+(a, d+l) \leq f'_+(a, d) \text{ and } f'_-(a, d+l) \leq f'_-(a, d),$$

ii) The same results hold when $l \in \text{ri}(K(a))$ *and* $d \in \text{cl}(K(a))$.
iii) The same results hold when $d \in \text{ri}(K(a))$ *and* $l \in \text{cl}(K(a))$.
iv) For all $d \in \text{ri}(K(a))$ *and* $l \in \text{cl}(K(a)) \cap -\text{cl}(K(a))$ *one has*

$$f'_+(a, d+l) = f'_+(a, d) \text{ and } f'_-(a, d+l) = f'_-(a, d).$$

Proof: Without loss of generality, we assume that $f(a) = 0$.
i) If $f'_+(a, d) = +\infty$ ($f'_-(a, d) = +\infty$), the property is trivial. If $f'_+(a, d) = -\infty$ ($f'_-(a, d) = -\infty$), then $d \in K(a)$. Hence

$$d + l \in \text{ri}(K(a)) = \text{int}(K(a))$$

and by Corollary 1,

$$f'_+(a, d+l) = -\infty \quad (f'_-(a, d+l) = -\infty).$$

We are left with the cases $-\infty < f'_+(a, d) < \infty$ or $-\infty < f'_-(a, d) < \infty$. Since $K(a)$ is a convex cone and $l \in \text{int}(K(a))$, $\bar{\alpha} \geq 1$ exists such that

$$d + \alpha l \in \text{int}(K(a)) \text{ for all } \alpha \geq \bar{\alpha}.$$

Let $\alpha > \bar{\alpha}$. Then there exists $\theta > 0$ such that $f(a + \theta d + \alpha\theta l) < 0$.

Firstly, we work on the lower Dini-derivative. Let λ and μ such that $-\infty < \lambda < f'_-(a,d) < \mu < \infty$. By definition of $f'_-(a,d)$, there exists $\bar{t} > 0$ such that for all $t \in (0,\bar{t})$, there is $\xi(t) \in (0,t)$ such that

$$f(a + \theta d + \alpha\theta l) < \lambda\xi(t) < f(a + \xi(t)d) < \mu\xi(t).$$

Notice that

$$a + \frac{\alpha\theta\xi(t)}{\alpha\theta - \theta + \xi(t)}(d + l) =$$

$$\frac{\xi(t)}{\alpha\theta - \theta + \xi(t)}(a + \theta d + \alpha\theta l) + \frac{\alpha\theta - \theta}{\alpha\theta - \theta + \xi(t)}(a + \xi(t)d).$$

Hence by quasiconvexity of f and since $\alpha > 1$,

$$f(a + \frac{\alpha\theta\xi(t)}{\alpha\theta - \theta + \xi(t)}(d + l) \geq f(a + \xi(t)d).$$

Dividing by $\xi(t)$ and passing to the limit as $t \longrightarrow 0$, we obtain

$$f'_-(a, \frac{\alpha}{\alpha - 1}(d + l)) = \frac{\alpha}{\alpha - 1}f'_-(a, d + l) \leq \mu.$$

This result holds for all $\alpha > \bar{\alpha}$ and all $\mu > f'_-(a,d)$, hence

$$f'_-(a, d + l) \leq f'_-(a, d).$$

Next, let us consider the upper Dini-derivative. Since $f(a+\theta d+\alpha\theta l) < 0 = f(a)$, there exists $\bar{t} > 0$ so that for all $t \in (0,\bar{t})$

$$f(a + \theta d + \alpha\theta l) \leq f(a + td).$$

Since

$$a + \frac{\alpha\theta t}{\alpha\theta - \theta + t}(d + l) = \frac{t}{\alpha\theta - \theta + t}(a + \theta d + \alpha\theta l) + \frac{\alpha\theta - \theta}{\alpha\theta - \theta + t}(a + td),$$

and f is quasiconvex, one has

$$f(a + \frac{\alpha\theta t}{\alpha\theta - \theta + t}(d + l)) - f(a) \leq f(a + td) - f(a).$$

Then, divide by t and pass to the limit when $t \longrightarrow 0$, it follows that

$$f'_+(a, \frac{\alpha}{\alpha - 1}(d + l)) \leq f'_+(a, d),$$

and finally

$$f'_+(a, d + l) \leq f'_+(a, d).$$

ii) Since $d \in \mathrm{cl}(K(a))$ and $l \in \mathrm{ri}(K(a))$, $\bar{\alpha} \geq 1$ exists such that

$$d + \alpha l \in \mathrm{int}(K(a)) \text{ for all } \alpha \geq \bar{\alpha}.$$

Proceed as in i).
iii) If $l \in \mathrm{cl}(K(a))$ and $d \in \mathrm{ri}(K(a))$, then $d + \alpha l \in \mathrm{ri}(K(a))$ for all $\alpha \geq 1$. Take $\bar{\alpha} = 1$ and proceed as in i).
iv) This is a direct consequence of iii). Indeed $d + l \in \mathrm{ri}(K(a))$ and $d = (d + l) - l$. ◇

When the closure of $K(a)$ is a half space, then Theorem 2 has strong implications as seen below.

THEOREM 3. *Assume that* $\mathrm{cl}(K(a))$ *is a half space and let* $x^\star \in K^\circ(a)$ *such that* $\|x^\star\| = 1$. *Then there exist* $\alpha_-, \alpha_+, \beta_-$ *and* β_+ *such that*
i) $0 \leq \alpha_+ \leq \alpha_- \leq +\infty, 0 \leq \beta_- \leq \beta_+ \leq +\infty$.
ii) if $\langle d, x^\star \rangle < 0$ *then*

$$f'_-(a,d) = \alpha_- \langle d, x^\star \rangle \text{ and } f'_+(a,d) = \alpha_+ \langle d, x^\star \rangle,$$

iii) if $\langle d, x^\star \rangle > 0$ *then*

$$f'_-(a,d) = \beta_- \langle d, x^\star \rangle \text{ and } f'_+(a,d) = \beta_+ \langle d, x^\star \rangle.$$

Moreover, if α_- *and* β_- *are finite then*

$$f'_-(a,d) = 0 \text{ for all } d \text{ such that } \langle d, x^\star \rangle = 0. \tag{1}$$

Finally, if α_+ *and* β_+ *are finite then*

$$f'_+(a,d) = 0 \text{ for all } d \text{ such that } \langle d, x^\star \rangle = 0. \tag{2}$$

Proof: Notice that $\mathrm{int}(K(a)) = \{d : \langle d, x^\star \rangle < 0\}$. Set $y^\star = d - \langle d, x^\star \rangle x^\star$, then $\langle y^\star, x^\star \rangle = 0$ and for all $\epsilon > 0$ both $y^\star - \epsilon x^\star$ and $-y^\star - \epsilon x^\star$ lie in $\mathrm{int}(K(a))$. On the other hand

$$d = (\langle d, x^\star \rangle + \epsilon) x^\star - \epsilon x^\star + y^\star,$$

and

$$(\langle d, x^\star \rangle - \epsilon) x^\star = d - \epsilon x^\star - y^\star.$$

Then, by Theorem 2

$$f'_-(a, (\langle d, x^\star \rangle - \epsilon) x^\star) \leq f'_-(a,d) \leq f'_-(a, (\langle d, x^\star \rangle + \epsilon) x^\star)$$

Assume that $\langle d, x^\star \rangle > 0$. Pass to the limit when $\epsilon \to 0$, then

$$\langle d, x^\star \rangle f'_-(a, x^\star) \leq f'_-(a,d) \leq \langle d, x^\star \rangle f'_-(a, x^\star).$$

Similarly, assume that $\langle d, x^\star \rangle < 0$. Then,

$$-\langle d, x^\star \rangle f'_-(a, -x^\star) \leq f'_-(a, d) \leq -\langle d, x^\star \rangle f'_-(a, -x^\star).$$

Proceed in the same way for the upper Dini-derivative. Then i) to iii) hold with $\alpha_- = -f'_-(a, -x^\star)$, $\alpha_+ = -f'_+(a, -x^\star)$, $\beta_- = f'_-(a, x^\star)$, and $\beta_+ = f'_+(a, x^\star)$.

Finally, assume that $\langle d, x^\star \rangle = 0$. Then for all $\epsilon > 0$

$$\epsilon f'_-(a, -x^\star) = f'_-(a, -\epsilon x^\star) \leq f'_-(a, d) \leq f'_-(a, \epsilon x^\star) = \epsilon f'_-(a, x^\star)$$

Hence $f'_-(a, d) = 0$ when both $f'_-(a, -x^\star)$ and $f'_-(a, x^\star)$ are finite. Proceed similarly for $f'_+(a, d)$. $\diamond$

Remark: When (α_- and β_-) or (α_+ and β_+) are not finite then (i) or (ii) are false as shown below:

Example: Take $n = 2$, $a = (0, 0)$ and set

$$f(x_1, x_2) = \begin{cases} -\pi/2 & \text{if } x_2 < 0, \\ \pi/2 & \text{if } x_2 > 0, \\ tan^{-1}(x_1) & \text{if } x_2 = 0. \end{cases}$$

Then,

$$K(a) = \{(d_1, d_2) : d_2 < 0 \text{ or } (d_2 = 0 \text{ and } d_1 < 0)\}.$$

Here $\mathrm{cl}(K(a))$ is a halfspace and $x^\star = (0, 1)$. Take $d = (1, 0)$, then $\langle d, x^\star \rangle = 0$ but $f'_-(a, d) = f'_+(a, d) = 1$.

As a consequence ot the theorem, we recover a necessary and sufficient condition for the differentiability of quasiconvex functions already stated in Crouzeix [2] but with a different proof.

THEOREM 4. *Assume that* $d \in \mathrm{int}(K(a))$. *Then* f *is differentiable at* a *if and only if* $\mathrm{cl}(K(a))$ *is a half space and the function* $\theta(t) = f(a+td)$ *is differentiable at* $t = 0$.

Proof: If f is differentiable at a, then $\mathrm{cl}(K(a))$ is a half space. The differentiability of θ at 0 is equivalent to the assertion that

$$-\infty < f'_-(a, d) = f'_+(a, d) = -f'_-(a, -d) = -f'_+(a, -d) < +\infty.$$

According to Theorem 3, this assertion is equivalent to the statement that the directional derivative $f'(a, d)$ is linear in d, i.e., f is Gâteaux-differentiable at a. But Gâteaux-differentiability and Fréchet-differentiability of quasiconvex functions coincide (Crouzeix [3]). $\diamond$

4. Continuity of the Dini-derivatives in the direction

The convexity structure of $f'_+(a,.)$ induces on this function continuity properties . Working on the lower Dini-derivative is more difficult, Proposition 8 of [2] gives some information, but its proof is rather complex. Here we derive some results using the monotonicity properties of the Dini-derivatives. This approach is much simpler and has the advantage of being the same for both lower and upper Dini-derivatives.

THEOREM 5. *Assume that* $\text{int}(K(a))$ *is nonempty. Then,*
i) $f'_-(a,.)$ *and* $f'_+(a,.)$ *are continuous on* $\text{int}(K(a))$ *and* $-\text{int}(K(a))$.
ii) If $f'_-(a,d) = 0$ ($f'_+(a,d) = 0$) *for some* $d \in -\text{int}(K(a))$, *then* $f'_-(a,h) = 0$ ($f'_+(a,h) = 0$) *for all* $h \notin K(a)$.

Proof: First, we show that $f'_-(a,.)$ is continuous at any $h \in \text{int}(K(a))$. For $\epsilon \in (0,1)$, let us consider the neighborhood of h

$$V_\epsilon = ((1-\epsilon)h + \text{int}(K(a))) \cap ((1+\epsilon)h - \text{int}(K(a))).$$

It results from Theorem 2 that for any $d \in V_\epsilon$

$$f'_-(a,(1+\epsilon)h) \leq f'_-(a,d) \leq f'_-(a,(1-\epsilon)h).$$

Hence,

$$(1+\epsilon)f'_-(a,h) \leq f'_-(a,d) \leq (1-\epsilon)f'_-(a,h),$$

from what the continuity of $f'_-(a,.)$ at h follows. The proof of the continuity of $f'_-(a,.)$ at $h \in -\text{int}(K(a))$ is obtained in the same manner by considering

$$V_\epsilon = ((1-\epsilon)h - \text{int}(K(a))) \cap ((1+\epsilon)h + \text{int}(K(a))).$$

Here the inequalities become

$$(1-\epsilon)f'_-(a,h) \leq f'_-(a,d) \leq (1+\epsilon)f'_-(a,h).$$

Assume now that $f'_-(a,d) = 0$ for some $d \in -\text{int}(K(a))$. For fixed $h \notin K(a)$, the quantity $f'_-(a,h)$ is nonnegative. On the other hand, there exists some $\alpha > 0$ such that $h \in \alpha d + \text{int}(K(a))$. Hence

$$0 \leq f'_-(a,h) \leq f'_-(a,\alpha d) = 0.$$

The proofs for $f'_+(a,.)$ are exactly the same. ◇

These results do not recover exactly those given in [2] where, for instance, the continuity of $f'_-(a,.)$ was proved at any h such that

$f'_-(a,h) > 0$ but only under the condition that there is some d such that $0 < f'_+(a,d) < \infty$.

What happens on the boundary of $K(a)$? Examples in [2] show that $f'_-(a,.)$ and $f'_+(a,.)$ are not necessarily continuous at points of this boundary. Nevertheless, we have the following property.

THEOREM 6. *Assume that $K(a)$ is nonempty. Then for all $h \in Bd(K(a))$,*

$$f'_-(a,h) \geq \liminf_{d \to h,\ d \in \mathrm{ri}(K(a))} f'_-(a,d),$$

and

$$f'_+(a,h) \geq \liminf_{d \to h,\ d \in \mathrm{ri}(K(a))} f'_+(a,d).$$

Proof: Firstly, assume that $\mathrm{int}(K(a))$ is nonempty. Take $k \in \mathrm{int}(K(a))$ and set

$$d_i = h + \frac{1}{i}h, \quad i = 1, 2, \cdots .$$

By Theorem 2, $f'_-(a,d_i) \leq f'_-(a,h)$ and $f'_+(a,d_i) \leq f'_+(a,h)$ for all i. On the other hand $d_i \in \mathrm{int}(K(a))$. The result follows.

When $\mathrm{int}(K(a)) = \emptyset$, it suffices to consider the restriction of f to the affine space $a + K(a) - K(a)$. ◇

5. A one dimensional condition for Lipschitz continuity

If there exists some $h \in \mathrm{int}(K(a))$ such that the function $\theta(t) = f(a+th)$ is continuous at $t = 0$, then f is continuous at a. See, for instance, Crouzeix [4]. Now we prove that in order to have a Lipschitz condition on quasiconvex functions it is enough to have a one dimensional condition.

THEOREM 7. *Assume that f is quasiconvex on an open convex set C. Asume in addition that there exist a convex cone K, $h \in \mathrm{int}(K)$ and a constant L such that*
a) $K \subseteq K(x)$ for all $x \in C$,
b) $|f(x+th) - f(x)| \leq L|t|$ for all t and x such that both x and $x+th$ lie in C.
Then there exists a constant $\hat{L}$ such that

$$|f(x) - f(y)| \leq \hat{L}\|x-y\| \text{ for all } x, y \in C.$$

Proof: Without loss of generality, we assume that $f(x) > f(y)$. For all $t \in (0,1)$ define

$$\theta(t) = f(ty + (1-t)x) - tf(y) - (1-t)f(x).$$

It follows from the assumptions that f is continuous at any $a \in C$. Hence θ is continuous on $[0,1]$. We choose $\bar{t} \in [0,1]$ so that

$$\theta(\bar{t}) = max(\theta(t) : t \in [0,1]).$$

Then

$$f'_-(\bar{t}y + (1-\bar{t})x, y-x) + f(x) - f(y) = \theta'_-(\bar{t}, 1) \leq 0.$$

Next take $\rho = max(r : B(h,r) \subseteq K)$. By Theorem 1 and since $\rho \leq \rho(\bar{t}y + (1-\bar{t})x, y-x)$

$$f'_-(\bar{t}y+(1-\bar{t})x, y-x) \geq \frac{\|y-x\|}{\rho} f'_\cdot(\bar{t}y+(1-\bar{t})x, y-x) \geq -\frac{L\|y-x\|}{\rho}.$$

The result follows with $\hat{L} = \frac{L}{\rho}$. ◇

Remark: The existence of such a cone K in a neighborhood of a point a is, as already said, ensured as soon as $f(a)$ is finite and λ exists so that $a \notin \mathrm{cl}(S_\lambda(f))$ and $\mathrm{int}(S_\lambda(f)) \neq \emptyset$ [1]. In economics, utility functions are usually assumed to be monotone with respect to the order induced by the nonegative orthant, set K equals this orthant in the theorem.

6. Towards a new subdifferential for quasiconvex and pseudoconvex functions

A function f is said to be *lower subdifferentiable* at a if there exists $x^\star$ such that

$$f(x) \geq f(a) + \langle x^\star, x-a\rangle \text{ whenever } f(x) < f(a)$$

Then $x^\star$ is said to be an element of $\partial_- f(a)$, the *lower subdifferential* of f at a. This concept introduced by Plastria [16] and studied by some other authors (see for instance [14]) has been, in particular, used for solving some quasiconvex problems by cutting plane methods.

Because this lower subdifferential is not local and because the optimality conditions for quasiconvex and pseudoconvex programs are essentially local, Martinez-Legaz [15] has introduced the concept of *weak lower subdifferentiability* by relaxing Plastria's definition.

A function f is said to be *weak lower subdifferentiable* at a if there exist $x^\star$ and a strictly positive and positively homogeneous function of degree zero ϵ such that

$$f(x) \geq f(a) + \langle x^\star, x-a\rangle$$

whenever

$$f(x) < f(a) \quad \text{and} \quad \langle x^\star, x - a\rangle + \epsilon(x - a) > 0\,.$$

Then $x^\star$ is said to be an element of $\partial^w f(a)$, the *weak lower subdifferential* of f at a. Weak lower subdifferentiable functions are quasiconvex and conversely if f is strictly quasiconvex and locally Lipschitz at a then f is weak lower subdifferentiable at this point.

For any function f it is clear that $x^\star \in \partial^w f(a)$ if and only if for any unit vector d there exists $\xi(d) > 0$ such that

$$f(a + td) \geq f(a) + t\langle x^\star, d\rangle$$

whenever

$$f(a + td) < f(a) \quad \text{and} \quad \xi(d) + t\langle x^\star, d\rangle > 0\,.$$

In the case where f is quasiconvex, we can restrict our attention to directions $d \in K(a)$. Thus $x^\star \in \partial^w f(a)$ if and only if for any unit vector $d \in K(a)$ one has $\langle x^\star, d\rangle < 0$ and the existence of some $\hat{\xi}(d) > 0$ such that

$$\frac{f(a + td) - f(a)}{t} \geq \langle x^\star, d\rangle \text{ for all } t \in (0, \hat{\xi}(d)).$$

From this we deduce that

$$x^\star \in \partial^w f(a) \text{ and } d \in K(a) \Rightarrow \langle x^\star, d\rangle < 0 \text{ and } f'_-(a, d) \geq \langle x^\star, d\rangle$$

and

$$\langle x^\star, d\rangle < 0 \text{ and } f'_-(a, d) > \langle x^\star, d\rangle \text{ for all } d \in K(a) \Rightarrow x^\star \in \partial^w f(a)\,.$$

This suggests to replace $\partial^w f(a)$ by

$$\partial^{\overline{w}} f(a) = \{x^\star : f'_-(a, d) \geq \langle x^\star, d\rangle \text{ for all } d \in K(a)\}\,,$$

whose expression is directly related to the lower Dini-derivatives. Clearly, $\partial^w f(a) \subseteq \partial^{\overline{w}} f(a)$. This new subdifferential has the advantage on the former one of causing no trouble in formulae like c) of corollary 2.16 or (3.7) or (3.8) of observation 3.5 in [15]. As $\partial^w f(a)$, $\partial^{\overline{w}} f(a)$ enjoys nice properties. In particular, it is closed and convex ($\partial^w f(a)$ is only convex) and

$$\partial^{\overline{w}} f(a) = \bigcup_{\lambda \geq 1} \lambda \partial^{\overline{w}} f(a)\,.$$

As said in the introduction, optimality conditions in quasiconvex programming involve the normal cone $K^\circ(a)$ at a to $\tilde{S}(a)$. Therefore, to be useful, a subdifferential must allow to recover this normal cone $K^\circ(a)$. This is the case for $\partial^{\overline{w}} f$ as shown below.

THEOREM 8. *If there exists some* $h \in \text{ri}(K(a))$ *such that* $-\infty < f'_-(a,h) \leq 0$, *then* $\partial^{\bar{w}} f(a)$ *is nonempty. Moreover,* $K^\circ(a)$ *is the closure of the cone generated by* $\partial^{\bar{w}} f(a)$.

Proof: If $f'_-(a,h) = 0$, then $f'_-(a,d) = 0$ for all $d \in K(a)$ and $\partial^{\bar{w}} f(a)$ is exactly $K^\circ(a)$.
Assume now that $-\infty < f'_-(a,h) < 0$. Firstly, we shall prove that for all $x^\star \in \text{ri}(K^\circ(a))$ there is some $\alpha > 0$ such that $\alpha x^\star \in \partial^{\bar{w}} f(a)$. For this consider L the lineality space of $K(a)$, i.e. , $L = K(a) \cap -K(a)$. We consider three cases:

i) $L = \{0\}$. Then $\text{int}(K^\circ(a)) = \text{ri}(K^\circ(a)) \neq \emptyset$ and $\langle d, x^\star \rangle < 0$ for all $d \in \text{cl}(K(a))$. Thus there exists some $\beta > 0$ such that

$$\beta \langle d, x^\star \rangle \leq -1 \text{ for all } d \in K(a) \text{ such that } \|d\| = 1.$$

Now, take

$$\alpha = -\frac{\beta f'_-(a,h)}{\rho(a,h)},$$

and apply Theorem 1. Then

$$f'_-(a,d) \geq \alpha \langle d, x^\star \rangle \text{ for all } d \in K(a).$$

ii) $L \neq \{0\}$ and $\text{int}(K(a)) \neq \emptyset$. In view of Theorem 6, it suffices to prove the existence of some $\alpha > 0$ such that

$$f'_-(a,d) \geq \alpha \langle d, x^\star \rangle \text{ for all } d \in \text{int}(K(a)).$$

But $\langle l, x^\star \rangle = 0$ for all $l \in L$. Hence by Theorem 2

$$f'_-(a, d+l) = f'_-(a,d) \text{ for all } d \in \text{int}(K(a)), l \in L.$$

Then, it is enough to prove the existence of some $\alpha > 0$ such that

$$f'_-(a,d) \geq \alpha \langle d, x^\star \rangle \text{ for all } d \in \text{int}(K(a)) \cap L^\perp.$$

For this consider the restriction g of f to the affine space $a + L^\perp$. Then g is quasiconvex and the lineality space of the set $\{d : g(a+td) < g(a) \text{ for some } t > 0\}$ is reduced to $\{0\}$. Apply i) to g and go back to f.

iii) $L \neq \{0\}$ and $\text{int}(K(a)) = \emptyset$. Consider the restriction g of f to the affine space $a + K(a) - K(a)$. Apply ii) to g and go back to f.

In the three cases, we have proved that $\text{ri}(K^\circ(a))$ is contained in the cone generated by $\partial^{\bar{w}} f(a)$. Hence $K^\circ(a)$ is contained in the closure of this cone. On the other hand, $f'_-(a,h) \leq 0$ for all $h \in K(a)$. Hence $\partial^{\bar{w}} f(a) \subseteq K^\circ(a)$. ◇

We already said on the inclusion between $\partial^w f(a)$ and $\partial^{\bar{w}} f(a)$. In view of the inequality between the lower and the upper Dini-derivatives we have also $\partial^{\bar{w}} f(a) \subseteq \partial^- f(a)$ where $\partial^- f(a)$ is the subdifferential introduced in [2] from the upper Dini-derivative,

$$\partial^- f(a) = \{x^\star : f'_+(a,d) \geq \langle x^\star, d\rangle \text{ for all } d \in K(a)\}.$$

In view of Corollary 1, we have the following condition.

COROLLARY 3. *The subdifferential* $\partial^{\bar{w}} f(a)$ *is nonempty if and only if* $-\infty < f'_-(a,h)$ *for all* h. *In this case,* $K^\circ(a)$ *is the closure of the cone generated by* $\partial^{\bar{w}} f(a)$.

In general, $K^\circ(a)$ is not generated by $\partial^{\bar{w}} f(a)$ as seen from the example below.
Example: Take $n = 2,\ a = (0,0)$ and set

$$f(x_1, x_2) = \begin{cases} -\sqrt{x_1 x_2} & \text{if } x_1 \leq 0, x_2 \leq 0, \\ \max(x_1, x_2) & \text{otherwise.} \end{cases}$$

Then $\partial^{\bar{w}} f(a) = \{(d_1, d_2) : d_1 > 0, d_2 > 0, d_1 d_2 \geq 1\}$ while $K^\circ(a)$ is the nonnegative orthant.

If f is convex and f is finite in a neighborhood of a, then $\partial^{\bar{w}} f(a) = \bigcup_{\lambda \geq 1} \lambda \partial f(a)$ where by $\partial f(a)$ we denote the Fenchel-Moreau subdifferential of f at a. It follows that $K^\circ(a)$ is exactly the cone generated by $\partial^{\bar{w}} f(a)$. This is also the case if $f'_-(a,.)$ is defined on the whole space $\Re^n$ and is convex. Such functions are called tangentially convex at a in [15]. More generally, we shall say that f is *regularly subdifferentiable* at a if $K^\circ(a)$ is the cone generated by $\partial^{\bar{w}} f(a)$. We can now establish a calculus rule for subdifferentials.

PROPOSITION 1. *Assume that* $g = \max_{i=1,\cdots,p} g_i$, g *is finite in* a *and* $I(a) = \{i : g(a) = g_i(a)\}$. *Then*
i) $co(\bigcup_{i \in I(a)} \partial^{\bar{w}} g_i(a) \subseteq \partial^{\bar{w}} g(a)$.
ii) If functions $g_i s$ *for* $i \notin I(a)$ *are upper semicontinuous at a, functions* $g_i s$ *for* $i \in I(a)s$ *are quasiconvex and regularly subdifferentiable at* a *and* $\text{int}(\{x : g(x) < g(a)\}) \neq \emptyset$, *then the equality holds. Moreover* g *is regularly subdifferentiable at a.*

Proof: Clearly, $g'_-(a,d) = \max_{i \in I(a)} (g_i)'_-(a,d)$. Hence i) holds. On the other hand, $T(a) = \bigcap_{i \in I(a)} T_i(a)$ and $\text{int}(T(a)) \neq \emptyset$ where $T(a)$ and $T_i(a)$ denote the cones $T(a) = \{d : g(a + td) < g(a) \text{ for some } t > 0\}$, and $T_i(a) = \{d : g_i(a + td) < g_i(a) \text{ for some } t > 0\}$. It follows that $T^\circ(a) = \sum_{i \in I(a)} T_i^\circ(a)$ and the equality follows. ◇

7. Optimality conditions

We go back to the optimization problem $\inf[f(x) : x \in C]$ where f is quasiconvex and C is convex. We assume that a belongs to the boundary of C and $\tilde{S}(a)$ is nonempty. A necessary and sufficient condition for optimality at a is that $\tilde{S}(a) \cap C = \emptyset$. Hence another necessary condition is the existence of some $x^\star \neq 0, x^\star \in K^\circ(a) \cap -N_C(a)$. If f is regularly subdifferentiable at a, the condition becomes

$$0 \in \partial^{\overline{w}} f(a) + N_C(a).$$

If, in addition f is pseudoconvex at a, then the condition is also sufficient. Indeed, if $x^\star \in \partial^{\overline{w}} f(a)$, then

$$f(a+td) < f(a) \Rightarrow \langle d, x^\star \rangle < 0 \Rightarrow x^\star \notin -N_C(a).$$

The particular case where $C = \{x : g_i(x) \leq 0\}$ can be treated by writing the expression of the normal cone of C in terms of the normal cone to the set $\{x : g(x) = \max_{i=1,\cdots,p} g_i(x) \leq 0\}$ at a, then in terms of the subdifferentials of g_is at a when Proposition 1 applies.

References

1. Y. Chabrillac and J. P. Crouzeix, Continuity and differentiability properties of monotone real functions of several variables, *Mathematical Programming Study*, 30, 1987, pp. 1–16.
2. J. P. Crouzeix, Some differentiability properties of quasiconvex functions on $\Re^n$, *Optimization and Optimal Control*, Springer-Verlag, Lecture Notes in Control and Information Sciences, Vol. 30, 1981, pp. 9–20.
3. J. P. Crouzeix, About differentiability properties of quasiconvex functions, *Journal of Optimization Theory and Applications*, 36, 1982, pp. 367–385.
4. J. P. Crouzeix, Continuity and differentiability properties of quasiconvex functions, *Generalized Concavity in Optimization and Economics*, Schaible and Ziemba editors, Academic Press, New-York, 1981, pp. 109–130.
5. W. E. Diewert, Alternative characterization of six kinds of quasiconvex functions, *Generalized Concavity in Optimization and Economics*, Schaible and Ziemba editors, Academic Press, New-York, 1981, pp. 51–93 .
6. G. Giorgi and S. Komlosi, Dini-derivatives in optimization-Part 1, *Rivista A.M.A.S.E.S.* 15 $n^0 1$, 1992, pp. 3–30.
7. G. Giorgi and S. Komlosi, Dini-derivatives in optimization-Part 2, *Rivista A.M.A.S.E.S.* 15 $n^0 2$, 1992, pp. 3–24.
8. G. Giorgi and S. Komlosi, Dini-derivatives in optimization-Part 3, *Rivista A.M.A.S.E.S.* 18 $n^0 3$, 1995, pp. 47–63.
9. J. A. Gromicho, *Quasiconvex Optimization and Location Theory.*, Dissertation Thesis, Timbergen Institute Research Series, nr.90 , 1995.
10. S. Komlosi, Some properties of nondifferentiable pseudoconvex functions, *Mathematical Programming* 26, 1983, pp. 232–237.
11. S. Komlosi, On a possible generalization of Pshenichnyi's differentiability, *Optimization*, 21, 1990, pp. 189–201.

12. S. Komlosi, On generalized upper quasidifferentiability, *Nonsmooth optimization: methods and applications*, F. Gianessi editor, Gordon and Breach, London, 1992, pp. 189–201.
13. S. Komlosi, Quasiconvex first-order approximations and generalized Kuhn-Tucker conditions, *European Journal on Operations Research*, 65, 1993, pp. 327–335.
14. J. E. Martinez-Legaz, On lower subdifferentiable functions, *Trends in Mathematical functions*, Hoffman, Hiriart-Urruty, Lemaréchal and Zowe editors, Birkhäuser-Verlag, Basel, 1988, pp. 197–232.
15. J. E. Martinez-Legaz, Weak lower subdifferential and applications, *Optimization*, 21, 1990, pp. 321–341.
16. F. Plastria, Lower subdifferentiable functions and their minimization by cutting planes, *Journal of Optimization Theory and Applications*, 46, 1985, pp. 37–53

Necessary conditions for two-function minimax inequalities

Ferenc Forgó
Budapest University of Economic Sciences,
Department of Operations Research,
H-1828 Budapest, Hungary.
E-mail: fforgo@ursus.bke.hu

István Joó
Eötvös Loránd University,
H-1088 Budapest, Hungary

Abstract. Necessary conditions are given for the two-function minimax inequality to hold for a large family of subsets of the sets involved. The functions should belong to a class consisting of certain generalizations of convex (concave)-like functions. Most of the analysis takes place in topological vector spaces but the results are also extended to pseudoconvex spaces.

1. Introduction

Given two sets, X and Y and two functions $f, g : X \times Y \longrightarrow \mathbf{R}$, a two-function minimax inequality is

$$\max_X \min_Y g \geq \min_Y \max_X f ,$$

where $f \leq g$ is usually assumed. Being a direct generalization of classical minimax equalities ($f = g$) and because of its close relation to variational inequalities, two-function minimax inequalities have drawn considerable attention lately and a host of existence theorems have been proved under different conditions imposed on X, Y, f, g. For a good review see Simons [9], [10]. Much less attention has been given to necessary conditions. For the case $f = g$ Joó [5], Kindler [6] and Simons [8] addressed the problem but we were unable to find results concerning the general case.

In this paper, we give necessary conditions of the kind introduced in mathematical programming by Martos [7] and applied to the classical minimax case by Forgó and Joó [3]. The basic idea is to characterize the functions for which maxmin $\geq$ minmax holds for a (possibly large) family of subsets of X and Y. Our results are also extended to pseudoconvex spaces introduced and studied by Joó [5]. A particular generalization of convex (concave)-like functions and "to within ε" conditions have a key role in our analysis.

F. Giannessi et al. (eds.), New Trends in Mathematical Programming, 59–64.

2. The main results

DEFINITION 1. $\varphi : \mathbf{R}^2 \longrightarrow \mathbf{R}$ is called a *submaximum function*, if for any $u, v \in \mathbf{R}$, $\varphi(u,v) \leq u \vee v$.

Also, $\psi : \mathbf{R}^2 \longrightarrow \mathbf{R}$ is called a *superminimum function*, if for any $u, v \in \mathbf{R}$, $\psi(u,v) \geq u \wedge v$.

Let X and Y be nonempty sets and $f, g : X \times Y \longrightarrow \mathbf{R}$.

DEFINITION 2. Given a submaximum function φ, f is said to be *φ-concave-like* with respect to g, if for all $\lambda > 0$ and $x_1, x_2 \in X$, there exists $x_3 \in X$ such that

$$y \in Y \Longrightarrow g(x_3, y) \geq \varphi[f(x_1, y), f(x_2, y)] - \lambda\,.$$

Symmetrically, given a superminimum function ψ, g is said to be *ψ-convex-like* with respect to f, if for all $\lambda > 0$ and $y_1, y_2 \in Y$, there exists $y_3 \in Y$ such that

$$x \in X \Longrightarrow f(x, y_3) \leq \psi[g(x, y_1), g(x, y_2)] + \lambda\,.$$

Note that x_3 depends on $\lambda, \varphi, x_1, x_2$, while y_3 depends on λ, ψ, y_1, y_2.

Given a set D and a function $w : D \longrightarrow \mathbf{R}$, we will denote

$$\arg\max_{d \in D} w(d) := \{d' \in D : w(d') \geq w(d'') \text{ for all } d'' \in D\}\,.$$

In the following, it will be assumed that X and Y are convex, compact sets in locally convex Hausdorff topological vector spaces, f and g are continuous and φ is a continuous submaximum function. For any fixed $x_1, x_2 \in X$ and $\lambda > 0$, define $Z : X \longrightarrow P(Y)$ and for any $\widehat{x} \in X$, $h_{\widehat{x}} : X \longrightarrow \mathbf{R}$

$$\begin{aligned} Z(x) &:= \{y \in Y : g(x,y) \leq \varphi[f(x_1, y), f(x_2, y)] - \lambda\} \\ h_{\widehat{x}}(x) &:= \min_{y \in Z(\widehat{x})} g(x,y)\,. \end{aligned}$$

THEOREM 1. *If*

(*i*) *for any closed set $K \subset Y$,*

$$\max_{x \in X} \min_{y \in K} g(x,y) \geq \min_{y \in K} \max_{x \in X} f(x,y)\,;$$

(*ii*) *for any closed set $K \subset Y$, the set of maximizers*

$$\arg\max_{x \in X} \{\min_{y \in K} g(x,y)\}$$

is convex, then f is φ-concave-like with respect to g.

Proof. Suppose, on the contrary, that f is not φ-concave-like with respect to g. Then there exist $\lambda > 0$ and $x_1, x_2 \in X$ such that for any $x \in X$, the set $Z(x)$ is not empty. By the continuity of f, g and φ, $Z(\widehat{x})$ is compact for any $\widehat{x} \in X$ and thus $h_{\widehat{x}}$ is continuous. Therefore, for any $\widehat{x} \in X$, the set-valued mapping $\widehat{x} \longrightarrow S(\widehat{x})$ where

$$S(\widehat{x}) := \{x \in X : x \in \arg\max_{x \in X} h_{\widehat{x}}(x)\}$$

is upper semicontinuous. By assumption (ii), the map is also convex, thus by Kakutani's fixed-point theorem, there exists $x^* \in X$ such that $x^* \in S(x^*)$, i.e.

$$h_{x^*}(x^*) = \min_{y \in Z(x^*)} g(x^*, y) = \max_{x \in X} \min_{y \in Z(x^*)} g(x, y).$$

Then, for any $y \in Z(x^*)$,

$$\max_{x \in X} \min_{y \in Z(x^*)} g(x, y) \leq \varphi[f(x_1, y), f(x_2, y)] - \lambda.$$

Since φ is a submaximum function,

$$\varphi[f(x_1, y), f(x_2, y)] \leq f(x_1, y) \vee f(x_2, y)$$

implying

$$\max_{x \in X} \min_{y \in Z(x^*)} g(x, y) < \min_{y \in Z(x^*)} \max_{x \in X} f(x, y),$$

which is a contradiction to assumption (i). □

Another necessary condition can be derived, if $Z(x)$ is convex for all $x \in X$. Such is the case e.g., when $f(x, .), g(x, .)$ are linear on Y and φ is also linear on $\mathbf{R}^2$.

THEOREM 2. *If*

(i) $f(x, .)$ and $g(x, .)$ are linear on Y;

(ii) φ is linear on $\mathbf{R}^2$;

(iii) for any closed, convex set $C \subset Y$,

$$\max_{x \in X} \min_{y \in C} g(x, y) \geq \min_{y \in C} \max_{x \in X} f(x, y);$$

(iv) for any closed, convex set $C \subset Y$, the set of maximizers

$$\arg\max_{x \in X} \left\{ \min_{y \in C} g(x, y) \right\}$$

is convex, then f is φ-concave-like with respect to g.

Proof. Same as for Theorem 1. □

Analogous theorems can be stated for ψ-convex-like functions.

Theorems 1 and 2 can also be stated without assuming any linear structure on X and Y, if vector space is replaced by *pseudoconvex space* introduced and studied by Joó [5].

We recall briefly the notion of a pseudoconvex space. A *convex hull operation* is a mapping on X

$$\langle\ .\ \rangle : P(X) \longrightarrow P(X)$$

defined by the following properties:

a) $\langle \emptyset \rangle = \emptyset\,, \quad \langle \{x\} \rangle = \{x\}\,,$ for all $\; x \in X$;

b) $\langle A \rangle = \cup\{\langle F \rangle : F \subset A$ is finite $\}$, for all $\; A \subset X$;

c) $\langle \langle A \rangle \rangle = \langle A \rangle$.

A set $A \subset X$ is called convex, if $A = \langle A \rangle$.

A pseudoconvex space is a triplet $(X, \langle . \rangle\,, \phi)$ where

(i) X is a topological space and $\langle . \rangle$ is a convex hull operation on X;

(ii) $\phi := \{\varphi_F : F \subset X$ is finite $\}$ where

$$\varphi_F : \Delta^n \longrightarrow \langle \{x_0, \dots, x_n\} \rangle\,, \qquad n = \mathrm{card}(F) - 1$$

is a continuous mapping of Δ^n onto $\langle F \rangle = \langle \{x_0, \dots, x_n\} \rangle$ where Δ^n denotes the unit simplex of $\mathbf{R}^n$. Further φ_F is convex hull preserving, i.e. for all subsimplices $(e_{i_0}, \dots, e_{i_k}) \subset (e_0, \dots, e_n)$ of Δ^n,

$$\varphi_F(\mathrm{co}(e_{i_0}, \dots, e_{i_k})) = \langle \{x_{i_0}, \dots, x_{i_k}\} \rangle\ .$$

(Here $e_0, \dots, e_n$ denote the vertices of the unit simplex in $\mathbf{R}^n$). In Joó [5], the following generalization of Browder's fixed point theorem [2] is proved:

THEOREM A. *Let $(X, \langle\ .\ \rangle\,, \phi)$ be a compact Hausdorff pseudoconvex space and $T : X \longrightarrow P(X)$ a mapping for which*

(*a*) $\quad T(x) \neq \emptyset$ *and convex, for all* $\; x \in X$;

(*b*) $\quad T^{-1}(y) = \{x \in X\,,\ y \in T(x)\}$ *is open in* X, *for all* $\; y \in X$.

Then T has a fixed point, i.e. there exists $\; x_0 \in X$ with $x_0 \in T(x_0)$.

Keeping the notation of Theorem 1 we state:

THEOREM 1' *Let $(X, \langle\ .\ \rangle\,, \phi)$ be a compact Hausdorff pseudoconvex T_2 space, Y be a compact topological space and suppose that*

(*i*) *for any closed set $K \subset Y$,*

$$\max_{x\in X} \min_{y\in K} g(x,y) \geq \min_{y\in K} \max_{x\in X} f(x,y)\,;$$

(ii) the sets

$$\left\{x \in X : \min_{y \in K} g(x,y) > \max_{x \in X} \min_{y \in K} g(x,y) - \mu\right\}$$

are convex for any $\mu > 0$ and every closed set $K \subset Y$;

(iii) f and g are continuous on $X \times Y$, and φ is continuous on $\mathbf{R}^2$;

(iv) the mapping $x \longrightarrow Z(x)$ as defined before Theorem 1 is continuous.

Then f is φ-concave-like with respect to g.

Proof. The proof is indirect again. As in Theorem 1, we have that there exists $\lambda > 0$ and $x_1, x_2 \in X$ such that $Z(x)$ is not empty for all $x \in X$. Define, for any $\widehat{x} \in X$, the set-valued mapping $\widehat{x} \longrightarrow T(\widehat{x})$,

$$T(\widehat{x}) := \left\{x \in X : \; h_{\widehat{x}}(x) > \max_{x \in X} h_{\widehat{x}}(x) - \frac{\lambda}{2}\right\},$$

which is nonempty and convex by (ii). Taking (iii) and (iv) into account, we see that

$$T^{-1}(x) := \{\widehat{x} : x \in T(\widehat{x})\} = \left\{\widehat{x} : h_{\widehat{x}}(x) > \max_{x \in X} h_{\widehat{x}}(x) - \frac{\lambda}{2}\right\}$$

is open (since $\widehat{x} \longrightarrow h_{\widehat{x}}(x)$ and $\widehat{x} \longrightarrow \max\limits_{x \in X} h_{\widehat{x}}(x)$ are continuous).

Then, by Browder's theorem (Theorem A), there exists $x^* \in X$ with $x^* \in T(x^*)$, i.e.

$$h_{x^*}(x^*) > \max_{x \in X} h_{x^*}(x) - \frac{\lambda}{2},$$

which means that

$$\max_{x \in X} \min_{y \in Z(x^*)} g(x,y) = \max_{x \in X} h_{x^*}(x) < h_{x^*}(x^*) + \frac{\lambda}{2} = \min_{y \in Z(x^*)} g(x^*,y) + \frac{\lambda}{2} \leq$$

$$\varphi[f(x_1,y), f(x_2,y)] - \lambda + \frac{\lambda}{2} \leq f(x_1,y) \vee f(x_2,y) - \frac{\lambda}{2}$$

for all $y \in Z(x^*)$. Consequently, as in Theorem 1,

$$\max_{x \in X} \min_{Z(x^*)} g(x,y) \leq \min_{y \in Z(x^*)} [f(x_1,y) \vee f(x_2,y)] - \frac{\lambda}{2} \leq$$

$$\min_{y \in Z(x^*)} \max_{x \in X} f(x,y) - \frac{\lambda}{2},$$

a contradiction to (i). □

THEOREM 2'. *In addition to the conditions in Theorem 1', we assume the following to hold:*

a) Y *is a compact, convex subset of a topological vector space and* (i), (ii) *are required only for closed, convex subsets* $C \subset Y$*;*

b) φ *is linear on* $\mathbf{R}^2$*;*

c) $f(x,.)$ *and* $g(x,.)$ *are linear on* Y*.*

Then f *is* φ*-concave-like with respect to* g*.*

Proof. Along the lines of the proof of Theorem 1'. □

Note that the assumption $f \leq g$ was not explicitly used anywhere. It is an open question what assumptions (if any) are needed in addition to φ-concave and ψ-convex-likeness to ensure the validity of a maxmin $g \geq$ minmax f type inequality.

Acknowledgement

We are indebted to an anonymous referee whose constructive criticism greatly improved the paper. Research was funded by grant MM 480.

References

1. Berge,C., *Topological spaces*, Macmillan, New York, (1963).
2. Browder, F. E., *The fixed point theory of multivalued mappings in topological vector spaces*, Math. Ann. **177** (1968), 283–302.
3. Forgó, F. and Joó, I., *Necessary conditions for maxmin=minmax*, to appear in Acta Math. Acad. Sci. Hung. (1996).
4. Joó, I. and Kassay, G., *Convexity, minimax theorems and their applications*, Annales Univ. Sci. Budapest Sect. Math. **38** (1995), 71–93.
5. Joó, I., *On some convexities*, Acta Math. Hung. **54** (1989), 163–172.
6. Kindler, J., *Intersection theorems and minimax theorems based on connectedness*, J. Math. Anal. Appl. **178** (1993) 529–546.
7. Martos, B., *Nonlinear programming theory and methods*, Akadémiai Kiadó, Budapest, (1975).
8. Simons, S., *Criteres de faible compacite en termes du théoreme de minimax*, Seminaire Choquet 1970/71, no. 23, 8 pages.
9. Simons, S., *Two-function minimax theorems and variational inequalities for functions on compact and noncompact sets, with some comments on fixed-point theorems*, Proceedings of Symposia in Pure Mathematics, **45** (1986), Part 2, 377–392.
10. Simons, S., *Minimax theorems and their proofs*, in: Minimax and Applications ed. Ding-Zhu Du and Panos M. Pardalos, Kluwer Academic Publishers, Boston, 1995, 1–23.

Fitting Circles and Spheres to Coordinate Measuring Machine Data

Saul I. Gass
University of Maryland

Christoph Witzgall
National Institute of Standards and Technology

Howard H. Harary
National Institute of Standards and Technology

Abstract. This work addresses the problem of enclosing given data points between two concentric circles (spheres) of minimum distance whose associated annulus measures the out-of-roundness (OOR) tolerance. The problem arises in analyzing coordinate measuring machine (CMM) data taken against circular (spherical) features of manufactured parts. It also can be interpreted as the "geometric" Chebychev problem of fitting a circle (sphere) to data so as to minimize the maximum distance deviation. A related formulation, the "algebraic" Chebychev, determines the equation of a circle (sphere) so as to minimize the maximum violation of this equation by the data points. The algebraic Chebychev amounts to a straightforward linear-programming problem. In this paper, we compare the algebraic Chebychev against the popular algebraic least-squares solutions for various data sets. In most of these examples, the algebraic and geometric Chebychev solutions coincide, which appears to be the case for most real applications. Such solutions yield concentric circles whose separation is less than that of the corresponding least-squares solution. It is suggested that the linear-programming approach be considered as an alternate solution method for determining OOR annuluses for CMM data sets.

Key words: Coordinate Measuring Machine, Linear Programming, Quality Control, Chebychev Minmax, Circle and Sphere Out-of-Roundness Tolerance

1. Introduction

Mechanical parts are manufactured to meet stated specifications. A surface must be level, a drilled hole circular, and a ball-bearing spherical. But, exactly how level, how circular and how spherical are the manufactured parts? In practice, we need to answer the question: does an object meet the tolerances required by its specifications? The answer to this question is often obtained by measuring the part using a coordinate measuring machine (CMM). A small number of measurements are made which are then analyzed to determine an associated fitted reference object (circle or sphere). If the reference object meets the stated tolerance, the part is said to meet specifications. The tolerance is given in terms of a band width. For example, a drilled hole might have a specified tolerance band of 0.4 mm. CMM measurements

F. Giannessi et al. (eds.), New Trends in Mathematical Programming, 65–91.

of the hole, as analyzed by a tolerance procedure (as described below), yield a computed reference circle with radius r_0 and center (x_0, y_0). The hole is within specifications if all the measured points (x_i, y_i) fall within the annulus formed by two concentric circles, the inscribed and circumscribed circles with center at (x_0, y_0), such that the difference between the radii of the inscribed and circumscribed circles is less than or equal to the stated tolerance. A similar discussion of tolerance holds for a sphere in which an "annular shell" is determined. [See ASME (1972) and Phillips, Borchardt and Caskey (1993) for a discussion on types of CMMs and measurement uncertainty in CMMs.]

In this paper, we are concerned with developing a method for determining reference circles and spheres and related tolerance annuluses based on CMM measurements of circular surfaces (holes or discs) and spherical parts. A given set of CMM measurements has an associated "out-of-roundness" value (OOR). The OOR value is determined by selecting a center point for the reference circle or sphere, with the value being the difference between the largest and smallest radii to the measured points, ASME (1972). The center is calculated by one of the following four alternative methods for determining the circularity or sphericalness of an object, ASME (1972), Chou, Woo and Pollock (1993):

i. **Minimum-Radial-Separation (MRS)**. This center is that for which the radial difference between two concentric circles (spheres) that just contains the measured points is minimum (often referred to as the minimum zone). That is, it is the center which defines an annulus (shell) of minimum width, where the annulus is defined by two concentric circles (spheres), a circumscribed circle (sphere) and an inscribed circle (sphere). Note that there is an associated reference concentric circle (sphere) with radius between that of the inscribed and circumscribed circles (spheres) that, when found, can be used to generate the radii of the inscribed and circumscribed circles (spheres) and the width of the annulus. The finding of this latter reference circle (shell) that generates the minimum annulus is the main topic of this paper. The mathematical problem of finding the minimum annulus can be stated as a Chebychev problem in which one wants to minimize the maximum deviation of all the measured points from the circumference of the reference circle. This is a rather complicated nonlinear problem and exact algorithms for solving it are not readily available (Anthony et al. 1993). In this paper we show how a reference circle can be determined using standard linear-programming procedures that find an annulus whose width is very close to the width of the true Chebychev annulus, and may in fact be identical.

ii. **Least-Squares-Center (LSC).** This center is that of a circle (sphere) from which the sum of the squares of the radial ordinates between this circle (sphere) and the measured points is a minimum. An annulus, similar to the one found by MRS, is formed by the minimum inscribed circle (sphere) and the maximum circumscribed circle (sphere) having the least-squares center. Just about all CMMs compute the OOR value from the LSC.

iii. **Maximum-Inscribed-Circle (Sphere) (MIC/MIS).** This center is that of the largest circle (sphere) which can be inscribed within the set of measured points. The annulus of interest is formed by this inscribed circle (sphere) and the concentric circumscribed circle (sphere).

iv. **Minimum-Circumscribed-Circle (Sphere) (MCC/MCS).** This center is that of the smallest circle (sphere) which just contains all the measured points. The annulus of interest is formed by this circumscribed circle (sphere) and the concentric inscribed circle (sphere).

Algorithms for calculating the above centers exist, with varying computational complexity, Chou, Woo and Pollock (1993), Etesami and Qiao (1988). For circles, a Voronoi diagram approach can be used to find the MRS center; for LCS, a nonlinear program can be solved; for MIC, a Voronoi diagram method is available; and for MCC, a linear-programming model or a Voronoi diagram algorithm can be used, Hopp and Reeve (1986). Except for least-squares, it is not clear whether the other techniques can be readily adapted for spheres, for example, the Voronoi diagram method does not carry over readily to higher dimensions, Elzinga and Hearn (1972), Jones and Schnabel (1986), Preparata and Shamos (1985). Least-squares seems to be the method of choice, although this requires the solution of a set of nonlinear simultaneous equations. An approximate method for calculating the least-squares center and radius is given in ASME (1972).

Anthony et al. (1993) note that the least-squares center (LSC) is appropriate where measurement errors predominate and, currently, it is almost universally used. They further note that the minimum zone, that is, the MRS center, is appropriate in the more usual case where measurement errors are small compared to form errors. A rationale for our approximating the MRS center can be seen from the above comments and the following quote (Anthony et al., 1993, p. 2): "With increasing accuracy of CMMs, the minimum zone criterion for determining a substitute element (geometric form that best fits the measured points) becomes more appropriate. However, the design of Chebychev algorithms is a far more challenging task than the design of the corresponding least-squares algorithms. So, software for the Chebychev assessment of form is likely to be less satisfactory than that originally available for Gaussian assessment. In fact, very little such software is currently commercially available." Also, from Carr and Ferreira (1994), we quote: "Current CMM verification algorithms are based on least-squares solution, which minimizes the sum of the squared errors, resulting in a possible over estimation of the form tolerance. Therefore, while CMM algorithms successfully reject bad parts, they may also reject some good parts."

As stated in ASME (1972), the preferred center from which the OOR value is to be determined is the MRS center. We next describe a method for approximating the MRS center and examine its efficacy by applying it to a number of CMM sets of measurements for circles and spheres.

2. Fitting a Circle by an Approximate Chebychev Procedure

Chebychev MinMax Circle Problem: For a set of n observations (x_i, y_i), find a center of a circle (x_0, y_0) and radius r_0 such that the maximum radial deviation of the measured points from the determined circumference is minimized.

2.1. The Approximate Model and Optimality Criterion

The basic model is derived from the usual algebraic expression of a circle, that is, $r_0^2 = (x_i - x_0)^2 + (y_i - y_0)^2$. For a calculated center (x_0, y_0) and associated radius r_0, we would not expect $r_i^2 = (x_i - x_0)^2 + (y_i - y_0)^2$ to be equal to r_0 for the measured (x_i, y_i). However, we would like the (x_0, y_0, r_0) to be determined in a manner that minimizes the maximum of the absolute differences $|r_i - r_0|$. This form of the problem requires the application of rather complex nonlinear programming or geometric based algorithms (Anthony et al. 1993, Hearn and Vijay 1982, Butler et al. 1994) or the analysis of related Voronoi diagrams (Chou, Woo and Pollock 1993, Etesami and Qiao 1988). In this paper, we investigate the effectiveness of the surrogate measure of minimizing the maximum of the absolute differences of the squared terms $|r_i^2 - r_0^2|$. We show below that a solution to the *squared difference problem* by a linear-programming model yields an excellent approximate solution to the unsquared problem for our CMM measured data. For future reference, we note that $|r_i^2 - r_0^2| = |(r_i - r_0)(r_i + r_0)|$, that is, for the determined (x_0, y_0, r_0), the unsquared absolute difference can be obtained from the squared absolute differences by applying the positive weights $1/(r_i + r_0)$ to the associated squared absolute terms. [See Jones and Schnabel

(1986) on applying this relationship to the least squares approximation to a sphere.]

Gander, Golub and Strebel (1994) describe the fitting of circles and ellipses to given points in the plane in two different senses: *geometric* in which the sum of squares of the distances to the given points is minimal, and *algebraic* in which the parameters of the underlying algebraic equation $F(x) = 0$ are determined by least-squares. From this perspective, the method we describe here is algebraic.

The squared difference problem, minimizing the maximum absolute deviation between the set of r_i^2 and r_0^2, is as follows:

$$\text{Minimize } (x_0, y_0, r_0) \quad \max_{i=1,\ldots,n} \{ |(x_i - x_0)^2 + (y_i - y_0)^2 - r_0^2| \}$$

where (x_0, y_0) is a variable point and the (x_i, y_i) are the measured points, $i = 1, \ldots, n$.

Let $R_i = (x_i - x_0)^2 + (y_i - y_0)^2 - r_0^2$. Expanding the squares, we have

$$R_i = (x_i^2 + y_i^2) - 2x_i x_0 - 2y_i y_0 - \rho_0 ,$$

where $\rho_0 = (r_0^2 - x_0^2 - y_0^2)$.

Let $t \geq |R_i|$ for all i. The problem is transformed into a linear-programming problem P0 as follows:

$(P0)$ Minimize t

subject to

$$t \geq R_i$$
$$-t \leq R_i$$

or

Minimize t

subject to

$$t \geq (x_i^2 + y_i^2) - 2x_i x_0 - 2y_i y_0 - \rho_0$$
$$-t \leq (x_i^2 + y_i^2) - 2x_i x_0 - 2y_i y_0 - \rho_0$$

for all $i, t \geq 0$ and x_0, y_0, and ρ_0 unrestricted (free) variables.

Rearranging terms, the problem P1 is

$(P1)$ Minimize t

subject to

$$t + 2x_i x_0 + 2y_i y_0 + \rho_0 \geq x_i^2 + y_i^2 \qquad (1)$$
$$t - 2x_i x_0 - 2y_i y_0 - \rho_0 \geq -x_i^2 - y_i^2 \qquad (2)$$

for all $i, t \geq 0$ and x_0, y_0, and ρ_0 unrestricted variables.

Solving P1 finds a ρ_0 and center (x_0, y_0) that can be used to determine an OOR value [the width of the annulus between the inscribed and circumscribed circles with center at (x_0, y_0)] for the measured points. From $\rho_0 = (r_0^2 - x_0^2 - y_0^2)$, we determine the associated radius r_0. The value of t will be the largest absolute value of the differences (deviations) of the squared radii from the (x_i, y_i) to the calculated center (x_0, y_0) and the associated r_0^2. That is, given a circle at (x_0, y_0) with radius r_0, the squared distance from any observation (x_i, y_i) to the circumference of the circle will be less than or equal to t, as measured along the radius from (x_0, y_0) to (x_i, y_i). We also note, based on the relationship $|r_i^2 - r_0^2| = |(r_i - r_0)(r_i + r_0)|$, that for the circle at (x_0, y_0) with radius r_0 the observations (x_i, y_i) will be within t units of its circumference. [Carr and Ferreira (1994) give a model similar to P1 based on the linearization of a nonlinear vector formulation for finding the minimum zone cylinder, that is, for the problem of finding the minimum zone circumscribed and inscribed coaxial cylinders for a set of three dimensional data points. Their solution procedure is iterative and requires the solution of a set of linear programs that converge to the optimum.]

The linear-programming problem P1 will have $2n$ inequality constraints and 4 variables, whereas its dual problem will have 4 constraints and $2n$ variables. From a computational point of view, as our study involves problems with large n, and the solution time for solving a linear-programming problem is a function of the number of constraints of the problem, the dual is of interest. It is developed as follows.

Let u_i be the dual variable associated with the ith inequality of constraint set (1), and v_i be the dual variable associated the ith inequality of constraint set (2). The dual problem D1 is then

$$(D1) \qquad \text{Maximize } w = \sum_i (x_i^2 + y_i^2)u_i - \sum_i (x_i^2 + y_i^2)v_i$$

subject to

$$\sum_i u_i + \sum_i v_i \leq 1 \tag{3}$$

$$\sum_i 2x_i u_i - \sum_i 2x_i v_i = 0 \tag{4}$$

$$\sum_i 2y_i u_i - \sum_i 2y_i v_i = 0 \tag{5}$$

$$\sum_i u_i - \sum_i v_i = 0 \tag{6}$$

with $u_i \geq 0$ and $v_i \geq 0$. As shown in the Appendix, if an optimal solution to D1 has constraint (3) holding as an inequality, then there is an equivalent solution for which (3) holds as an equation. Thus, in what follows, we consider (3) to be an equation. The solution to P1 or D1 gives us the information to determine the minmax deviation circle. For D1, the dual variable for (3) is t, the dual variable for (4) is x_0, the dual variable for (5) is y_0 and the dual variable for (6) is ρ_0. As ρ_0 is an unrestricted variable and since $\rho_0 = r_0^2 - x_0^2 - y_0^2$, or $r_0^2 = \rho_0 + x_0^2 + y_0^2$, we need to ensure against the computed values of (ρ_0, x_0^2, y_0^2) yielding a negative value for r_0^2. A discussion of why that cannot happen is given in the Appendix.

Note that in D1 that we can divide both equations (4) and (5) by 2 and obtain an equivalent problem that uses the observations x_i and y_i as given. However, the returned dual variables for x_0 and y_0 would have to be divided by 2 to obtain the correct values associated with P1. (The division by 2 in D1 is equivalent to scaling the x_0 and y_0 columns of P1.) We use the form of D1 as given above, with (3) an equation, to avoid any confusion in interpreting the dual variables of D1.

We next present and discuss some computational examples.

Example 1 (EX1/P1)

We first solve a four point problem in primal format, and then compare its solution with its related dual. The four observations are:

i	(x_i, y_i)
1	(0, 0)
2	(4, 1)
3	(2, 3)
4	(0, 1)

The primal problem is given by:

$$\text{Minimize } T$$

subject to

$$\begin{aligned}
&2) \quad T + R \geq 0\\
&3) \quad T - R \geq 0\\
&4) \quad T + R + 8X + 2Y \geq 17\\
&5) \quad T - R - 8X - 2Y \geq -17\\
&6) \quad T + R + 4X + 6Y \geq 13\\
&7) \quad T - R - 4X - 6Y \geq -13\\
&8) \quad T + R + 2Y \geq 1\\
&9) \quad T - R - 2Y \geq -1\\
&T > 0 \qquad R, X, Y \text{ Free}
\end{aligned}$$

Here $T = t$, $R = \rho_0$, $X = x_0$, and $Y = y_0$. The above problem statement and its solution output, given below, were obtained by solving EX1/P1 using the LINDO linear-programming software (Schrage 1989).

SOLUTION: OBJECTIVE FUNCTION VALUE
OF) .33333330

VARIABLE	VALUE	REDUCED COST
T	.333333	.000000
R	-.333333	.000000
X	2.000000	.000000
Y	.833333	.000000

ROW	SLACK OR SURPLUS	DUAL PRICES
2)	.000000	−.333333
3)	.666667	.000000
4)	.666667	.000000
5)	.000000	−.083333
6)	.000000	−.166667
7)	.666667	.000000
8)	.666667	.000000
9)	.000000	−.416667

The center $(x_0, y_0) = (2, 0.833333)$, and with $R = \rho_0 = -0.333333$, $r_0 = 2.088327$. The minmax deviation $t = 0.333333$. For the given four points, the radii from (x_0, y_0) are the following:

Point	Radius (r_i)	$(r_i - r_0)$
$(0, 0)$	2.166667	0.078340
$(4, 1)$	2.006932	−0.081395
$(2, 3)$	2.166667	0.078340
$(0, 1)$	2.006932	−0.081395

The width of the annulus that contains these points is $\Delta = 2.166667 - 2.006932 = 0.159735$ (OOR value). With $r_0 = 2.088327$, two circles at center $(2, 0.833333)$ with radii $2.088327 + 0.078340$ and $2.088327 - 0.081395$ form, respectively, circumscribed and inscribed circles for the given points. Here the dual prices are the negative of the u_i and v_i corresponding to the primal equations, e.g., $u_1 = 0.333333$, $v_2 = 0.083333$, $u_3 = 0.166667$, and $v_4 = 0.416667$, as seen next in the solution of the dual problem.

Example 1 (EX1/D1)
The dual problem is given by:

$$\text{MAX } w = 17U_2 + 13U_3 + U_4 - 17V_2 - 13V_3 - V_4$$

SUBJECT TO

$$
\begin{array}{ll}
2) & U_1 + V_1 + U_2 + U_3 + U_4 + V_2 + V_3 + V_4 = 1 \\
3) & 8U_2 + 4U_3 - 8V_2 - 4V_3 = 0 \\
4) & 2U_2 + 6U_3 + 2U_4 - 2V_2 - 6V_3 - 2V_4 = 0 \\
5) & U_1 + U_2 + U_3 + U_4 - V_1 - V_2 - V_3 - V_4 = 0
\end{array}
$$

(All variables nonnegative)

SOLUTION: OBJECTIVE FUNCTION VALUE

OF) .33333330

VARIABLE	VALUE	REDUCED COST
U_2	.000000	.666667
U_3	.166667	.000000
U_4	.000000	.666667
V_2	.083333	.000000
V_3	.000000	.666666
V_4	.416667	.000000
U_1	.333333	.000000
V_1	.000000	.666667

ROW	SLACK OR SURPLUS	DUAL PRICES
2)	.000000	.333333
3)	.000000	2.000000
4)	.000000	.833333
5)	.000000	−.333333

The values of the variables of interest are contained in the DUAL PRICES column. The minmax deviation $= w = t = 0.333333$, $x_0 = 2.000000$, $y_0 = 0.83333$, $\rho_0 = -0.333333$, and $r_0 = 2.088327$.

Harary Data Set CB062R

The Harary data is a set of CMM measurements consisting of 256 (x_i, y_i) points taken of a hole drilled to have a 1" or 2.54 cm diameter, with an assumed center at $(0, 0)$. The measured points were taken at 256 points of the compass, starting with 0. Figure 1 shows a plot of the hole magnified 500 times, with a least-squares circle superimposed; the lobes are typical for such drilled elements. The hole and measurements were made by H. Harary for his study of probe metrology (Harary 1992).

2.2. Summary of MRS Computations for a Circle

Of the 22 circle problems solved and reported here, eight had four points selected so that the first problem 4.1, had points (1, 65, 129, 193). For problem 4.2, the numbers were incremented by nine, and so on for the other problems. Similarly, problem 6.1 had points (1, 49, 97, 145, 193, 241); problem 8.1 had points (1, 33, 65, 97, 129, 161, 193, 125); problem 16.1 points started with point 1 and selected every 16th point; problem 32 started with point 1 and selected every 8th point; finally, problem 256 used all 256 points.

The eight 4-point problems were solved on the CPLEX linear- programming system, as we found that greater accuracy was required in determining the centers and radii for these small problems. The six 6-point, the four 8-point, two 16-point, and the one 32-point problems were solved using the LINDO linear-programming software. As our version of LINDO can solve problems with up to 500 variables, the one 256-point, 512 variable dual linear-programming problem was solved using the XPRESS-MP LP optimizer. The results of these computations are shown in Table 1.

As described in Section 1, the MRS annulus for a circle is that for which the radial difference between two concentric circles that just contains the measured points is minimum. The Chebychev approximation described in Section 2 determines a minmax deviation t and a circle with center (x_0, y_0) and radius r_0 that is centered in an annulus, where the annulus is defined by two circles (an outer circumscribed circle and inner inscribed circle) that are concentric to this centered circle. The radius of the outer circle is equal to $[r_0 + \max(r_i - r_0)]$ and the radius of the inner circle is $[r_0 + \min(r_i - r_0)]$.

The Chebychev approximation model of Section 2 determines the minmax of $|r_i^2 - r_0^2|$, instead of the desired minmax of $|r_i - r_0|$, where r_i are the radii from the calculated center (x_0, y_0) to the measured (x_i, y_i). Although the MRS for a set of measured points is used to define out-of-roundness, it is usually not determined as existing algorithms tend to be rather complex and not available; an MRS algorithm is not a standard element of a CMM's software.

Recall that $|r_i^2 - r_0^2| = |(r_i - r_0)(r_i + r_0)|$. For the Chebychev determined (x_0, y_0, r_0), an approximation of the unsquared absolute differences for each i can be obtained from the squared absolute differences $|r_i^2 - r_0^2|$ by multiplying them by their associated positive weights $1/(r_i + r_0)$. These approximations are, of course, for the Chebychev determined center. The question that then needs to be pursued deals with how close is the Chebychev circle for minmax $|r_i^2 - r_0^2|$ to the true MRS circle determined by minmax $|r_i - r_0|$. As a test of the accuracy of the approximation, we adjusted the mathematical formulation of the primal form P0 by multiplying the right-hand-sides R_i of each constraint by the weights $1/(r_i + r_0)$ that were obtained from the unweighted solution, and resolved the problems. For the dual form, this meant that each variable u_i and v_i in equation (3) of problem D1 were multiplied by their corresponding weights. In the few cases tested, the width of the annuluses agreed to the fourth decimal place or better.

Further, using the nonlinear FFSQP software (Zhou, Tits, and Lawrence, 1995) the "true" Chebychev minmax $|r_i - r_0|$ were calculated for the set of 22 problems sampled from the Harray data set of 256 points. (The calculations were made by Albert Sang, University of Michigan.) We found that the radial differences between the true Chebychev FFSQP determined annuluses and our approximate Chebychev annuluses were of the order of +0.0000016825 (32-point problem) or less. That is, the more complex FFSQP algorithm's annuluses were all smaller in width, but not by much. Further, the regular least-squares solution to these problems were calculated by Marjorie McClain (National Institute of Standards and Technology) using the ODRPACK Software for Weighted Orthogonal Distance Regression, Boggs et al. (1992). For all problems, the approximate Chebychev annulus was smaller than the corresponding least-squares annulus, with the differences being of the order of +0.0032236811 or less (6-point problem 6.1). In addition, the true Chebychev minmax solutions were also calculated by Andrew Vakhutinsky (University of Maryland) using the Voronoi diagram algorithm of Barber et al. 1995. The Voronoi diagram solutions matched the approximate Chebychev and exact Chebychev centers and annuluses to six decimal places. For our problems, the width of the minimum zone (MRS) annulus is smallest for the true Chebychev, with the linear-programming approximate Chebychev having the next smallest widths, followed by the least-square widths.

One should not be too surprised by the closeness of the approximate and exact Chebychev solutions. Both approaches determine a center for an inscribed and circumscribed circle by selecting a subset of points from the given measured points. To determine the inscribed and circumscribed circles and their joint center, both methods need only four points to "pin" the circles down. That is, we can have the following situations: (1) one point (of the convex hull of the set of points) is on the circumscribed circle and three points are on the inscribed circle, (2) two points (on the convex hull) are on the circumscribed circle and two points are on the inscribed circle, with the lines joining the respective pairs intersecting, and (3) three points (on the convex hull) are on the circumscribed circle and one point is on the inscribed circle. The number of CMM measurements that are taken on a circle or sphere is usually small and, hence, the number of combinations of four points is also small. It would seem unusual for the set of four points selected by the approximate Chebychev to be different from the four points selected by the true Chebychev method (Voronoi or nonlinear optimization). This is borne out by our computations in that methods (approximate Chebychev, and true Chebychev by Voronoi or nonlinear optimization) yield the same centers within six decimal places. The resulting MRS annuluses are all smaller in width than the usual least-squares annuluses. (We assume here that we do not encounter degenerate situations in which, for example, four points are on the inscribed or circumscribed circles. The real-world of CMM measurements and computational accuracy would rule out such occurrences.)

Table 1. Summary results for 22 circle problems based on 256 point data set

Problem	LP center X	LP center Y	LP radius
4.1	0.00013702518820	-0.00030252878553	12.694789039208220
4.2	-0.00037412555692	0.00056930189391	12.699646441928690
4.3	9.83720302E-06	0.00055973984874	12.704806383806420
4.4	0.00030325619703	0.00042407248273	12.707061432526440
4.5	0.00039709601146	0.00028489565636	12.704754641470470
4.6	-0.00020550883202	-2.394184445E-05	12.700292998588720
4.7	-0.00054065846921	-0.00091320271486	12.695635372047520
4.8	-0.00024037105352	-0.00068312421648	12.693456572187050
6.1	-0.006745	-0.001292	12.699793193760640
6.2	-0.005487	-0.004082	12.699265599627920
6.3	0.001025	-0.003636	12.700075364781150
6.4	0.005689	-0.004042	12.699978295354880
6.5	0.004808	0.001782	12.701634788183290
6.6	-0.001968	0.002705	12.701142121480610
8.1	-0.000190	-0.000302	12.702255710199820
8.2	-0.000206	-0.000024	12.700291336934440
8.3	-0.002043	-0.000291	12.699015877560360
8.4	-0.001099	-0.000513	12.701228344966090
16.1	-0.000724	-0.000837	12.700401616671220
16.2	-0.001099	-0.000513	12.701228344966090
32	-0.001257	-0.000617	12.701126011528980
256	-0.000230	-0.000059	12.700432830000000

	LP Min Max	LP Min Max Annulus	Chebychev Annulus	Least Sq Annulus
4.1	.045240822747	.003563742663	.003563732906	.003563794595
4.2	.022653916865	.001783818349	.001783812150	.001783874493
4.3	.099668036321	.007844918982	.007844902272	.007845129748
4.4	.149703453996	.011781102281	.011781095782	.011781407103
4.5	.171342726265	.013486512471	.013486512026	.013486809856
4.6	.165606243154	.013039584637	.013039566415	.013039716150
4.7	.160967081232	.012678941607	.012678923697	.012679522550
4.8	.100574253403	.007923350736	.007923340410	.007923596010
6.1	.197378400000	.015542815941	.015541790391	.018766497039
6.2	.137462300000	.010825219818	.010824205562	.013365053103
6.3	.111512200000	.008781880547	.008780725034	.009606000778
6.4	.159221900000	.012537893908	.012537102976	.014527753040
6.5	.198202000000	.015604803987	.015604411145	.017028282496
6.6	.219871300000	.017312243664	.017311134082	.019200410775
8.1	.234851900000	.018489726392	.018489058218	.018898491801
8.2	.165607500000	.013040261775	.013039566415	.013474359934
8.3	.246788900000	.019434383436	.019433927052	.020794738582
8.4	.297924500000	.023457088719	.023456340963	.024361628076
16.1	.281968300000	.022202616654	.022201806881	.023180108382
16.2	.297924500000	.023457088719	.023456340963	.024271034773
32	.300458100000	.023657991917	.023656309416	.024829788246
256	.320809000000	.025260223387	.025259741289	.025456267023

Figures 2-4 illustrate the relationship between the minmax residuals $(r_i^2 - r_0^2)$ and the corresponding radius residuals $(r_i - r_0)$ for circle problems 6.1, 16.1 and 256. Note the cyclic nature of both sets of residuals.

We had hoped to be able to prove that the approximate Chebychev annulus would always be smaller than the least-squares determined annulus for any set of points in a plane. The following example five point counterexample shows that this is not the case.

Counterexample

i	(x_i, y_i)
1	$(1,0)$
2	$(12,0)$
3	$(0,12)$
4	$(-8,0)$
5	$(0,-12)$

The true Chebychev solution is $(x_0, y_0) = (5,0)$, with an annulus of 9; the approximate linear-programming Chebychev solution is $(x_0, y_0) =$

(0, 0), with an annulus of 11; and the least-squares solution is (4.2745, 0), with an annulus of 9.4641. We note that the counterexample is anomalous in terms of typifying CMM data.

3. Fitting a Sphere by an Approximate Chebychev Procedure

Fitting a sphere to measurements using Chebychev estimation can be done in a fashion similar to what was done above for the circle.

Chebychev Minmax Sphere Problem: For a set of observations (x_i, y_i, z_i), find a center of a sphere (x_0, y_0, z_0) and radius r_0 such that the maximum deviation of the observed points from the surface of the determined sphere is minimized.

The basic model is given by the sphere

$$(x_i - x_0)^2 + (y_i - y_0)^2 + (z_i - z_0)^2 = r_0^2 \,.$$

For the observed points (x_i, y_i, z_i), the approximate problem is to find the center (x_0, y_0, z_0) and radius r_0 that

$$\text{Minimize } (x_0, y_0, z_0) \quad \max_i \{|(x_i - x_0)^2 + (y_i - y_0)^2 + (z_i - z_0)^2 - r_0^2|\}$$

where (x_0, y_0, z_0) is a variable point and the (x_i, y_i, z_i) are the measured points.

Let $R_i = (x_i - x_0)^2 + (y_i - y_0)^2 + (z_i - z_0)^2 - r_0^2$. Expanding the squares, we have $R_i = (x_i^2 + y_i^2 + z_i^2) - 2x_ix_0 - 2y_iy_0 - 2z_iz_0 - \rho_0$, where $\rho_0 = (r_0^2 - x_0^2 - y_0^2 - z_0^2)$.

Let $t \geq |R_i|$. The problem is transformed into a linear-programming problem as follows:

$$\text{Minimize } t$$

subject to

$$t \geq R_i$$

$$-t \leq R_i$$

or

$$\text{Minimize } t$$

subject to

$$t \geq (x_i^2 + y_i^2 + z_i^2) - 2x_ix_0 - 2y_iy_0 - 2z_iz_0 - \rho_0$$

$$-t \leq (x_i^2 + y_i^2 + z_i^2) - 2x_ix_0 - 2y_iy_0 - 2z_iz_0 - \rho_0$$

for all $i, t \geq 0$, and x_0, y_0, z_0 and ρ_0 unrestricted variables. Rearranging terms, the problem P2 is

$$(P2) \qquad \text{Minimize } t$$

subject to

$$t + 2x_i x_0 + 2y_i y_0 + 2z_i z_0 + \rho_0 \geq x_i^2 + y_i^2 + z_i^2 \tag{7}$$

$$t - 2x_i x_0 - 2y_i y_0 - 2z_i z_0 - \rho_0 \geq -x_i^2 - y_i^2 - z_i^2 \tag{8}$$

for all $i, t \geq 0$, and x_0, y_0, z_0 and ρ_0 unrestricted (free) variables.

Given a set of observations, we can solve P2 and find ρ_0 and the center (x_0, y_0, z_0). From $\rho_0 = (r_0^2 - x_0^2 - y_0^2 - z_0^2)$, we determine the radius r_0. The value of t will be the largest absolute value of the differences (deviations) of the squared distances of the observations from r_0^2. That is, given a sphere at (x_0, y_0, z_0) with radius r_0, any observation (x_i, y_i, z_0) will be within t units of its surface, as measured along the radius from (x_0, y_0, z_0) to (x_i, y_i, z_0).

As with the circle problem P1, it is convenient to take the dual to the sphere problem P2. The dual problem D2 is developed as follows.

Let u_i be the dual variable associated with the ith inequality of constraint set (7), and v_i be the dual variable associated the ith inequality of constraint set (8). The dual problem is then

$$(D2) \qquad \text{Maximize } w = \sum_i (x_i^2 + y_i^2 + z_i^2)u_i - (x_i^2 + y_i^2 + z_i^2)v_i$$

subject to

$$\sum_i u_i + \sum_i v_i \leq 1 \tag{9}$$

$$\sum_i 2x_i u_i - \sum_i 2x_i v_i = 0 \tag{10}$$

$$\sum_i 2y_i u_i - \sum_i 2y_i v_i = 0 \tag{11}$$

$$\sum_i 2z_i u_i - \sum_i 2z_i v_i = 0 \tag{12}$$

$$\sum_i u_i - \sum_i v_i = 0 \tag{13}$$

with $u_i \geq 0$ and $v_i \geq 0$. Again, as described in the Appendix, we can take inequality (9) to be an equation. The solution to P2 or D2 gives us the information to determine the minmax deviation sphere.

Example 2 (D2)

We have five observations:

i	(x_i, y_i, z_i)
1	$(0.98, 0.00, 0.00)$
2	$(0.00, 1.01, 0.00)$
3	$(-1.02, 0.00, 0.00)$
4	$(0.00, 0.99, 0.00)$
5	$(0.00, 0.00, 1.10)$

The D2 minmax problem is given by:

$$\text{MAX } w = 0.9604U_1 + 1.0201U_2 + 1.0404U_3 + 0.9801U_4 + 1.21U_5 - 0.9604V_1 - 1.0201V_2 - 1.0404V_3 - 0.9801V_4 - 1.21V_5$$

SUBJECT TO

$$\begin{array}{ll} 2) & U_1 + U_2 + U_3 + U_4 + U_5 + V_1 + V_2 + V_3 + V_4 + V_5 = 1 \\ 3) & 1.96U_1 - 2.04U_3 - 1.96V_1 + 2.04V_3 = 0 \\ 4) & 2.02U_2 + 1.98U_4 - 2.02V_2 - 1.98V_4 = 0 \\ 5) & 2.2U_5 - 2.2V_5 = 0 \\ 6) & U_1 + U_2 + U_3 + U_4 + U_5 - V_1 - V_2 - V_3 - V_4 - V_5 = 0 \end{array}$$

END

SOLUTION: OBJECTIVE FUNCTION VALUE

1) .19797040E-01

VARIABLE	VALUE	REDUCED COST
$U1$	.005050	.000000
$U2$	.490099	.000000
$U3$	.004851	.000000
$U4$	.000000	.039594
$U5$	.000000	.039594
$V1$	.000000	.039594
$V2$	.000000	.039594
$V3$	.000000	.039594
$V4$	.500000	.000000
$V5$	.000000	.000000

ROW	SLACK OR SURPLUS	DUAL PRICES
2)	.000000	.019797
3)	.000000	−.020000
4)	.000000	.010149
5)	.000000	.113634
6)	.000000	.979803

From the dual prices column we have the minmax $w = t = 0.19797$, $x_0 = -0.020000$, $y_0 = 0.010149$, $z_0 = 0.113634$, $\rho_0 = 0.979803$. From these values, we have $r_0 = 0.996604$.

Point	Radius (r_i)	($r_i - r_0$)
(0.98, 0.00, 0.00)	1.006618	0.010014
(0.00, 1.01, 0.00)	0.997546	0.000942
(-1.02, 0.00, 0.00)	1.006618	0.010014
(0.00, 0.99, 0.00)	0.977683	-0.018921
(0.00, 0.00, 1.10)	0.986755	-0.009849

The width of the annulus that contains these points is $\Delta = 1.006618 - 0.977683 = 0.028935$ (OOR value). With $r_0 = 0.996604$, two spheres at center $(-0.020000, 0.010149, 0.113634)$ with radii $r_0 + 0.010014$ and $r_0 - 0.018921$ form, respectively, circumscribed and inscribed spheres for the given points.

Example 3 (61 Point Sphere)

The data for this problem were obtained from measurements of a sphere by a coordinate measuring machine (CMM) at the National Institute of Standards and Technology (NIST). The sphere was first measured at the North pole and then 12 equally spaced measurements were made at the five latitudes (polar angles) of 30°, 60°, 90°, 110°, 130°.

For this problem, we have the minmax $w = 0.076068$, $x_0 = 0.000502$, $y_0 = 0.001064$, $z_0 = -0.006078$, $\rho_0 = 63.893250$, with $r_0 = 7.9933277$. The annulus width is 0.00951673. The radii of the circumscribed and inscribed spheres are given by $r_0 + 0.00475650$ and $r_0 - 0.00476023$, respectively.

The least-squares solution to the 61-point set of data yielded the following solution for the center and radius: $x_0 = 0.00098$, $y_0 = 0.00194$, $z_0 = -0.00426$, with $r_0 = 7.9928633$. The corresponding annulus width is 0.01044755, with the circumscribed and inscribed circles having radii of $r_0 + 0.00562304$ and $r_0 - 0.00482451$, respectively. Note that the linear-programming minmax annulus has a smaller width than that of the least-squares annulus.

Example 4 (13 Points from a 361 Point Sphere)

The thirteen points for this problem were obtained by 12 equally spaced CMM readings of a sphere taken at latitude 90°, plus one measurement at the pole. The minmax solution is $x_0 = 0.000295$, $y_0 = 0.002238$, $z_0 = 0.002842$, and $r_0 = 7.991314233$. The corresponding annulus width is 0.00429858; the circumscribed and inscribed circles radii are $r_0 + 0.00214867$ and $r_0 - 0.00214991$.

The least-squares solution is $x_0 = 0.000820881$, $y_0 = 0.0016959$, $z_0 = 0.00059597$, and $r_0 = 7.991141$. The corresponding annulus width is 0.00492597; the circumscribed and inscribed circles radii are $r_0 + 0.00199971$ and $r_0 - 0.00292626$. The width of the minmax annulus is less than the least-squares annulus. Figure 5 shows a plot of the minmax radius residuals and the radius residuals for the 13 points.

Example 5 (361 Point Sphere)

The data for this problem were obtained by CMM readings of a sphere at ten latitudes (15°, 30°, 45°, 60°, 75°, 90°, 100°, 110°, 120°, 130°), with 36 equally spaced measurements made at each latitude, plus one polar measurement, for a total of 361 points. The minmax linear-programming problem was solved using XPRESS-MP LP. The minmax solution is $x_0 = 0.003948$, $y_0 = 0.00024977$, $z_0 = -0.014052$, with $r_0 = 7.97940148$. The MRS annulus width is 0.02417821, with the circumscribed and inscribed circles have radii of $r_0 + 0.01209343$ and $r_0 - 0.01208478$.

The corresponding results for the least square solution is $x_0 = 0.005617$, $y_0 = -0.0010268$, $z_0 = -0.013924$, with $r_0 = 7.98109556$. The least-squares annulus width is 0.02541751, with the circumscribed and inscribed circles have radii of $r_0 + 0.00996937$ and $r_0 - 0.01544814$. Note that the width of the minmax annulus is less than the least-squares annulus.

4. Appendix: Proof of equality formulation of the dual problem and related conditions of the optimal solution

To show that an optimal solution to Problem D1 exists for which inequality (3) holds as an equation, we note the following.

$$(D1) \qquad \text{Maximize } w = \sum_i (x_i^2 + y_i^2)u_i - \sum_i (x_i^2 + y_i^2)v_i$$

subject to

$$\sum_i u_i + \sum_i v_i \leq 1 \tag{3}$$

$$\sum_i 2x_i u_i - \sum_i 2x_i v_i = 0 \tag{4}$$

$$\sum_i 2y_i u_i - \sum_i 2y_i v_i = 0 \tag{5}$$

$$\sum_i u_i - \sum_i v_i = 0 \tag{6}$$

Denote an optimal solution to Problem D1 by u_i^* and v_i^* such that constraint (3) is an inequality. Let $s^* > 0$ be the value of the optimal slack variable for (3), that is

$$\sum_i u_i^* + \sum_i v_i^* + s^* = 1 \tag{3'}$$

Let $s' = s^*/2n$, where n is the total number of observations. Define $u_i' = u_i^* + s'$ and $v_i' = v_i^* + s'$. It is straightforward to demonstrate that the vector $(u_i', v_i') \geq 0$ is an optimal feasible solution to D1 by direct substitution in to constraints (1), (2), (3) and (4). We see that all the constraints are satisfied and constraint (1) holds as an equation. Also, by direct substitution, the objective function value for the vector (u_i', v_i') is the same as for the given optimal solution (u_i^*, v_i^*).

With constraint (3) being an equation, solving D1 by the simplex method forces exactly four variables from the set (u_i, v_i) to be in any basic solution. Also, the matrix of the system (3), (4), (5), and (6) is generated by columns of the form:

Column Type 1 (u_i)	Column Type 2 (v_i)	Dual Variable	Equation number
1	1	t	(3)
$2x_i$	$-2x_i$	x_0	(4)
$2y_i$	$-2y_i$	y_0	(5)
1	-1	ρ_0	(6)

For n measured points (x_i, y_i), there are n type 1 columns associated with the u_i and n type 2 columns associated with the v_i. A feasible (4×4) basis will have four such columns from the available $2n$. Note that the four columns cannot be all of type 1 or type 2. If that were the case, then we would not have a basis as the first row and fourth row would be identical (all 1's if all type 1) or the negative of each other (the first row all 1's and the fourth row all -1's for all type 2).

Recall that we need only four points to determine concentric inscribed and circumscribed circles for a set of points where the four points can be arranged in three ways: (1) one point (of the convex hull of the set of points) is on the circumscribed circle and three points are on the inscribed circle, (2) two points (on the convex hull) are on the circumscribed circle and two points are on the inscribed circle, and (3) three points (on the convex hull) are on the circumscribed circle and one point is on the inscribed circle. For the dual, if a u_i is in the solution then the corresponding measured point (x_i, y_i) is on the circumscribed circle; if a v_i is in the solution then the (x_i, y_i) is on the inscribed circle. The four basic variables correspond to points that define the inscribed and circumscribed circles. Further, we note that for the same i, both u_i and v_i cannot be in an optimal basis. This statement leads to the conclusion that the solution to problem D1 will yield values of (ρ_0, x_0, y_0) such that $r_0^2 = (\rho_0 + x_0^2 + y_0^2) > 0$. The validity of these last two statements is discussed next.

Recall that ρ_0 is an unrestricted variable. Since $\rho_0 = (r_0^2 - x_0^2 - y_0^2)$, or $r_0^2 = (\rho_0 + x_0^2 + y_0^2)$, we need to ensure against the computed values of (ρ_0, x_0^2, y_0^2) yielding a negative value for r_0^2. Geometrically, this means that the linear-programming solution must select four points such that three of them are on the circumscribed (inscribed) circle and one on the inscribed (circumscribed) circle, or two are on the circumscribed circle and two on the inscribed circle. We have already seen that the solution to the dual cannot select all four points on the circumscribed (all u_i in the basis) or inscribed (all v_i in the basis) circle. What we now need to show is that we must have three u_i (v_i) and one v_i (u_i) or two u_i and two v_i for four different i. This is equivalent to showing that both a u_i and v_i for the same i cannot be in the optimal basic feasible solution.

If for some i, both u_i and v_i were in the optimal solution, then the simplex optimality criterion requires the following conditions to hold simultaneously for the u_i and v_i, respectively:

$$t + 2x_i x_0 + 2y_i y_0 + \rho_0 = (x_i^2 + y_i^2) \qquad (u_i)$$

$$t - 2x_i x_0 - 2y_i y_0 - \rho_0 = -(x_i^2 + y_i^2) \qquad (v_i)$$

Together, these conditions state that

$$t = (x_i^2 + y_i^2) - 2x_i x_0 - 2y_i y_0 - \rho_0 \qquad (u_i)$$

$$t = -(x_i^2 + y_i^2) + 2x_i x_0 + 2y_i y_0 + \rho_0 \qquad (v_i)$$

which can hold only if $t = 0$. As t is a measure of the width of the annulus separating the circumscribed and inscribed circles, it will be equal to zero only in the geometrically degenerate case of the circumscribed and inscribed circles coinciding. For CMM data sets, the probability of this occurring is zero. Thus, the linear-programming solution to problem D1 will consist of three u_i and one v_i, three v_i and one u_i, or two u_i and two v_i, with all indices i being different. Given the corresponding four points, then circumscribed and inscribed circles can be found such that the width of the annulus separating them is nonzero. This implies that the four points in the linear-programming solution correspond to computed values of (ρ_0, x_0^2, y_0^2) such that $r_0^2 = (\rho_0 + x_0^2 + y_0^2) > 0$.

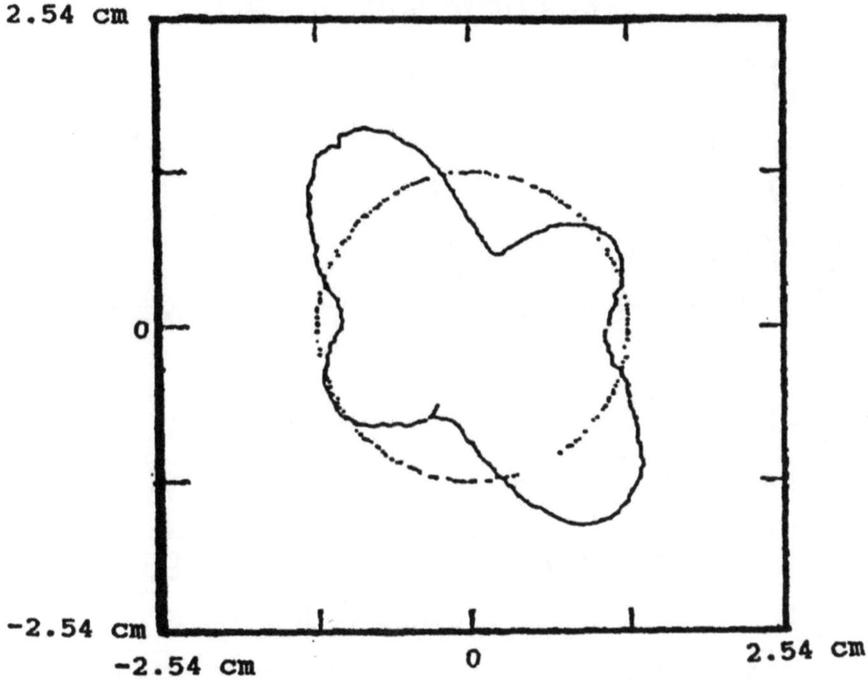

Figure 1. Profile of 256 point drilled and counter-bored hole and its least-squares circle approximation

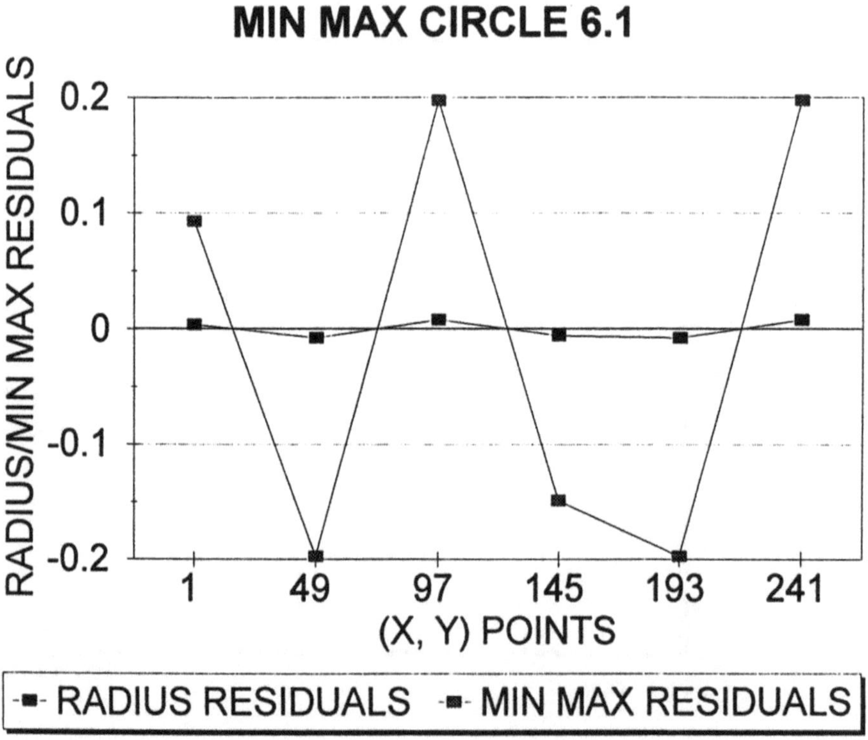

Figure 2. Circle 6.1 residuals: radius res.= $(r_i - r_0)$; minmax res.= $(r_i^2 - r_0^2)$

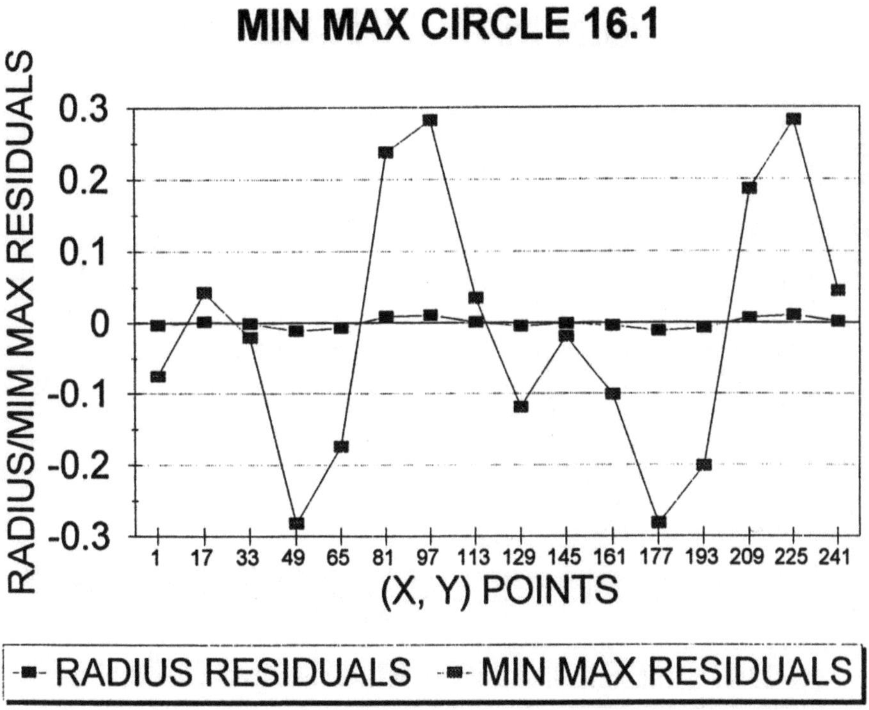

Figure 3. Circle 16.1 residuals: radius res.= $(r_i - r_0)$; minmax res.= $(r_i^2 - r_0^2)$

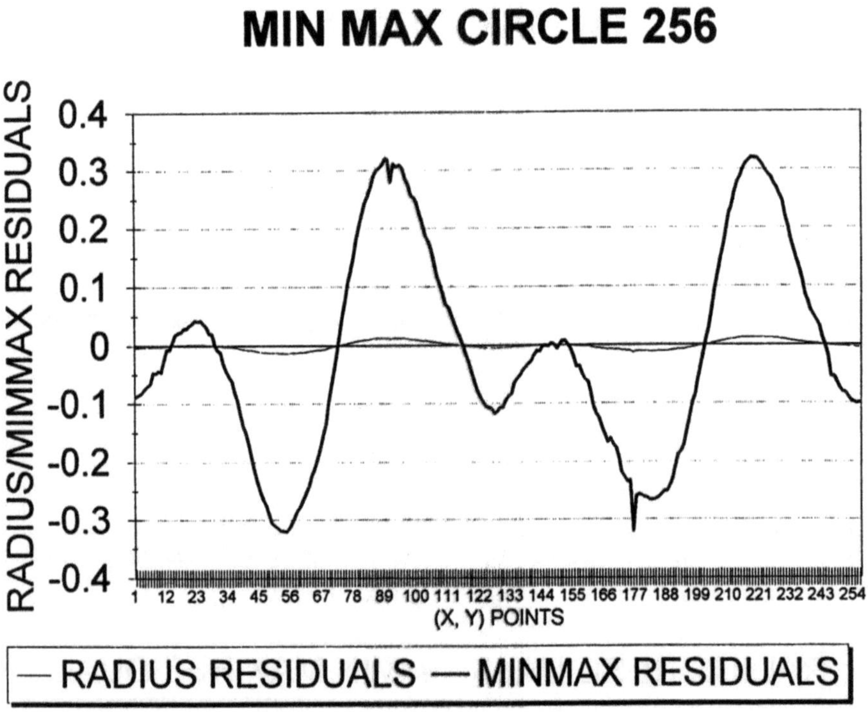

Figure 4. Circle 256 residuals: radius res.= $(r_i - r_0)$; minmax res.= $(r_i^2 - r_0^2)$

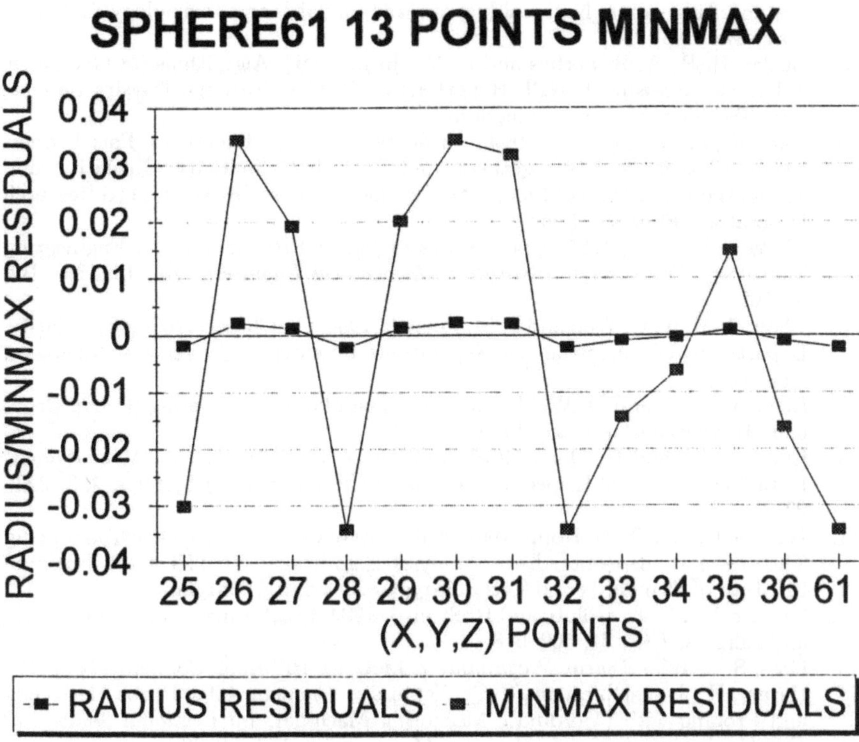

Figure 5. Sphere 61/13 residuals: radius res.$= (r_i - r_0)$; minmax res.$= (r_i^2 - r_0^2)$

References

1. American Society of Mechanical Engineers, 1972. *Measurement of Out-Of-Roundness*, ANSI B89.3.1-1972, NY, NY.
2. Anthony, G. T., H. M. Anthony, B. Bittner, H. P. Butler, M. G. Cox, R. Drieschner, R. Elligsen, A. B. Forbes, H. Gross, S. A. Hannaby, P. M. Harris and J. Kok, 1993. Chebychev Best-Fit Geometric Elements, NPL Report DITC 221/93. National Physical Laboratory, Teddington, United Kingdom.
3. Barber, C. B., D. P. Dobkin and H. T. Huhdanpaa, 1995, The Quickhull Algorithm for Convex Hulls, University of Minnesota.
4. Boggs, P. T., R. H. Byrd, J. E. Rogers and R. B. Schnabel, 1992. User's Reference Guide for ODRPACK Version 2.1 – Software for Weighted Orthogonal Distance Regression, National Institute of Standards and Technology, Gaithersburg, Maryland.
5. Butler, B. P., A. B. Forbes and P. M. Harris, 1994. Algorithms for Geometric Tolerance Assessment, NPL Report DITC 228/94, National Physics Laboratory, Teddington, United Kingdom.
6. Carr, K., and P. Ferreira, 1994. Verification of Form Tolerances, Part I: Basic Issues, Flatness and Straightness, and Part II: Cylindricity, Cirularity, and Straightness of a Median Line, Ford Technical Report SR-94-14, Ford Research Laboratory, Flint, Michigan.
7. Chetwynd, D. G., 1985. Applications of Linear Programming to Engineering Metrology, *Proceedings, Institute of Mechanical Engineers*, Vol. 199, No. B2, 93–100.
8. Chou, S.-Y., T. C. Woo and S. M. Pollock, 1994. On Characterizing Circularity, Department of Industrial and Operations Research, University of Michigan, Ann Arbor.
9. Elzinga, D. J., and D. W. Hearn, 1972. The Minimum Covering Sphere Problem, *Management Science*, 19, 1, 96–104.
10. Etesami, F., and H. Qiao, 1988. Analysis of Two-Dimensional Measurement Data for Automated Inspection, *Journal of Manufacturing Systems*, 7, 3, 223–232.
11. Feng, S. C., and T. H. Hopp, 1991. A Review of Current Geometric Tolerancing Theories and Inspection Data Analysis Algorithms, NISTIR 4509, National Institute of Standards and Technology, Gaithersburg, Maryland.
12. Gander, W., G. H. Golub, and R. Strebel, 1994. Least-Squares Fitting of Circle and Ellipses, *BIT*, 24, 560–578.
13. Gass, S. I., 1985. *Linear Programming*, McGraw-Hill Book Company, NY, NY.
14. Harary, H., J.-O. Dufraigne, P. Chollet and A. Clement, 1993. Probe Metrology and Probing with Coordinate Measuring Machines, *International Journal of Flexible Automation and Integrated Management*, 1, 1, 59–70.
15. Hearn, D. W., and J. Vijay, 1982. Efficient Algorithms for the (Weighted) Minimum Circle Problem, *Operations Research*, 30, 4, 777–795.
16. Hopp, T. H. and C. A. Reeve, 1996. An Algorithm for Computing the Minimum Covering Sphere in any Dimension, NISTIR 5831, National Institute of Standards and Technology, Gaithersburg, MD.
17. Jones, B. A., and R. B. Schnabel, 1986. A Comparison of Two Sphere Fitting Methods, *Proceedings of the Instrumentation and Measurement Technology*, Subgroup of the IEEE, Boulder, Colorado.
18. Le, V.-B., and D. T. Lee,1991. Out-of-Roundness Problem Revisited, IEEE Transactions on Pattern Analysis and Machine Intelligence, 13, 3, 217–223.
19. Lee, D. T., and R. L. Drysdale, III, 1081. Generalization of Vornoi Diagrams in the Plane, *Siam Journal of Computing*, 10, 1, 73–87.
20. Megiddo, N., 1983. Linear-Time Algorithms for Linear Programming in R^3 and Related Problems, *Siam Journal of Computing*, 12, 4, 759-776.

21. Preparata, F. P. and M. I. Shamos, 1985. *Computational Geometry*, Springer-Verlag, New York.
22. Phillips, S. D., 1995. Performance Evaluation, Chapter 7 in *Coordinate Measuring Machines and Systems*, edited by J. Bosch, Marcel Dekker, 137–220.
23. Phillips, S. D., B. Borchardt, and G. Caskey, 1993. Measurement Uncertainty Considerations for Coordinate Measuring Machines, NISTIR 5170, National Institute of Standards and Technology, Gaithersburg, Maryland.
24. Shannos, M. I, and D. Hoey, 1975. Closest-Point Problems, *Proceedings of the 16th Annual Symposium on Foundations of Computer Science*, IEEE Computer Society, Long Beach, California, 151–162.
25. Schrage, L.,1989. *User's Manual: Linear, Integer and Quadratic Programming with Lindo*, 4th Edition, Scientific Press.
26. Zhou, J. L., A. L. Tits and C. T. Lawrence, 1995. FFSQ Software, Electrical Engineering Department, University of Maryland, College Park, Maryland.

On Minty Variational Principle

Franco Giannessi
Department of Mathematics, Università di Pisa
Via Buonarroti 2, 56127 Pisa, Italy

Abstract. The well known Minty Variational Inequality is extended to the vector case. Such an inequality is shown to recover Vector Optimization problems in the convex case.

Key words: Variational and Vector Variational Inequalities; Minty Lemma; Minty Variational Principle; Vector Optimization

1. Introduction

Variational Inequalities have shown to be important mathematical models in the study of many real problems, especially equilibrium problems [1,5]. Sometimes the problems have received a "scalar" formulation, even if reality had a "vector nature". For this reason in [3] the proposal was done to introduce a vector version of a Variational Inequality. Then several studies have been done on this subject. However, the relationships between this theory and Vector Optimization need further investigation. The present note aims to contribute to this topic. The classic Minty Inequality and Minty Lemma [1,5] have shown to be an important tool in the variational field when the operator is monotone and the domain is convex. They offer a variational principle and, when the operator is a gradient, also a minimum principle for convex optimization problems [4]. Minty Variational Inequality is here formulated for the vector case in the finite dimensional case and in the simplest format. It is shown that the resulting inequality collapses to a Vector Optimization Problem, when the domain is convex and the operator is a monotone gradient.

In the sequel we will adopt the following ordering relations. Consider the cone $C := \mathbf{R}^n_+ \setminus \{0\}$, namely the Pareto case [6,7]; given a set S, $\forall x, y \in S$ we will have:

$$\begin{aligned} x \geq_C y \quad &\Longleftrightarrow \quad x - y \in C; \\ x \not\geq_C y \quad &\Longleftrightarrow \quad x - y \notin C; \end{aligned}$$

analogous definitions can be introduced for the opposite inequalities.

F. Giannessi et al. (eds.), New Trends in Mathematical Programming, 93–99.

2. Minty Vector Variational Inequality

Let the positive integers ℓ and n be given, as well as the non–empty domain $K \subseteq \mathbb{R}^n$ and the matrix–valued function $F : \mathbb{R}^n \to \mathbb{R}^{\ell \times n}$. Consider the following Vector Variational Inequality: find $y \in K$ such that

$$\langle F(x), y - x \rangle_\ell \not\geq_C 0 \quad , \quad \forall x \in K, \tag{2.1}$$

where $\langle \cdot, \cdot \rangle_\ell$ denotes a vector of ℓ scalar products in $\mathbb{R}^n$; obviously, $\langle \cdot, \cdot \rangle_1 = \langle \cdot, \cdot \rangle$, and $\langle \cdot, \cdot \rangle_\ell$ is a Jacobian product when F is a gradient. At $\ell = 1$ (2.1) becomes:

$$\langle F(x), y - x \rangle \leq 0 \quad , \quad \forall x \in K, \tag{2.2}$$

and is the classic Minty Variational Inequality [1,5]. Thus, (2.1) can be called Minty Vector Variational Inequality.

Now, consider the vector–valued function $f : \mathbb{R}^n \to \mathbb{R}^\ell$, and the Vector–Minimization Problem:

$$\min{}_C f(x) \quad , \quad x \in K, \tag{2.3}$$

where $\min_C$ marks vector–minimum with respect to the cone C: y is a vector–minimum point of (2.3) iff

$$f(y) \not\geq_C f(x) \quad , \quad \forall x \in K. \tag{2.4}$$

When $\ell = 1$, (2.4) becomes obviously $f(y) \leq f(x)$, $\forall x \in K$.

PROPOSITION 1. *Let K be closed, convex and with non–empty interior; f convex, differentiable on an open set containing K, and such that $f'(x) = F(x)$. Then y is a vector–minimum point of (2.3) iff it is a solution of (2.1).*

Proof. "If". Let y be a solution of (2.1). Ab absurdo, suppose that $\exists z \in K$ such that:

$$f(y) \geq_C f(z). \tag{2.5}$$

Because of the convexity of K, we have $z(\alpha) := \alpha y + (1 - \alpha) z \in K$ $\forall \alpha \in [0, 1]$. Because of the convexity of f and of the Lagrange Mean Value Theorem, $\exists \hat{\alpha} \in]0, 1[$, such that [1] :

$$\frac{d}{d\alpha} f(z(\hat{\alpha})) \geq f(y) - f(z). \tag{2.6}$$

[1] Without any fear of confusion, we write $\geq$ instead of $\geq_{\mathbb{R}^\ell}$; the same will happen in the sequel.

From (2.5) and (2.6) we obtain:

$$\langle F(\hat{x}), y - z \rangle_\ell \geq_C 0 \quad , \quad \hat{x} := z(\hat{\alpha}) \in K.$$

Multiplying both sides of this inequality by $1 - \hat{\alpha}$, we find:

$$\langle F(\hat{x}), y - \hat{x} \rangle_\ell \geq_C 0 \quad , \quad \hat{x} \in K,$$

which contradicts (2.1). "Only if". Let y be a solution of (2.3). Ab absurdo, suppose that $\exists \hat{x} \in K$, such that:

$$\langle F(\hat{x}), y - \hat{x} \rangle_\ell \geq_C 0.$$

From the convexity of f we have:

$$f(y) - f(\hat{x}) \geq \langle F(\hat{x}), y - \hat{x} \rangle_\ell.$$

From the two above inequalities we deduce:

$$f(y) - f(\hat{x}) \geq_C 0 \quad , \quad \hat{x} \in K,$$

which contradicts (2.4). This completes the proof. □

Suppose that there exists f such that $F(x) = f'(x)$ with f convex and K convex. At $\ell = 1$ (2.2) represents the non–negativity of the directional derivative of f from every $x \in K$ in the direction $y - x$, and gives a Minimum Principle for scalar minimization problems [4]. Therefore, when $\ell > 1$, it is natural to interpret (2.1) in a quite analogous way and call it *Minty Vector Minimum Principle.*

Now consider a perturbed Vector Variational Inequality: find $y \in K$ for which $\exists \bar{\epsilon} \in]0, 1[$, such that

$$\langle F(y + \epsilon(x - y)), x - y \rangle_\ell \not\leq_C 0 \;, \; \forall x \in K, \; \forall \epsilon \in]0, \bar{\epsilon}]. \tag{2.7}$$

PROPOSITION 2. *Let K be non–empty, closed and convex, and F monotone. Then y is a solution of (2.7) iff it is a solution of (2.1).*

Proof. "If". Let y be a solution of (2.1). The convexity of K implies $x_\epsilon := y + \epsilon(z - y) \in K$, $\forall z \in K$, $\forall \epsilon \in [0, 1]$. At $x = x_\epsilon$, with $\bar{\epsilon} = 1$ and $\epsilon \in]0, \bar{\epsilon}]$, (2.1) becomes equivalent to (2.7). "Only if". Let y be a solution of (2.7). The convexity of K implies $x_\epsilon := y + \epsilon(x - y) \in K$, $\forall x \in K$, $\forall \epsilon \in [0, 1]$. Inequality (2.7), multiplying both its sides by ϵ, becomes:

$$\langle F(x_\epsilon), y - x_\epsilon \rangle_\ell \not\geq_C 0 \;, \; \forall x \in K, \; \forall \epsilon \in]0, \bar{\epsilon}].$$

Taking into account this inequality, the monotonicity of F, and choosing any $\epsilon \in]0, \bar{\epsilon}]$, we find:

$$\langle F(x), y - x\rangle_\ell = \frac{1}{1-\epsilon}\langle F(x), x_\epsilon - x\rangle_\ell \leq \frac{1}{1-\epsilon}\langle F(x_\epsilon), x_\epsilon - x\rangle_\ell =$$

$$= \frac{1}{\epsilon}\langle F(x_\epsilon), y - x_\epsilon\rangle_\ell \not\geq_C 0 \quad , \quad \forall x \in K.$$

Hence (2.1) follows. This completes the proof. □

The above proposition shows a "perturbation nature" of Minty Variational Inequality.

3. The weak case

The extension of the (simplest form of) Stampacchia Variational Inequality to the vector field leads to consider the following Vector Variational Inequality [3]: find $y \in K$ such that

$$\langle F(y), x - y\rangle_\ell \not\leq_C 0 \, , \quad \forall x \in K. \tag{3.1}$$

PROPOSITION 3. *Let K be convex and F monotone. $y \in K$ is a solution of (2.1) if it is a solution of (3.1).*

Proof. Let y be a solution of (3.1). From the monotonicity of F and (3.1) we have:

$$\langle F(x), y - x\rangle_\ell \leq \langle F(y), y - x\rangle_\ell \not\geq_C 0,$$

and hence (2.1) follows. This completes the proof. □

The above proposition cannot be inverted, as simple examples show; take, for instance, $n = 1$, $\ell = 2$, $K = [-1, 0]$, $F(x) = (1, 2x)$; $y = 0$ is not a solution of (3.1) while it is a solution of (2.1). When $F(x) = f'(x)$, a consequence of Propositions 1 and 3 is that (3.1) is a sufficient (but not necessary) condition for y to be a vector–minimum point of problem (2.3). See [2] for a wide analysis of this subject. The above lack of necessity is overcome, if the inequalities are weakened. This leads to replace (3.1) with the following Weak Vector Variational Inequality (of Stampacchia kind): find $y \in K$ such that

$$\langle F(y), x - y\rangle_\ell \not<_C 0 \quad , \quad \forall x \in K, \tag{3.2}$$

and to replace (2.1) with the following Weak Vector Variational Inequality (of Minty kind): find $y \in K$ such that

$$\langle F(x), y - x \rangle_\ell \not\geq_C 0 \quad , \quad \forall x \in K. \tag{3.3}$$

PROPOSITION 4. *If K is convex, F monotone and continuous, then y is a solution of (3.2) iff it is a solution of (3.3).*

Proof. "Only if". Let y be a solution of (3.2). The monotonicity of F and (3.2) imply

$$\langle F(x), y - x \rangle_\ell \leq \langle F(y), y - x \rangle_\ell \not\geq_C 0 \, , \; \forall x \in K,$$

and hence (3.3) follows. "If". Let y be a solution of (3.3). Consider any $z \in K$ and any sequence $\{\alpha_r\|_{r=1}^{\infty} \downarrow 0$ with $\alpha_r \in]0,1]$, $n \in \mathbb{N} := \{1, 2, \ldots\}$. The convexity of K implies $z_r := (1 - \alpha_r)y + \alpha_r z \in K$. At $x = z_r$ (3.3) becomes:

$$\alpha_r \langle F(z_r), y - z \rangle_\ell \not\geq_C 0 \, , \; \forall z \in K.$$

By dividing both sides by α_r, the above inequality becomes:

$$\langle F(z_r), y - z \rangle_\ell \not\geq_C 0 \, , \; \forall z \in K.$$

Passing to the limit, as $n \to +\infty$, in this inequality and taking into account the continuity of F, we find:

$$\langle F(y), y - z \rangle_\ell \not\geq_C 0 \, , \; \forall z \in K,$$

which is equivalent to (3.2). This completes the proof. □

Consider again the function f of Sect.2, and the Weak Vector–Minimization Problem:

$$CW\!-\!\min f(x) \quad , \quad x \in K, \tag{3.4}$$

where $CW-\min$ marks weak vector–minimum with respect to the cone C: y is a weak vector–minimum point of (3.4) iff

$$f(y) \not\geq_C f(x) \quad , \quad \forall x \in K. \tag{3.5}$$

When $\ell = 1$, (3.5) does not differ from (2.4).

PROPOSITION 5. *Let K be closed, convex and with non–empty interior; f convex, differentiable on an open set containing K and such that $f'(x) = F(x)$. Then y is a weak vector–minimum point of (3.4) iff it is a solution of (3.2).*

Proof. “If”. Let y be a solution of (3.2). Ab absurdo, suppose that $\exists z \in K$ such that:

$$f(y) \; >_C \; f(z). \tag{3.6}$$

Because of the convexity of f we have:

$$f(z) - f(y) \geq \langle f'(y), z - y \rangle_\ell \; , \; \forall z \in K.$$

This inequality and (3.6) imply

$$\langle f'(y), z - y \rangle_\ell <_C 0 \; , \; z \in K,$$

which contradict (3.2). “Only if”. Let y be a solution of (3.4). Ab absurdo, suppose that y be not solution of (3.2). Hence, because of Proposition 4, $\exists \tilde{x} \in K$, such that:

$$\langle F(\tilde{x}), y - \tilde{x} \rangle_\ell >_C 0.$$

The convexity of f implies:

$$f(y) - f(\tilde{x}) \geq \langle F(\tilde{x}), y - \tilde{x} \rangle_\ell.$$

From the two above inequalities we deduce:

$$f(y) - f(\tilde{x}) >_C 0, \quad \tilde{x} \in K,$$

which contradicts (3.5). This completes the proof. □

These propositions show that, in the weak case, the inequalities of Stampacchia kind, of Minty kind, and Vector–Optimization problems agree.

References

1. C. Baiocchi and A. Capelo, “Variational and Quasi–Variational Inequalities. Applications to free–boundary problems”. J. Wiley, 1994.
2. G.–Y. Chen and X. Q. Yang, “Vector Variational Inequality”. Monograph. To appear
3. F. Giannessi, “Theorems of the alternative, quadratic programs and complementarity problems”. In “Variational Inequalities and Complementarity problems”, R. W. Cottle et al. (Eds.), J. Wiley, 1980, pp. 151–186.

4. F. Giannessi, "On connections among separation, penalization and regularization for Variational Inequalities with point–to–set operators". Rendiconti del Circolo Matematico di Palermo, 1997.
5. D. Kinderlehrer and G. Stampacchia, "An introduction to Variational Inequalities". Academic Press, New York, 1980.
6. Dinh The Luc, "Theory of Vector Optimization". Springer-Verlag, 1989.
7. M. Pappalardo, "Stationarity in Vector Optimization". To appear

Single-facility location problems with arbitrary weights

Galina Jalal and Jakob Krarup
DIKU (Dept. of Computer Science,
University of Copenhagen)
Universitetsparken 1, DK-2100 Copenhagen, Denmark
E-mail: gala@diku.dk, krarup@diku.dk

Abstract. The literature on locational decision problems with *positive* weights representing demands is huge in volume and spans about four centuries of development. Recently, also *negative* weights have been considered, notably within the context of locating so-called *obnoxious* facilities. Investigations of models comprising both positive and negative weights, however, seem to be rather virgin territory. Embarking from a problem traditionally attributed to Pierre de Fermat (1601-1665), we summarize the findings of two papers dealing with single-facility location problems. The first one discusses a geometrical solution to a *Weber Problem* with weights $(-1, +1, +1)$, formulated – but incorrectly solved – by Courant and Robbins in 1941 whereas the second one outlines an exact algorithm for a specially structured *1-median problem* with arbitrary weights. Along the way, we address a series of open questions and propose directions for future research.

Key words: Locational decisions, Fermat, Weber, positive and negative weights

1. Introduction

Single-commodity facility location problems deal typically with the location of one or more *facilities* each providing the same kind of service to the *clients* allocated to it. We can view the facilities as *suppliers* of a certain commodity demanded by the clients. In general, *demands* are expressed by positive *weights* assigned to the clients or consumers of that commodity. The facilities to be located are normally regarded as "friendly" in the sense that "closeness" is viewed as an attractive property. Location theory, however, does also encompass the counterpart: the location of so-called *obnoxious* facilities where one frequently used criterion is the *maximum* distance between a facility and the closest client. For example, disregarding its workers and subcontractors, who else likes an obnoxious power plant in their backyard? Note that even a friendly facility may well become obnoxious unless "closeness" is taken with a grain of salt. Thus, optimal closeness to a noisy elementary school is rather "reachable within a few minutes' walk" than "next door".

Hybrid models with truly antagonistic criteria capturing both the friendly and the obnoxious aspect of a locational decision problem have

F. Giannessi et al. (eds.), New Trends in Mathematical Programming, 101–114.

not, to the best of our knowledge, been widely studied. Our point of embarkation is the following problem, allegedly first formulated by *Pierre de Fermat* (1601-1665) early in the 17th century. At the end of his celebrated essay on maxima and minima, Fermat threw out the challenge: *"Let he who does not approve of my method attempt the solution of the following problem: Given three points in a plane, find a fourth point such that the sum of its distances to the three given points is a minimum!"*

Fermat's problem was brought to the attention of another eminent mathematician, *Evangelista Torricelli* (1608-1647) who around 1640 devised a geometrical construction for its solution. An even simpler construction was devised by Simpson in 1750.

The achievements of Fermat, Torricelli, and Simpson are accounted for by Courant and Robbins in their famous book *"What is Mathematics?"* (Courant and Robbins, 1941). They have furthermore formulated – but incorrectly solved – a related, so-called *Complementary Problem* which essentially is a *Weber Problem* with both positive and negative weights.

The correct solution to the Complementary Problem is sketched in Section 2. Section 3 outlines the main ideas upon which an exact algorithm for a specially structured *1-median problem* with arbitrary weights is based. Other aspects as to the relevance of both positive and negative weights are discussed in the concluding section.

2. The Complementary Problem

For a given triangle ABC and for any point X in the plane, let $f(X) = AX + BX + CX$. A formal statement of the Fermat problem is then

$FERMAT(ABC)$: Find a point T such that

$$f(T) = AT + BT + CT = \min\{f(X) : X \in \mathbf{R}^2\}$$

$FERMAT(ABC)$ was, as mentioned above, solved geometrically by Torricelli around 1640. The Simpson variant (Simpson, 1750) of Torricelli's construction of T is illustrated in fig. 1.

ABC is the given triangle. ABC_1, AB_1C, A_1BC are the *outward* equilateral triangles. The three so-called *Simpson lines* AA_1, BB_1, CC_1 do all intersect at T and are all at the same length $f(T)$. The angle between any pair of neighbouring Simpson lines is 60°. T is accordingly the point from which each of the three sides AB, AC, BC subtends an angle of 120°. For a proof of these and related properties, see for example (Krarup and Vajda, 1996).

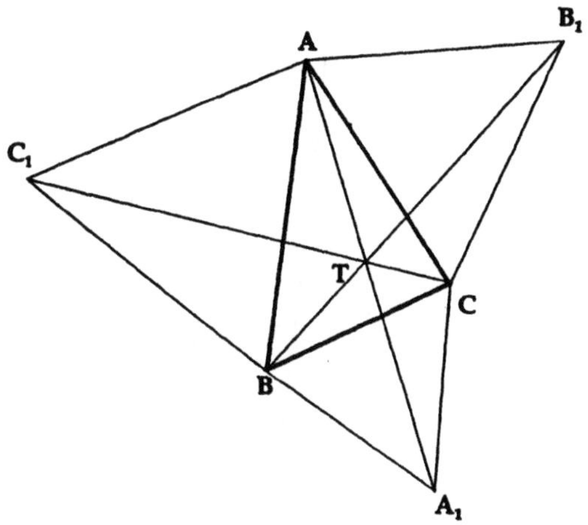

Figure 1. Simpson's construction of the fourth point T

Simpson's construction, however, does only provide the desired result if triangle ABC has no angle exceeding 120°. If one angle, say $\angle A$, equals 120°, the Simpson lines will intersect at A and the Torricelli point solving $FERMAT(ABC)$ will accordingly be A itself.

It is perhaps less obvious that A will solve $FERMAT(ABC)$ as it actually does when $\angle A$ is strictly greater than 120°, a result noted by Torricelli but first proved in (Heinen, 1834). A good question is then: what can in this case be deducted from the Torricelli-Simpson construction? Fig 2. shows the given triangle ABC with $\angle A > 120°$. We have furthermore the equilateral triangles ABC_1, BCA_1, and CAB_1 together with the Simpson lines, BB_1 and CC_1 which intersect at T'.

T' does certainly not solve $FERMAT(ABC)$ whereas A itself does, cf. the remarks above. Instead, T' is seen to represent the optimal solution to $FERMAT(AB_1C_1)$ since T' is the intersection of the Simpson lines BB_1 and CC_1.

Section 3 of §5 of Ch. VI in (Courant and Robbins, 1941) is titled "A Complementary Problem". With the points labelled as in fig. 4, the problem is introduced as follows:

"The formal methods of mathematics sometimes reach out beyond one's original intention. For example, if the angle A is greater than 120° the procedure of geometrical construction produces, instead of the point T (which in this case is the point A itself) another point T', from which the larger side BC of the triangle ABC appears under an angle of 120°, and the two smaller sides under the angle of 60°. Certainly, T' does not solve our minimum problem, but we may suspect that it has some relation to it. The answer is that T' solves the following problem:

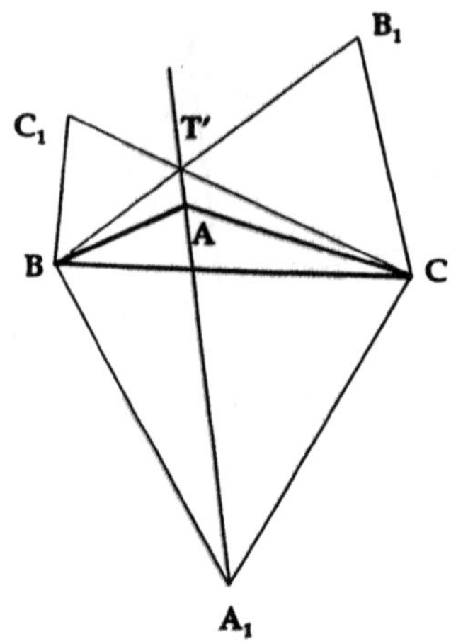

Figure 2. The Simpson construction for triangles $ABC, \triangle A > 120$

to minimize the expression $b+c-a$. *The proof [...] is left as an exercise for the reader.".*

As above, a, b, c denote the distances from the point sought (and claimed to be T') to A, B, C respectively. However, as shown in (Krarup, 1996), readers engaging themselves in the proposed exercise will become disappointed since the interpretation of T' is wrong.

For a given triangle ABC and for any point X in the plane, let $g(X) = BX + CX - AX$. The *Complementary Problem*, henceforth referred to as $CP(ABC)$, is now the problem of identifying a point V minimizing $g(X)$, that is,

CP(ABC): Find V such that

$$g(V) = BV + CV - AV = \min\{g(X) : X \in \mathbf{R}^2\}$$

To see that T' in fig. 2 does *not* solve $CP(ABC)$ is fairly straightforward. Disregarding the degenerate cases where A, B, C are collinear, T' and A must always be on the same side of the line containing B and C. If T'' is the mirror image of T' with respect to that line, $BT' + CT' = BT'' + CT''$ whereas $AT'' > AT'$ or $g(T'') < g(T')$. T'' is thus a better solution to $CP(ABC)$ than T'.

Where should the point V minimizing $g(X)$ then be found? Part of the answer can be derived from what may be termed the Restricted Complementary Problem or $RCP(ABC)$ for short. Without loss of generality we can assume the given triangle to be labelled such that $\angle B \geq C$ or $AB \leq AC$. $RCP(ABC)$ is then identical to $CP(ABC)$ with the additional constraint $AX \leq AC$. Let W be the point sought for:

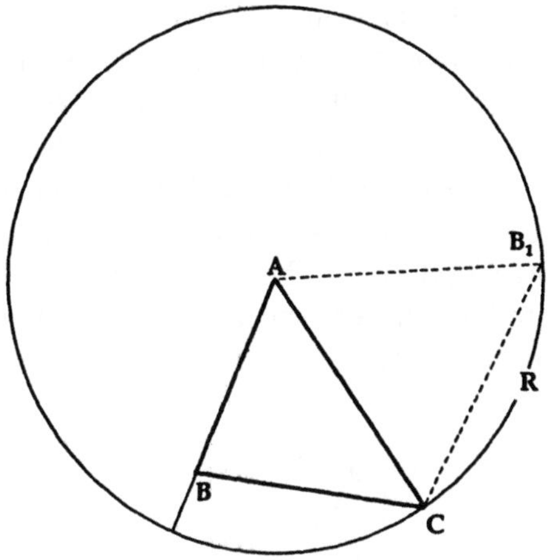

Figure 3. An instance $ABC, \Delta B \geq \Delta C$, of $RCP(ABC)$

RCP(ABC): Find W such that

$$g(W) = BW + CW - AW = \min\{g(X) : X \in \mathbf{R}^2, AX \leq AC\}$$

Fig. 3 shows a triangle ABC with $\angle B \geq \angle C$. Furthermore, a circle R having A as its centre and with radius $r = AC$ is drawn. B must be either within or on that circle. The same applies for T' since T' must be somewhere on the line segment BB_1 where B_1 is the third point on the equilateral triangle ACB_1.

For any point X in the plane it follows from the triangle inequality applied to ΔBXC that $BX + CX \geq BC$. BC is thus a *lower bound* on the first two terms of the objective function $g(X)$. Similarly, since $AX \leq r$, an *upper bound* on AX, the amount substracted in the third term of $g(X)$ must be AC itself. For any X on the line segment $BC, BX + CX$ will equal the lower bound BC. For $X = C$, *both* the lower bound BC on $BX + CX$ and the upper bound r on AX will be reached. If $\angle B > \angle C, B$ will be an interior point in the circle and $X = C$ is the uniquely determined optimal solution to $RCP(ABC)$ as asserted. For $\angle B = \angle C$, both B and C are optimal solutions.

It can be shown that $CP(ABC)$ and $RCP(ACB)$ have identical solutions if $\angle B \geq \angle C$ and $\angle C \leq 60°$. However, as will be commented on shortly, there *do* exist triangles ABC having one or infinitely many optimal solutions to $CP(ABC)$ *outside* the circle R.

The contents of (Krarup, 1996) are: If a point V is an optimal solution to $CP(ABC)$ then V will also solve $FERMAT(A'BC)$ for any point A' on the line containing A and V such that V belongs to the line segment AA'. It is furthermore noted that whereas an optimal solution to the Fermat problem is uniquely determined unless A, B, C are collinear, $CP(ABC)$ may have one, two, or infinitely many optimal solutions. Finally, to solve $CP(ABC)$, in general a shortcut is to identify a set of points A' which, together with B and C, defines a set of "companion instances" (not *equivalent* instances) of $FERMAT(A'BC)$ such that an optimal solution to $CP(ABC)$ also solves any $FERMAT(A'BC)$. The solution properties of the latter are thereafter exploited to obtain the final result. The validity of these observations extends easily to the degenerate case where A, B, C are collinear.

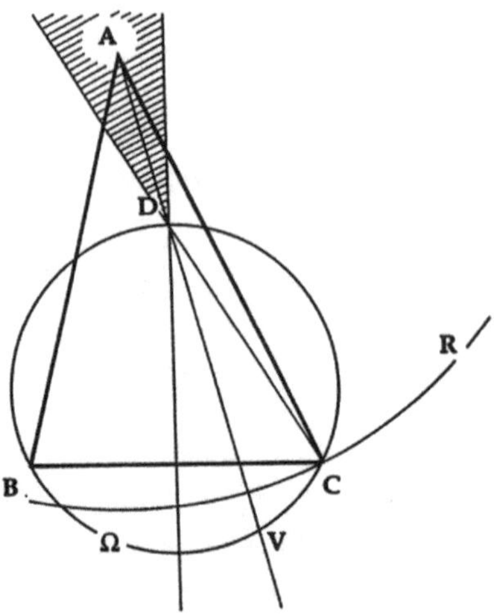

Figure 4. The geometrical solution to CP(ABC) e. g. with $\angle B > \angle C, \angle C > 60^0$

Without dwelling on details, the geometrical construction of V is shown in fig. 4 for triangles ABC with $\angle B \geq \angle C$ and $\angle C > 60^0$. A is then any point in the shaded area. R is the same circle as in fig. 3. $\triangle BCD$ is equilateral so $\angle BYC = 120°$ for all points Y on the arc Ω. From the connection between $CP(ABC)$ and $FERMAT(A'CB)$ quoted above, it can be deducted that A, D, V, A' are all collinear. V is thus the intersection of the line containing these four points and the arc Ω.

All of the above does in principle also apply when ABC is *equilateral*, the only exception being that the solution to $FERMAT(A'BC)$ still is uniquely determined whereas $CP(ABC)$ in this case has infinitely many optimal solutions.

Although $FERMAT(ABC)$ and $CP(ABC)$ appear to be almost identical, there is one important feature which indeed separates them. $FERMAT(ABC)$ is a *convex* problem which means that there is no *local minimum* different from the *global* one. $CP(ABC)$, on the other hand, is *nonconvex*. For example, if the given triangle is *isosceles* with $AB = AC$, an optimal solution can, as mentioned above, be found at both B and C. Now, if ABC is *almost* isosceles, a gradient technique, say, may take us to a local minimum which is not global.

2.1. FERMAT, WEBER, p-MEDIAN

The *Weber Problem*, known under an abundance of other names including the more suggestive *The Euclidean Minimum Distance Facility Location Problem* asks for a point minimizing the sum of the weighted Euclidean distances from itself to n given points $A_1, \ldots, A_n$, each associated with a given weight. We note in passing that *WEBER* $((A_1, w_1), \ldots, (A_n, w_n))$ with weights $w_1, \ldots, w_n$ is optimally solvable via a mechanical analogue device named the *Varignon frame* after its originator, *Pierre Varignon* (1654-1722), and described in his treatise *"Projet d'une nouvelle mécanique"*, published in 1687.

It is seen that *WEBER* $((A_1, 1), (A_2, 1), (A_3, 1))$ is identical to *FERMAT*(*ABC*) with $A = A_1, B = A_2, C = A_3$, that is, *FERMAT*(*ABC*) can be considered as the *unweighted* case of *WEBER*( ) for $n = 3$. Likewise, for $CP(ABC)$ we can define an equivalent Weber problem *WEBER* $((A, -1), (B, 1), (C, 1))$ where A, B, C represent the $n = 3$ clients to be served and $-1, 1, 1$ are the corresponding weights.

The (Euclidean or continuous) *p-median* problem is to establish p facilities and to supply each client from a subset of these such that the demands (represented by the weights) of all clients are met and such that the total costs thereby incurred are minimized. With distances replaced by costs, *WEBER* $((A_1, w_1), \ldots, (A_n, w_n))$ can thus be viewed as a $1 - median$ problem as well.

3. The "Pos/neg-weighted" 1-median Problem on a Cactus

$FERMAT(ABC)$ and $CP(ABC)$ are both examples of *continuous* location problems in the sense that the optimizing point sought for can be placed anywhere in the plane. Focussing on continuous models: how to proceed to more complex cases?

One possibility is to seek the geometrical constructions extended to cases where "more than three clients are to be served" for various combinations of positive and negative weights. The road ahead could then be to move from the triangles hitherto considered to a square, a rectangle, a general quadrilateral, a pentagon, and so forth. This aspect was discussed with R.E. Burkard after a talk on $CP(ABC)$ delivered at University of Technology, Graz, Austria, in October 1995. However, as Burkard pointed out, it is likely that we then may repeat the efforts made by some German geometer a hundred years ago.

Instead it was agreed in the first round to look at *discrete* formulations where the set of potential sites for the facilities to be placed is finite and often represented by the *vertices of a network.*

The so-called *NP-complete* problems are equivalent in the sense that either all or none of these problems can be solved in *polynomial time.* At present no polynomial-time algorithm is known for solving any *NP-complete* problem to optimality and it has furthermore been widely conjectured that no such algorithm can ever be devised. Actually, to prove that this conjecture is either true or false is viewed by many to be today's most famous unsolved problem within theoretical computer science.

In terms of *computational complexity,* almost all problems of interest within discrete location theory – or, for that matter, within combinatorial optimization as such – are in general *NP-hard.* This means that the existence of a polynomial-time algorithm for solving one of these would imply that all *NP-complete* problems are solvable in polynomial time.

How to approach an *NP-hard* problem? We need first to distinguish between *exact algorithms* guaranteeing both feasibility and optimality and *heuristics* guaranteeing nothing in general although feasbility in many cases is achieved. In principle, we can design a *search* algorithm based on, for example, the *branch-and-bound* principle for the optimal solution of any *NP-hard* problem. Such an effort, however, would be of no interest unless it was accompanied by extensive numerical experiments documenting its superiority over other algorithms hitherto proposed for the same problem. Alternatively, one may develop heuristics, possibly complemented by analyses of their worst-case (or average-case) behaviour.

Yet another possibility is to identify what is termed *well-solved special cases.* For example, if a problem defined on a network is known to be *NP-hard* in general, it may be polynomially solvable (and hence well-solved) if the underlying network exhibits a special structure. Thus, the network could be, say, a simple path, a tree, or a cycle, or it could be identified in terms of more complex forbidden subgraphs.

This is the line of approach taken in (Burkard and Krarup, 1996) for the *1-median problem in a network* with both positive and negative weights: find a vertex α minimizing the sum of the weighted shortest path distances from itself to all other vertices.

Let $N = (V, E, w, c)$ be a given connected network with vertex set V and edge set E. With each vertex $x \epsilon V$ is associated a certain *weight* w_x. Likewise, with each edge $(x, y) \epsilon E$ we associate a certain length c_{xy}. A cycle is a sequence $(x_1, \ldots, x_k, x_{k+1} = x_1)$ of k distinct vertices supplemented by $x_{k+1} = x_1$ such that (x_j, x_{j+1}) is an edge, $j = 1, ..., k$, $k \geq 3$. N is assumed to be a *cactus*, that is, no two cycles of N have more than one vertex in common. The weights w_x can be any real numbers and likewise for the lengths c_{xy}, the only exception being that the length of an edge included in a cycle must be nonnegative. For all pairs x, y of distinct vertices, let d_{xy} be the length of a *shortest path* (or the *shortest path distance*) between x and y. d_{xx}, all x, is assumed to equal zero throughout. The objective function to be minimized is defined for each vertex x and expresses the sum of the weighted shortest path distances between x and all other vertices, that is,

$$f(x) = \sum_{y \in V} w_y d_{xy}$$

For given $N = (V, E, w, c)$ where N is a cactus, the *1-median* problem (or *1-MP* for short) can now be stated

1-MP: Find a vertex α such that

$$f(\alpha) = \min\{f(x) : x \in V\}$$

For an ordinary *1-median* problem on a network where all weights are positive, the so-called *Vertex Optimality Property* asserts the existence of an optimal solution at a vertex. This property does not apply in the present case. We need therefore explicitly to request that a vertex solution is required as done in the formulation above.

Not much sophistication is needed to construct small data instances of 1-MP demonstrating the existence of local minima different from the global one. 1-MP and CP(ABC) do thus share the computationally complicating (but challenging!) property of being nonconvex, cf. the remarks made in the previous section.

The main ideas of the exact algorithm devised in (Burkard and Krarup, 1996) for solving 1-MP defined on a cactus and with arbitrary weights will now be explained with reference to the small example shown in fig. 5.

For ease of exposition we will first consider an instance of 1-MP where $N = (V, E, w, c)$ is a tree rooted at some vertex, here at vertex a. The path between a pair x, y of vertices is then uniquely determined and of length d_{xy}. The circled numbers shown are the weights.

Suppose we were to evaluate the objective function $f(x)$, for example, at vertex e. We find: $f(e) =$ the contribution from vertex a + the contribution from vertex $b + \ldots$:

$$f(e) = 5c_{ae} + (-3)(c_{ba} + c_{ae}) + \ldots$$

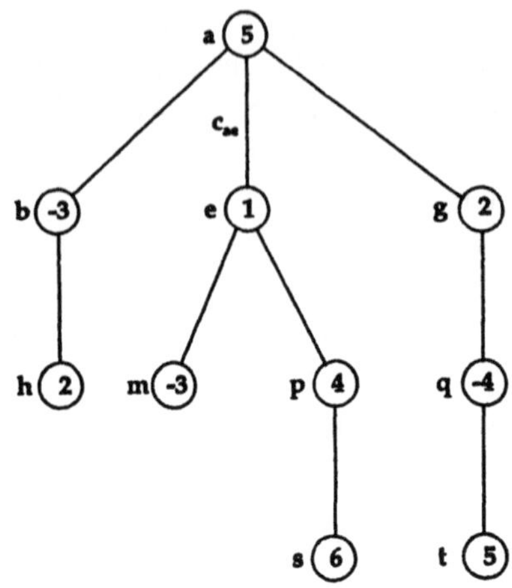

Figure 5. An instance $N = (V, E, w, c)$ of 1-MP where N is a tree

Let $n = \mid V \mid$ be the total number of vertices. The sum on the right hand side in the expression above will thus comprise $n-1$ terms. With a data structure allowing for the evaluation of each term in constant time, the time complexity of evaluating $f(x)$ for a specific vertex is $O(n)$. To solve 1-MP by *complete enumeration* means to repeat this exercise for all n vertices and then pick the best. The overall complexity of such a primitive approach is accordingly $O(n^2)$.

It appears, however, that the *difference* $\Delta_{xy} = f(y) - f(x)$ between the objective function values for a pair (x, y) of adjacent vertices can be calculated in *constant* time. Assume, for example, that we want to determine $\Delta_{ea} = f(a) - f(e)$ in fig. 5. We obtain,

$$f(a) = f(e) + (w_e + w_m + w_p + w_s)c_{ae} - (w_a + w_b + \ldots + w_t)c_{ae}$$

or

$$\Delta_{ea} = f(a) - f(e) = (W_E - W_A)c_{ae}$$

where the definition of W_A and W_E follows from the grouping of the two sets of terms.

In general, the removal of an edge (x, y) from a tree can be viewed as a *partition* (X, Y) of the vertex set V. Upon removal of (x, y), X is the nonempty subset of vertices reachable from x and $Y = V \backslash X$ is the equally nonemptive complement. W_X and W_Y are the total weights of all vertices in X and Y respectively.

Let $\mathbf{W} = W_X + W_Y$ be the total weight of all vertices. Hence,

$$\Delta_{xy} = f(y) - f(x) = (\mathbf{W} - 2W_X)c_{xy}$$

If **W** is known and x is an *endvertex* (i.e. a vertex of degree 1), W_X reduces to the weight w_x of x only. Δ_{xy} can thereafter be determined in constant time.

Initially, all vertices are *live*. In each step we choose a live endvertex x together with its *companion* vertex y, the only vertex adjacent to x. x is *killed* by evaluating Δ_{xy} and passing all relevant information on to y. Among the things recorded is the (locally) *optimizing vertex*: if $\Delta_{xy} = f(y) - f(x) < 0, y$ is a better solution than x; otherwise, x is noted as "the best solution so far". Ties can be resolved arbitrarily.

The live endvertices in fig.5 are killed in the following order: t, s, q, p, m, h, g, b. The only remaining live vertices are then e and a which we call *junctions*.

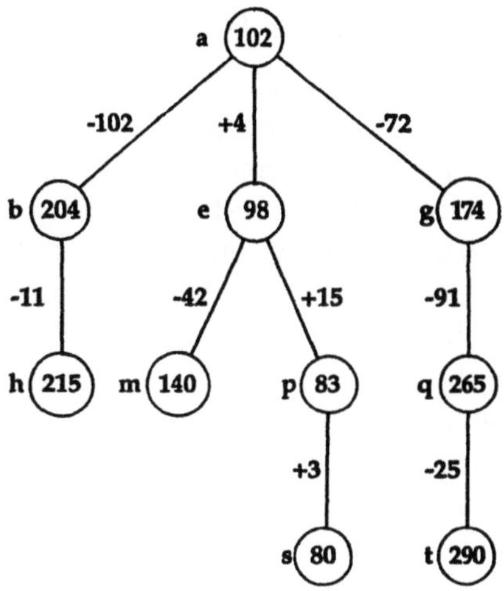

Figure 6. How to handle junctions. Circled numbers: $f(x)$, other numbers: Δ_{xy}

Fig. 6 shows how junctions can be handled. The numbers associated with the edges are the differences $\Delta_{xy} = f(y) - f(x)$. For comparison, the objective function values themselves, calculated "manually" based on edge lengths c_{xy} not shown, are given as the numbers in the circles representing the vertices.

To explain what happens at junction e after vertices m, p, s have been killed, we note that

$$f(p) - f(s) = 3$$

(s is the optimizing vertex along the path $s - p$);

$$f(e) - f(p) = 15$$

(s is still the optimizing vertex along the path $s - p - e$), or

$$f(e) = f(s) + 3 + 15 = f(s) + 18 \ .$$

Furthermore, for the other path $m-e$ starting at an endvertex (here m) and terminating at a junction (here e), we obtain

$$f(e) - f(m) = -42$$

(e is recorded as the optimizing vertex).

The value of $f(e)$ is unknown at this stage. What we do know, however, is that the two equations involving $f(e)$ must be satisfied simultaneously, that is, $f(e) = f(s) + 18, f(e) = f(m) - 42$. Hence, s is the locally optimal solution among vertices s, p, m, e.

With s, p, m killed, e is now an endvertex which is killed in the final step. Three different paths do now terminate at junction a and three equations involving $f(a)$ must accordingly be satisfied:

$$f(a) = f(b) - 102, \qquad f(a) = f(s) + 22, \qquad f(a) = f(g) - 72 .$$

Since $\max\{-102, 22, -72\} = 22$ and since a is the root of the tree, we can conclude that s is the *globally* optimal solution whereby we are done. Furthermore, since each vertex is considered only once and can be killed in constant time, the overall complexity is $O(n)$.

The difficulty with a *cycle* as opposed to a *tree* is that any pair of vertices is connected by two edge-disjoint paths. For a network with certain lenghts associated with its edges, a *shortest path tree* for a given vertex specifies the shortest paths from that vertex to any other vertex. For any vertex of a tree, the shortest path tree is the tree itself. For a cycle, however, where we in a well-defined way can talk of shortest paths as being either *clockwise* or *counterclockwise*, the shortest path tree may vary fron one vertex to another. This fact must obviously complicate the updating whenever a vertex is being killed. A cycle, on the other hand, has no junctions which indeed simplifies the bookkeeping. In spite of the difficulties encountered, it is nevertheless possible also in this case to devise a data structure such that the total number of computations is a linear function of the number of vertices.

We shall finally allow for *combinations* of cycles and trees. If a single cycle and a single tree can be processed in linear time such that all information so to speak is collected in a single vertex, then the same procedure will work for networks with two or more cycles provided that no two cycles have more than one vertex in common. In other words, the linear algorithm can handle instances $N = (V, E, w, c)$ where N is a cactus.

4. Suggestions for further research

Among the myriads of formulations considered within the general area of *discrete location theory*, only four of these, *the p-median problem (p-*

MP), the p-center problem (p-CP), the uncapacitated facility location problem (UFLP), and the quadratic assignment problem (QAP) – at times referred to as prototype location problems – have played a particularly dominant role. Besides offering a wealth of theoretical challenges, these models have, despite the seeming simplicity of their underlying assumptions, provided important quantitative bases for the investigation of numerous practical location decision problems. For an overview of these four "families of prototype location problems" as they actually are, see e.g. (Krarup and Pruzan, 1990).

For given finite sets of clients and of potential sites for locating facilities, p-MP is the problem of locating exactly p facilities and to allocate each client to one of them such that all demands are satisfied at least total cost. $UFLP$ differs from p-MP in that certain *fixed costs* are incurred whenever facilities are established and p, the number of facilities to be located, is not prespecified. p-CP, however, differs significantly from both p-MP and $UFLP$ in several respects, primarily with respect to the criterion used for assessing the quality of a feasible solution. Whereas p-MP and $UFLP$ are *minisum* problems, p-CP has a *minimax* objective: open p facilities and assign each client to one of them such that the *maximum distance* (unweighted case) or *maximum weighted distance* from any open facility to any of the clients assigned to it is a minimum. Finally, for QAP, the clients are the facilities themselves with "demands" replaced by interaction between pairs of facilities.

It is tempting to seek the linear algorithm discussed in the previous section, or the principles upon which it is based, extended to more complex models like those considered above. For example, is it possible to identify a 1-center, the optimal solution to 1-CP, in a similar way?

Below is an instance of 1-CP where the underlying network is a simple path with five vertices a–e. The numbers above the vertices are the positive or negative weights whereas the numbers between a pair of vertices, here restricted to be positive, are the edge lengths.

The instance shown is here solved by complete enumeration and by killing the vertices one by one from left to right.

The entry in, for example, "row e" and "column b" is the weighted (shortest path) distance between a client with weight w_e at vertex e and the facility located at vertex $b : w_e d_{eb} = 3(3 + 1 + 4) = 24$.

Suppose now that vertices a and b are killed. With c added to the set, the information available so far is contained in the topmost left 3×3 submatrix. The maximum values in the three columns are 20, 12, 10 respectively. Since $\min\{20, 12, 10\} = 10$, vertex c should be chosen as the site for the facility. c is in other words the *local* 1-center for the path a–b–c. Upon the inclusion of the remaining two vertices d, e, however, d appears to be the *global* optimum.

The example shows furthermore that the inclusion of a vertex with negative weight does not affect the current position of the 1-center. This suggests that a different criterion should be employed in the presence of negative weights. With the idea of obnoxious facilities in mind, an obvious possibility is to introduce the notion of a Γ-*center*: find a vertex such that the absolute value of the weighted distance to the closest vertex with negative weight is maximum.

For the minisum problem 1-MP, an optimal solution is represented by a single vertex seen in relation to all other vertices. For the minimax problem 1-CP, the solution is represented by a *pair* of vertices: the vertex at which the 1-center is found and the vertex from which the weighted distance to that 1-center is maximum. It is thus very unlikely that we can devise an algorithm for 1-CP which looks at each vertex only once and can eliminate a vertex in constant time. The same applies for p-MP, $p > 1$.

The motivation for this study was the interest taken in $FERMAT$ (ABC) and the subsequent confrontation with the incorrect solution to the *Complementary Problem*. As regards realistic applications, however, we believe that models capable of handling arbitrary weights may well provide useful decision support, for example, to companies operating in a competitive environment or to monopolistic companies wishing to increase or decrease the number of existing facilities.

References

1. Burkard, R. E., and J. Krarup, *"A linear algorithm for the 'pos/neg-weighted' 1-median problem on a cactus"*, (1996), Bericht Nr. 90, Spezialforschungsbereich F 003, Optimierung und Kontrolle, Technische Universität Graz.
2. Courant, R., and H. Robbins, *What is Mathematics?*, Oxford University Press, 1941.
3. Heinen, F., *"Progr. Cleve"* (1834) 18-19.
4. Krarup, J. *"On 'A Complementary Problem' of Courant and Robbins"* (1996), Report 96/39, DIKU (Department of Computer Science, Univ. of Copenhagen), submitted for publication.
5. Krarup, J., and P. M. Pruzan, *"Ingredients of locational analysis"*, Ch. 1 in (P. B. Mirchandani and R. L. Francis, eds.) *Discrete Location Theory*, Wiley-Interscience, 1990.
6. Krarup, J., and S. Vajda, *"On Torricelli's geometrical solution to a problem of Fermat"* (1996), to appear in (S. Powell, and H. P. Williams, eds.) *"Special Edition of J. Mathematics in Business and Industry: In celebration of work of S. Vajda"*.
7. Simpson, T., *The Doctrine and Application of Fluxions*, London, 1750.

On Testing SLP Codes with SLP-IOR

P. Kall and J. Mayer
Institute for Operations Research
University of Zurich
Moussonstr. 15
CH-8044 ZURICH.

Abstract. This paper gives a summary of selected testing features of the model management system SLP-IOR. The pseudo random test problem generator GENSLP is described and numerical examples with randomly generated test problems are presented.

Key words: Stochastic Linear Programming, Test Problem Generators, Model Management Systems

1. SLP models in SLP-IOR

SLP-IOR is a model management system designed to facilitate the handling of stochastic linear programming (SLP) problems of various types as well as their solution by various solvers appropriate for the particular problem types, see Kall and Mayer [15],[16],[17]. So far we have implemented the following categories of SLP problems:

- *SLP with fixed recourse*, in particular
  - *arbitrary* fixed recourse,
  - *complete* fixed recourse, and
  - *simple* recourse;
- *SLP with joint chance constraints* (according to the state of the art allowing for random right hand sides with multivariate (nonsingular) normal distributions);
- *SLP with separate chance constraints* (at present restricted to random right hand sides with a variety of continuous type univariate probability distributions).

The model types mentioned above are briefly described as follows:

F. Giannessi et al. (eds.), New Trends in Mathematical Programming, 115–135.

SLP with fixed recourse

Here we consider problems given as

$$(1.1)\qquad \begin{cases} \min\{c^{\mathrm{T}}x + E_{\omega}Q(x,\omega)\} \\ \text{s.t. } Ax \propto b^{1} \\ \qquad x \in [l,u], \end{cases}$$

where A is a matrix of order $m_1 \times n_1$ and

$$(1.2)\qquad \begin{cases} Q(x,\omega) = \min q^{\mathrm{T}}(\omega)y \\ \text{s.t. } Wy \propto h(\omega) - T(\omega)x \\ \qquad y \geq 0, \end{cases}$$

with W and $T(\cdot)$ of order $m_2 \times n_2$ and $m_2 \times n_1$, respectively, and vectors $x,\ y$ as well as $c,\ b,\ l,\ u,\ q(\cdot),\ h(\cdot)$ of appropriate dimensions.
We distinguish

— *arbitrary fixed recourse*, i.e. W an arbitrary fixed matrix;

— *complete fixed recourse*, i.e. W fixed such that
$\{z \mid z = Wy, y \geq 0\} = \mathbb{R}^{m_2}$
(and $\{u \mid W^{\mathrm{T}}u \leq q(\omega)\} \neq \emptyset$ w.p. 1);

— *simple recourse*, i.e.
$W = (I, -I),\ T(\omega) \equiv T,\ q(\omega) \equiv q.$

SLP with joint chance constraints

These problems are stated as

$$(1.3)\qquad \begin{cases} \min c^{\mathrm{T}}x \\ P(\{\omega \mid Tx \geq h(\omega)\}) \geq \alpha \\ \qquad\qquad Ax \propto b \\ \qquad\qquad x \in [l,u], \end{cases}$$

[1] For $\propto$ any one of the relations $=, \leq, \geq$ row-wise.

with $0 \leq \alpha \leq 1$ being some (high) probability level.

SLP with separate chance constraints

In principal these problems can be formulated as

$$(1.4) \qquad \begin{cases} \min c^{\mathrm{T}} x \\ P(\{\omega \mid t_i^{\mathrm{T}}(\omega) x \geq h_i(\omega)\}) \geq \alpha_i, \ \forall i \\ \qquad\qquad Ax \propto b \\ \qquad\qquad\ x \in [l, u], \end{cases}$$

with $0 \leq \alpha_i \leq 1, \forall i$, given (high) probability levels. Here $t_i^{\mathrm{T}}(\omega)$ denotes the i-th row of $T(\omega)$, and row-wise independence is assumed. As mentioned above, at present we have implemented only the particular case $t_i^{\mathrm{T}}(\omega) \equiv t_i^{\mathrm{T}}$.

In the models above we assume that

— $\omega \in \Omega$, some given sample space,

— $(\Omega, \mathcal{F}, P)$ is a probability space,

— $q(\omega), h(\omega), t_i(\omega)$ are random vectors,

— $T(\omega)$ is a random matrix.

Furthermore, with the random vector $\xi(\omega) \in \mathbb{R}^k$, for the above random entries the following regression structure is assumed:

$$(1.5) \qquad \begin{cases} q(\omega) &= q^0 + \sum_{j=1}^k q^j \xi_j(\omega) \\ h(\omega) &= h^0 + \sum_{j=1}^k h^j \xi_j(\omega) \\ T(\omega) &= T^0 + \sum_{j=1}^k T^j \xi_j(\omega) \\ t_i(\omega) &= t_i^0 + \sum_{j=1}^k t_i^j \xi_j(\omega). \end{cases}$$

2. SLP solvers in SLP-IOR

For the various SLP models mentioned above a variety of solution methods was proposed in the literature, some of which have been implemented (and tested) and are available as source code. By a *solver* we understand

— a source code (which then can easily be linked to SLP-IOR), or

— an executable code linked to GAMS [3].

For *SLP with recourse* we distinguish two classes of solution methods ('+' for solvers included in SLP-IOR):

- Algorithms to solve the algebraic equivalent (for discrete distributions a large scale LP with dual decomposition structure), e.g.
  - — the *L*-shaped method of Van Slyke and Wets [32],
  - — the basis reduction method of Strazicky [28],
  - \+ the regularized decomposition method QDECOM of Ruszczyński [27],
  - \+ MSLiP of Gassmann [9],
  - — any general purpose LP method like the Simplex or Interior Point methods; available with us are
    - \+ MINOS5 (coming with GAMS),
    - \+ CONOPT (coming with GAMS),
    - \+ OSL (coming with GAMS),
    - \+ ZOOM (coming with GAMS),
    - \+ XMP (Marsten [23]),
    - \+ BPMPD (interior point method) by Cs. Mészáros [26],
    - \+ HOPDM (interior point method) by J. Gondzio [10],
    - \+ OB1 (experimental version from Marsten).
- Algorithms applied to the original problem formulation (1.1) like e.g.
  - \+ DAPPROX, a successive approximation for complete recourse problems of Kall and Frauendorfer [7], Frauendorfer [6], employing QDECOM
  - — SQG, the stochastic quasi-gradient method of Ermoliev [5] and Gaivoronski [8],
  - \+ SDECOM, the stochastic decomposition of Higle and Sen [11], (presently implemented without the stopping rule based on bootstrap methods),
  - \+ SRAPPROX, the successive approximation for simple recourse (using separability) of Kall and Stoyan [18].

For *SLP with joint chance constraints* the methods existing so far are adaptations of NLP procedures for multivariate normal RHS's; the adaptation relies on employing fast Monte Carlo simulation methods to evaluate (approximately) the probability function and its gradient. At present we have implemented

+ PCSPIOR using the method of Szántai [30], based on the *Supporting Hyperplane Method* proposed by Veinott,

+ PROBALL of J. Mayer, adapted from the *Central Cutting Plane Method* introduced by Elzinga and Moore,

+ PROCON of J. Mayer [24], based on the *Reduced Gradient Method* as discussed by Abadie and Carpentier.

Separate chance constraints are transformed into their deterministic equivalents (if possible).

For a detailed presentation of solution methods for the above SLP models see Mayer [25].

3. Some features of SLP-IOR

SLP-IOR is an interactive, menu driven system. The internal model representation consists of three elements,

— the *underlying algebraic structure* (see Fig. 1)

— the *underlying regression structure* (1.5)

— the *random variable structure*, i.e. the dependency structure and the joint distributions of groups of dependent random variables.

The system allows for various manipulations with a given model like

— editing and changing any of the above elements,

— extracting (and saving) as well as injecting any of the above structures,

— analyzing the actual model type,

— extracting, discretizing and then injecting again the distribution,

— export/import in other data formats (like S-MPS or GAMS),

— transforming into another model type (e.g. complete recourse into one of simple recourse, deterministic equivalent (if it exists), expected value problem or into a joint chance contrained problem).

In particular for testing purposes we have the features

— to generate randomly (GENSLP) single SLP models under various specifications (like model type, dimensions, stochastic parts, sparsity of arrays and magnitude of their NDef (i.e. nondefault) entries),

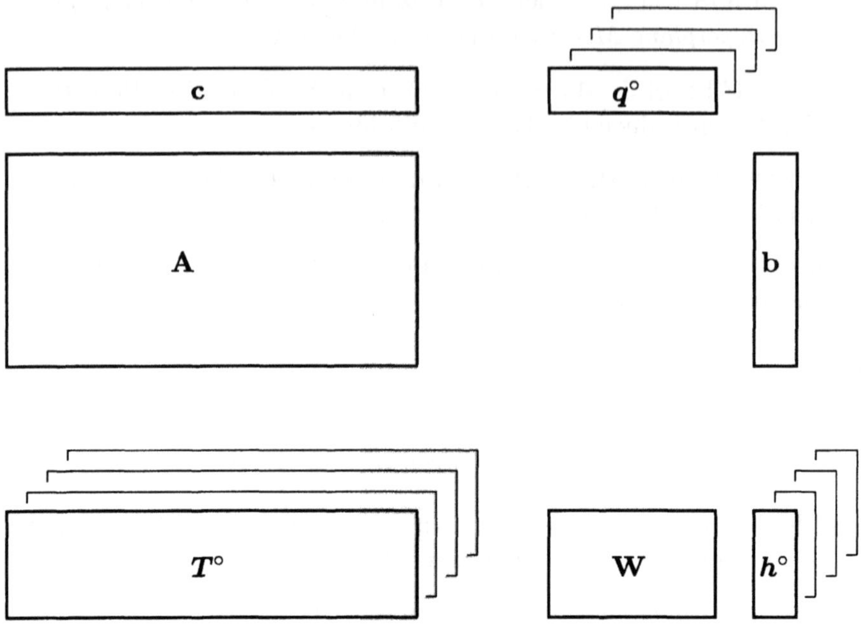

Figure 1. Two-stage model: Data arrays

— to generate randomly (GENSLP) test batteries of identical model type under the above specifications,

— to perturb data in the algebraic structure,

— to discretize various distributions and to inject distributions into single models and/or batteries.

4. The test problem generator GENSLP

In this section we give a brief description of GENSLP, a program for randomly generating test problem batteries for the following model classes:

- Deterministic LP problems;
- two stage problems with random, fixed, complete or simple recourse;
- jointly chance constrained problems with a single joint chance constraint and with only the RHS being stochastic, having a nondegenerate multinormal distribution.

The first version of GENSLP has been developed by P. Kall and E. Keller, see Keller [20], for generating two stage complete recourse test problems. This version was further developed by E. Keller and the authors by including sparse matrices, further specification facilities as well as a facility for generating jointly chance constrained test problems.

First of all let us discuss what we mean by "randomly generating" a test problem battery.

Let us begin the discussion with the case of a linear program in standard form, with technology matrix A, RHS b and objective c; let A be an $m \times n$ matrix. Let us consider a random linear program with all entries of A, b and c being independent random variables with a given distribution. In our case the distribution will be the uniform distribution with a prescribed range for the array entries. A randomly generated LP test problem battery arises by generating a sample for A, b and c, on the basis of pseudo random numbers.

For testing purposes these kinds of randomly generated problems have obvious limitations. On the one hand for large scale test problems the matrix A should be sparse and on the other hand we have good chance of obtaining test batteries which contain a large amount of infeasible and unbounded test problems. We performed an experiment with 100 test problems, randomly generated in the above sense. The dimensions were $m = 30$ and $n = 60$; the array elements range was $[-10, 10]$ and the density (sparsity) was 100%. It turned out that 47 test problems were infeasible and 47 unbounded, with the remaining 6 having an optimal solution. In a second experiment the objective coefficients were generated in the range $[0, 10]$ and the density was set to 20%. All of the other characteristics were the same as in the previous experiment. The battery now contained 44 infeasible problems and 56 problems with an optimal solution.

For including sparse matrices we define a random sparse matrix with a prescribed density of nonzero elements as follows. The row and column indices of the nonzero elements are independent discrete random variables, uniformly distributed in $[1, \ldots, m]$ and $[1, \ldots, n]$, respectively. The nonzero entries are themselves random variables, uniformly distributed in a prescribed range. Again, randomly generating an LP test problem battery with sparse A means generating a sample from the described population.

For generating test problems with existing optimal solutions we have to give up the aim of generating completely "random" problems. The matrix A is randomly generated in this case too. The RHS and objective function vectors however will be no more independent from the matrix entries. They are computed in such a way that feasibility and optimality is ensured, respectively. To achieve this goal a random solution point and Lagrange multiplier vector is generated. The components of both are uniformly distributed with the ranges chosen in such a way which ensures that b and c have entries in their prescribed ranges.

If further properties are required for A, e.g. that A has full row rank, then the independence of the matrix entries can obviously be no more ensured either.

For stochastic programming problems several further issues arise. The regression terms require no further specific discussion; they are generated as sparse arrays. Considering the remaining features we discuss the two stage and chance constrained cases separately.

For two stage problems we do not generate randomly a distribution; a one point distribution is returned. This immediately implies that for random and general fixed recourse problems there is no way to guarantee the existence of a solution. For these problem classes the various arrays appearing in the problem formulation are generated according to their prescribed sparsity and range.

Complete recourse problems and as a special case simple recourse problems have the property that the second stage subproblem has a feasible solution for any right hand side. Generating the second stage objetive q such that $q \geq 0$ holds ensures the existence of an optimal solution for the second stage, again independently of the right hand side. Now A, T and h are randomly generated according to range and sparsity. The RHS b is computed similarly to the LP case, on the basis of a randomly generated candidate solution such that feasibility is ensured. To guarantee the existence of an optimal solution c is computed such that the candidate solution minimizes $c^{\mathrm{T}}x$ subject to the first stage constraints. The crucial point in generating complete recourse problems is generating W which has the complete recourse property. The algorithm for generating such a matrix will be discussed later.

For chance constrained problems a joint chance constraint is generated with only the RHS being stochastic and having a nondegenerate multinormal distribution. The expected value and standard deviation vectors as well as the correlation matrix are randomly generated. The feasibility is guaranteed for the linear constraints in the way discussed above whereas for the joint chance constraint a linesearch has to be carried out w.r. to the stochastic constraint. This approach also ensures the existence of a Slater point. Optimality is achieved via Kuhn-Tucker conditions and GENSLP returns the optimal objective value as well as an optimal solution with the joint chance constraint being active at this solution point.

Below we give a more detailed description of the methods employed in GENSLP for randomly generating test problems. Let us begin with the elementary operations.

- *Generating a (pseudo) random entry* is performed by utilizing the uniform random number generator RND() of the Lahey FORTRAN F77L-EM/32 [21].
- *Generating a random index* in a range $[1, k]$ is carried out by generating a random number in $[0, 1]$ subdivided into k subintervals of equal length.

Generating matrices

General sparse matrix

A pair of random indices is generated. If the corresponding matrix entry is already nonzero then this procedure is repeated until a zero position is obtained. In this case a random entry is generated in the prescribed range. This procedure is repeated until the required target for the number of nonzeros is achieved.

Nonsingular sparse matrix

First an upper triangular matrix is generated in the same fashion as described above, with nonzeros along the diagonal. Afterwards a fixed number of steps of the "reverse" of a Gaussian elimination is carried out for achieving the prescribed number of nonzeros. This means that a random nonzero multiple of a randomly selected row/column is added to another randomly selected row/column, in a successive manner. This way a matrix arises which is no more triangular while the nonsingularity is preserved. Finally the matrix is "shaked" by performing a prescribed number of random row/column interchanges to achieve a "uniform" nonzero pattern.

Sparse matrix with full row/column rank

This is achieved by generating a nonsingular submatrix, followed by generating further nonzeros outside this submatrix, to achieve the prescribed number of nonzeros. Finally the matrix is "shaked".

Complete recourse W matrix

The procedure is based on the following theorem of Kall [13] (see also Kall and Wallace [19] and Vajda [31]).

PROPOSITION 1. *An $m \times n$ matrix W has the complete recourse property iff it has a full row rank and with B being an $m \times m$ regular submatrix the following condition holds: $\{ y \mid W \cdot y = -B \cdot \mathbf{1},\ y \geq 0 \} \neq \emptyset$, where $\mathbf{1}$ is a vector with all components equal to 1.*

The main steps in the algorithm are as follows.

1. Generate a nonsingular square submatrix by the procedure described above; let $W = (B, D)$ be a corresponding partition.
2. Compute $B\mathbf{1}$.
3. In each column of D generate a nonzero entry and afterwards fill up D in such a way that the required number of nonzeros is achieved and additionally each row of D also contains at least one nonzero.
4. Change the last nonzero entry in each row such that the condition in the above Proposition is fulfilled by $y = \mathbf{1}$.

Generating deterministic LP problems

For deterministic LP problems the generated test battery consists of LP problems of the minimization type having the following properties: The various arrays in the model have prescribed densities and entry ranges; the existence of an optimal solution is guaranteed. Optionally the technology matrix can be required to have full row rank. The techniques for achieving these goals have already been discussed above.

Generating two stage problems

The two stage problems generated are minimization problems.

The underlying algebraic structure

For random and general fixed recourse problems the underlying algebraic structure is generated analogously to the LP case just by fulfilling the sparsity and range targets for the various arrays of the model.

In the complete recourse case we also ensure the existence of an optimal solution. The procedure for complete recourse problems runs as follows.

1. Generate a complete recourse matrix W according to the algorithm described above.

2. Generate arrays T, h and q with $q \geq 0$, according to the prescribed entry ranges and densities.

3. Generate A according to density and range in such a way, that $(A^{\mathrm{T}}, T^{\mathrm{T}})$ does not contain a zero row. Generate a candidate feasible solution $x^\circ \geq 0$.

4. Generate b and c in such a way that x° is an optimal solution of $\min\{c^{\mathrm{T}}x \mid Ax = b, x \geq 0\}$.

Simple recourse problems are also generated according to the above scheme with the obvious difference that in this case W need not to be generated.

It is easy to see that the test problems generated in the above way always have an optimal solution, independently of the specification of the regression terms and probability distribution.

The underlying regression structure

The arrays in the affine sums are generated as sparse arrays. The regression structure serves also for pinpointing the random entries in the model. In this respect the following options have been included: The nonzero pattern in the terms corresponding to random variables is either a subset of the nonzero pattern of the constant term, or a superset of that nonzero pattern, or the nonzero pattern is arbitrary.

The underlying random variable structure

As already mentioned above GENSLP does not generate a distribution for the random variables; the one-point discrete distribution is returned for each of the random variables.

Generating jointly chance constrained problems

In the jointly chance constrained case the generated battery consists of chance constrained problems of the following type: Only the RHS is stochastic with a joint multinormal distribution. The problems are Slater regular; the optimal objective value and an optimal solution is known. A prescribed amount of the deterministic constraints is active at the optimal solution returned.

For ensuring the above mentioned properties the algebraic and random variable structures are generated in a closely connected manner; the regression terms are unit vectors. The procedure runs as follows:

1. Generate a positive definite correlation matrix. Two methods have been included. The first one is based on generating uniformly distributed points on the unit hypersphere, see Marsaglia and Olkin [22]. This method generates dense correlation matrices in general. The second method is based on Gershgorin's theorem which ensures positive definiteness via diagonal dominance. This approach allows also for generating sparse correlation matrices.

2. Generate matrices A and T according to their density and range target in such a way that none of them contains a zero row.

3. By solving an LP find a vector u such that $Tu > 0$ holds. Successively change negative entries of T to positive till a vector u with the above mentioned property can be found. Randomly generate a reference point x°.

4. Set the expected value of the random RHS to Tx° and generate standard deviations as a prescribed percentage of the expected values.

5. At this step the multivariate normal distribution is completely specified. Perform linesearch along the ray $x^\circ + \lambda u$ for finding a point x^* corresponding to the prescribed probability level α as well as a Slater point x^s. Due to the choice of u this can always be done.

6. Compute $b = Ax^*$ and fix the relation types such that the Slater point x^s becomes feasible. Relax the RHS according to the prescribed number of active constraints by randomly generated slack values. Randomly generate lower and upper bounds for the components of x such that x^* and x^s remain feasible thus ensuring compactness of the feasible domain.

7. Compute the gradient of the stochastic constraint function, and fix c in such a way that x^* becomes optimal. Select the direction of optimization such that the optimal objctive value becomes nonnegative.

For computing the probabilities in the linesearch as well as for computing the gradient of the stochastic constraint function Szántai's NORSUBS subroutine package [29] is employed.

GENSLP has been implemented in Fortran using the Lahey F77L-EM/32 compiler [21].

5. Numerical examples

Finally we present some test results. The test batteries (of 9 instances each) were generated with GENSLP with the following parameters:

Table I. Test problem characteristics

Name	A	T	W	Type	#rv's	#real.
UGSRD	30 × 70	8 × 70	8 × 16	SR	5	2'360
HGCRD	25 × 60	8 × 60	8 × 17	CR	3	693
SPACRD	25 × 60	8 × 60	8 × 18	CR	3	1'287
UGCC	20 × 35	4 × 35		CC	4	
UGCL	45 × 85	7 × 85		CC	7	

where SR = simple recourse, CR = complete recourse and CC = joint chance constraints.

For these computations we used a DOS PC with 32 MB storage and a Pentium-100 processor. The reports on the test runs yielded essentially the following: First, all the examples UGSRD* were solved to optimality. BDMPD deals with the deterministic equivalent LP, which in this case has 18'911 rows, 37'831 columns and 276'403 NDef's.

Table II. Elapsed time summary (sec)

	BPMPD	SRAPPROX
UGSRD1	164.83	3.13
UGSRD2	148.12	2.53
UGSRD3	100.52	2.96
UGSRD4	109.629	2.97
UGSRD5	202.3	2.86
UGSRD6	211.13	2.309
UGSRD7	148.13	2.97
UGSRD8	167.35	2.91
UGSRD9	118.8	2.8
min	100.52	2.309
max	211.13	3.13
mean	152.312	2.826
sdev	38.735	0.253

Table III. Objective values at termination

	BPMPD	SRAPPROX
UGSRD1	3883.695664	3883.692094
UGSRD2	3080.949255	3080.960586
UGSRD3	3980.167744	3980.17605
UGSRD4	3163.165678	3163.171884
UGSRD5	3830.549376	3830.527794
UGSRD6	5329.882368	5329.853737
UGSRD7	4993.811418	4993.823493
UGSRD8	3505.739463	3505.729738
UGSRD9	6148.742026	6148.73341

The following examples HGCRD* have deterministic equivalent LP's with 5'570 rows, 11'842 columns and 86'837 NDef's.

Table IV. Termination status

	BPMPD	DAPPROX	SDECOM
HGCRD1	Optimal	Optimal	Int NOpt[2]
HGCRD2	Optimal	Optimal	Int NOpt
HGCRD3	Optimal	Optimal	Int NOpt
HGCRD4	Optimal	Infeas[3]	Optimal
HGCRD5	Optimal	Optimal	Optimal
HGCRD6	Optimal	Optimal	Optimal
HGCRD7	Optimal	Optimal	Int NOpt
HGCRD8	Optimal	Optimal	Optimal
HGCRD9	Optimal	Optimal	Optimal

Table V. Elapsed time summary (sec)

	BPMPD	DAPPROX	SDECOM
HGCRD1	81.18	60.48	1320.3
HGCRD2	84.92	49.77	1329.69
HGCRD3	104.58	117.43	1159.91
HGCRD4	86.4	6.2	370.97
HGCRD5	89.31	68.33	112.33
HGCRD6	79.86	49.16	58.83
HGCRD7	75.57	52.4	1295.48
HGCRD8	91.39	48.67	242.83
HGCRD9	81.18	88.21	197.51
min	75.57	6.2	58.83
max	104.58	117.43	1329.69
mean	86.043	60.072	676.427
sdev	8.51	30.549	577.56

Although in SDECOM the runs for some examples were interrupted at the iterations limit of 1'000, the objective values are fairly close to the optimal values returned by the other two solvers.

[2] Interrupt after 1'000 iterations.
[3] Numerical difficulties to be overcome by change of parameters in QDECOM.

Table VI. Objective values at termination

	BPMPD	DAPPROX	SDECOM
HGCRD1	2307.837217	2306.435959	2297.081272
HGCRD2	695.917153	695.523632	721.669067
HGCRD3	887.693887	887.255218	890.267399
HGCRD4	361.81638	351.520862[4]	360.391923
HGCRD5	1242.49281	1241.956007	1198.233951
HGCRD6	693.739025	693.330419	687.184159
HGCRD7	1043.503202	1042.83476	1033.28592
HGCRD8	1548.714838	1547.804128	1471.62643
HGCRD9	988.165008	987.440983	985.122022

For problems SPACRD* we used the same matrix dimensions of A and T as in HGCRD* but generated them sparser. Using in turn a much finer discretization of the distribution lead to a deterministic equivalent LP with 10'322 rows, 23'227 columns and 101'814 NDef's.

Table VII. Termination status

	BPMPD	DAPPROX
SPACRD1	Optimal	Optimal
SPACRD2	Optimal	Optimal[5]
SPACRD3	Optimal	Optimal
SPACRD4	Optimal	Optimal
SPACRD5	Optimal	Optimal
SPACRD6	Optimal	Optimal
SPACRD7	Optimal	Optimal
SPACRD8	Optimal	Optimal
SPACRD9	Optimal	Optimal

[4] After parameter change in QDECOM: 359.735.

[5] After changing stopping tolerance from 0.1% to 1%.

Table VIII. Elapsed time summary (sec)

	BPMPD	DAPPROX
SPACRD1	89.25	40.919
SPACRD2	74.58	40.419
SPACRD3	71.58	30.31
SPACRD4	83.15	32.41
SPACRD5	69.98	19.0
SPACRD6	79.2	45.81
SPACRD7	79.71	38.39
SPACRD8	89.15	43.18
SPACRD9	91.34	21.42
min	69.98	19.0
max	91.34	45.81
mean	80.882	34.651
sdev	7.923	9.537

Table IX. Objective values at termination

	BPMPD	DAPPROX
SPACRD1	507.297789	506.19701
SPACRD2	499.479697	496.93794
SPACRD3	611.48373	609.521408
SPACRD4	200.652239	200.123386
SPACRD5	476.082548	475.307771
SPACRD6	368.996457	367.891609
SPACRD7	229.446129	228.818732
SPACRD8	2974.198764	2966.892509
SPACRD9	555.718935	554.320669

Finally we solved two batteries of chance constrained problems, namely UGCC* as well as UGCL*. UGCC* consisted of NLP's with 20 linear constraints and one nonlinear constraint in 35 variables, the latter one involving the 4–dimensional normal distribution function.

Table X. Elapsed time summary (sec)

	PCSPIOR	PROBALL	PROCON
UGCC1	1.809	1.37	1.32
UGCC2	1.37	1.42	1.32
UGCC3	1.2	1.27	0.93
UGCC4	1.48	1.54	0.93
UGCC5	10.44	4.34	0.77
UGCC6	8.4	4.07	2.03
UGCC7	11.87	7.19	3.46
UGCC8	1.159	1.48	1.26
UGCC9	1.21	1.42	0.77
min	1.159	1.27	0.77
max	11.87	7.19	3.46
mean	4.326	2.677	1.421
sdev	4.521	2.081	0.859

Table XI. Objective values at termination

	PCSPIOR	PROBALL	PROCON
UGCC1	3797.048097	3797.047809	3799.576499
UGCC2	2669.723895	2669.722285	2670.758955
UGCC3	475.239903	475.24104	463.080496
UGCC4	4463.264651	4463.270299	4470.097925
UGCC5	4396.108951	4396.106123	5038.796859
UGCC6	910.927364	910.92649	907.187266
UGCC7	294.62963	294.626084	294.631198
UGCC8	4625.189246	4625.196014	4625.600793
UGCC9	2276.493241	2276.490659	2277.766303

The problems in UGCL* involved 45 linear constraints and the nonlinear probability constraint defined by a 7–dimensional normal distribution function, depending on 85 variables.

Table XII. Elapsed time summary (sec)

	PCSPIOR	PROBALL	PROCON
UGCL1	154.12	78.98	4.45
UGCL2	96.01	102.32	2.69
UGCL3	200.36	62.4	7.14
UGCL4	318.95	63.76	3.13
UGCL5	222.72	71.51	25.65
UGCL6	154.289	89.97	3.79
UGCL7	147.58	77.06	27.41
UGCL8	151.479	31.75	1.98
UGCL9	166.7	114.35	5.88
min	96.01	31.75	1.98
max	318.95	114.35	27.41
mean	179.134	76.899	9.124
sdev	63.212	24.172	10.003

Table XIII. Objective values at termination

	PCSPIOR	PROBALL	PROCON
UGCL1	2731.039684	2731.037453	2730.057867
UGCL2	1040.214339	1040.213121	1040.048833
UGCL3	1182.963325	1180.784558	1180.608991
UGCL4	2744.325668	2737.525793	2737.2043
UGCL5	4288.637655	4288.639403	4288.229822
UGCL6	3832.980499	3832.981394	3831.127804
UGCL7	3141.512265	3141.509344	3141.494292
UGCL8	3962.939563	3962.976276	3962.683023
UGCL9	3825.559645	3825.562091	3825.564273

All problems of these last two batteries were solved to optimality without any fatal numerical problem.

References

1. J.F. Benders. Partitioning procedures for solving mixed-variables programming problems. *Numer. Math.*, 4:238–252, 1962.
2. J.R. Birge and F.V. Louveaux. A multicut algorithm for two stage linear programs. *Eur. J. Oper. Res.*, 34:384–392, 1988.
3. A. Brooke, D. Kendrick, and A. Meeraus. *GAMS. A User's Guide, Release 2.25*. The Scientific Press, San Francisco, 1992.
4. G.B. Dantzig and P.W. Glynn. Parallel processor for planning under uncertainty. In J.B. Rosen, editor, *Supercomputers and Large-Scale Optimization: Algorithms, Software, Applications*, pages 1–21. J.C. Baltzer AG, Basel, 1990. Ann. Oper. Res. 22.
5. Y. Ermoliev. Stochastic quasigradient methods and their application to systems optimization. *Stochastics*, 9:1–36, 1983.
6. K. Frauendorfer. Solving SLP recourse problems with arbitrary multivariate distributions—the dependent case. *Math. Oper. Res.*, 13:377–394, 1988.
7. K. Frauendorfer and P. Kall. A solution method for SLP recourse problems with arbitrary multivariate distributions – the independent case. *Probl. Contr. Inf. Theory*, 17:177–205, 1988.
8. A. Gaivoronski. Stochastic quasigradient methods and their implementation. In Y. Ermoliev and R.J-B. Wets, editors, *Numerical Techniques for Stochastic Optimization*, pages 313–351. Springer-Verlag, 1988.
9. H.I. Gassmann. MSLiP: A computer code for the multistage stochastic linear programming problem. *Math. Prog.*, 47:407–423, 1990.
10. J. Gondzio. HOPDM (version 2.12) – A fast LP solver based on a primal–dual interior point method. *Eur. J. Oper. Res.*, 85:221–225, 1995.
11. J.L. Higle and S. Sen. Stochastic decomposition: An algorithm for two stage stochastic linear programs with recourse. *Math. Oper. Res.*, 16:650–669, 1991.
12. G. Infanger. *Planning under Uncertainty: Solving Large-scale Stochastic Linear Programs*. Boyd & Fraser Publ. Co., Danvers, MA, 1994.
13. P. Kall. *Stochastic linear programming*. Springer Verlag, 1976.
14. P. Kall. Stochastic programs with recourse: An upper bound and the related moment problem. *ZOR*, 31:A119–A141, 1987.
15. P. Kall and J. Mayer. SLP-IOR: A model management system for stochastic linear programming — system design —. In A.J.M. Beulens and H.-J. Sebastian, editors, *Optimization-Based Computer-Aided Modelling and Design*, pages 139–157. Springer-Verlag, 1992.
16. P. Kall and J. Mayer. A model management system for stochastic linear programming. In P. Kall, editor, *System Modelling and Optimization*, pages 580–587. Springer-Verlag, 1992.
17. P. Kall and J. Mayer. SLP–IOR: An interactive model management system for stochastic linear programs. *Math. Programming*, 75:221–240, 1996.
18. P. Kall and D. Stoyan. Solving stochastic programming problems with recourse including error bounds. *Math. Operationsforsch. Statist., Ser. Opt.*, 13:431–447, 1982.
19. P. Kall and S. W. Wallace. *Stochastic programming*. John Wiley & Sons, 1994.
20. E. Keller. GENSLP: A program for generating input for stochastic linear programs with complete fixed recourse. Manuscript, IOR, University of Zurich, 1984.
21. Lahey Computer Systems Inc. Lahey FORTRAN F77L-EM/32. Language Reference, Lahey Computer Systems, Incline Village, NV, 1992.
22. G. Marsaglia and I. Olkin. Generating correlation matrices. *SIAM Journal on Scientific and Statistical Computations*, 5:470–475, 1984.
23. R.E. Marsten. The design of the XMP linear programming library. *ACM Transactions on Mathematical Software*, 7:481–497, 1981.

24. J. Mayer. Probabilistic constrained programming: A reduced gradient algorithm implemented on PC. Working Paper WP-88-39, IIASA, Laxenburg, 1988.
25. J. Mayer. Stochastic linear programming algorithms: A comparison based on a model management system, 1995. Habilitationsschrift, Wirtschaftswissenschaftliche Fakultät, Universität Zürich.
26. Cs. Mészáros. The augmented system variant of IPMs in two–stage stochastic linear programming computation. Working Paper WP-95-11, MTA SzTAKI, Budapest, 1995.
27. A. Ruszczyński. A regularized decomposition method for minimizing a sum of polyhedral functions. *Math. Prog.*, 35:309–333, 1986.
28. B. Strazicky. On an algorithm for solution of the two-stage stochastic programming problem. *Methods of Oper. Res.*, 19:142–156, 1974.
29. T. Szántai. Calculation of the multivariate probability distribution function values and their gradient vectors. Working Paper WP-87-82, IIASA, 1987.
30. T. Szántai. A computer code for solution of probabilistic-constrained stochastic programming problems. In Y. M. Ermoliev and R. J.-B. Wets, editors, *Numerical Techniques for Stochastic Optimization*, pages 229–235. Springer-Verlag, Berlin, 1988.
31. S. Vajda. *Probabilistic programming.* Academic Press, 1972.
32. R. Van Slyke and R. J-B. Wets. *L*-shaped linear programs with applications to optimal control and stochastic linear programs. *SIAM J. Appl. Math.*, 17:638–663, 1969.
33. R. Wets. Programming under uncertainty: The complete problem. *Z. Wahrsch. theorie u. verw. Geb.*, 4:316–339, 1966.

On Primal–Dual Path–Following Algorithms for Semidefinite Programming

E. de Klerk, C. Roos and T. Terlaky
Faculty of Technical Mathematics and Informatics,
Delft University of Technology,
P.O. Box 5031, 2600 GA Delft, The Netherlands.
e–mail: e.deklerk@twi.tudelft.nl, c.roos@twi.tudelft.nl, t.terlaky@twi.tudelft.nl

Abstract. Interior point methods for semidefinite programming have recently been studied intensively, due to their polynomial complexity and practical efficiency. Most of these methods are extensions of linear programming algorithms. The primal-dual central path following method for linear programming by Jansen et al. [6] has recently been extended to semidefinite programming by Jiang [7], utilizing the Nesterov-Todd direction and introducing a new distance measure. In this note we refine and extend this analysis: A weaker condition for a feasible full Newton step is established, and quadratic convergence to target points on the central path is shown. Moreover, we show how to compute large dynamic target updates which still allow full Newton steps.

1. Introduction

Primal-dual path following methods have emerged as the most successful interior point algorithms for linear programming (LP). Predictor-corrector methods are particularly in favour, following successful implementations [13].

The extension of algorithms from linear to semidefinite programming (SDP) has followed the same trends, and recent attention has focused on path following algorithms. Primal path following algorithms are studied in [3] (general analysis), [2] (long step method), and [5] (full Newton step method).

The primal-dual path following methods employ different search directions which arise from different linearization strategies of the relaxed optimality conditions. A comparison of the best known search directions can be found in [1] and [18].

One of the popular primal-dual directions is the so-called Nesterov-Todd (NT) direction, introduced in [15]. (Recently Kojima et al [10] showed the NT direction to be a special case of the primal–dual directions for monotone semidefinite complementarity problems introduced in [11].)

A long step primal-dual path following method using the NT direction was recently presented by Jiang [7]. This paper is an extension of the LP analysis by Jansen et al. [6]. This method targets a specific

F. Giannessi et al. (eds.), New Trends in Mathematical Programming, 137–157.

point on the primal-dual central path, which is then updated if the current iterates are 'close enough' to it. The novelty of this analysis lies in the use of a new centrality measure, which is analogous to the LP measure in [6].

In this paper we extend and refine this analysis – a weaker sufficient condition for full Newton steps is derived, and quadratic convergence to target points is proved. Moreover, a dynamic target updating scheme is discussed. The aim is to obtain a method which uses only full Newton steps, and uses large target updates to reduce the iteration count. A worst-case iteration complexity bound of $O(\sqrt{n})$ of the resulting algorithm is proved.

1.1. Preliminaries

The semidefinite programming problem is studied in the standard form. Thus the primal problem (P) is given by:

$$\min \operatorname{Tr}(CX)$$

subject to

$$\begin{aligned} \operatorname{Tr}(A_i X) &= b_i, \quad i = 1, \ldots, m \\ X &\succeq 0 \end{aligned}$$

and its dual problem (D) is:

$$\max b^T y$$

subject to

$$\begin{aligned} \sum_{i=1}^{m} y_i A_i + Z &= C \\ Z &\succeq 0 \end{aligned}$$

where C and the A_i's are symmetric $n \times n$ matrices, $b, y \in \mathbb{R}^m$ and '$X \succeq 0$' means X is symmetric positive semidefinite[1]. The notation 'Tr' denotes the trace operator.

The matrices A_i are assumed to be linearly independent. We will assume that a strictly feasible pair ($X \succ 0$, $Z \succ 0$) exists. This ensures the existence of an optimal primal-dual pair (X^*, Z^*) and a corresponding zero duality gap ($\operatorname{Tr}(X^* Z^*) = 0$).

[1] The symbol '$\succeq$' denotes the so-called Löwner partial order on the space of symmetric matrices. For symmetric matrices A and B the notation $A \succeq B$ therefore means $A - B$ is positive semidefinite, i.e. $A - B \succeq 0$.

The optimality conditions for the pair of dual problems are

$$\begin{aligned} \mathrm{Tr}(A_i X) &= b_i, \quad i = 1, \dots, m \\ \sum_{i=1}^{m} y_i A_i + Z &= C \\ XZ &= 0 \\ X, Z &\succeq 0. \end{aligned}$$

If these conditions are relaxed to

$$\begin{aligned} \mathrm{Tr}(A_i X) &= b_i, \quad i = 1, \dots, m \\ \sum_{i=1}^{m} y_i A_i + Z &= C \\ XZ &= \mu I \\ X, Z &\succ 0 \end{aligned}$$

for some $\mu > 0$, then a unique solution of the relaxed system exists, denoted by $\{X(\mu), y(\mu), Z(\mu)\}$. This solution gives a parametric representation of the central path as a function of μ.[2]

The existence and uniqueness of $\{X(\mu), y(\mu), Z(\mu)\}$ follow from the fact that $\{X(\mu), y(\mu), Z(\mu)\}$ corresponds to the unique minimum of the strictly convex primal–dual barrier function

$$f(X, Z, \mu) = \frac{1}{\mu}\mathrm{Tr}(XZ) - \ln\det(XZ) \tag{1.1}$$

defined on the primal–dual feasible region. Because of the two different associations, the parameter μ is called either the *barrier parameter*, or the *centering* parameter. We will also refer to μ as a target parameter, since successive points on the central path will be 'targeted'.

1.2. THE NESTEROV-TODD SEARCH DIRECTIONS

For a primal feasible $X \succ 0$ and dual feasible $Z \succ 0$, the scaling matrix

$$D := Z^{-\frac{1}{2}} \left(Z^{\frac{1}{2}} X Z^{\frac{1}{2}}\right)^{\frac{1}{2}} Z^{-\frac{1}{2}}, \tag{1.2}$$

satisfies $D^{-1}X = ZD$, or

$$L^{-1} X L^{-T} = L^T Z L := V,$$

[2] The central path is an analytic curve which converges to the analytic center of the optimal face as $\mu \to 0$; for details, see [4].

where $D = LL^T$ is a factorization of D. In other words, the matrix D may be used to scale the variables X and Z to the same symmetric positive definite matrix. The matrix V depends on the factorization of D.

This scaling was introduced by Nesterov and Todd [15], and its properties were further investigated by Sturm and Zhang [17] who used it to provide a framework for a number of central path following methods. The scaling seems to be a natural extension of that used in linear programming, and was used by De Klerk et al. [9] to generalize primal–dual affine scaling methods from LP to SDP.

Letting '$\sim$' denote the similarity relation, one has

$$V^2 = L^{-1}XZL \sim XZ,$$

i.e. V^2 has the same eigenvalues as XZ and is symmetric positive definite. As a consequence the duality gap is given by

$$\mathrm{Tr}(XZ) = \mathrm{Tr}(V^2) = \|V\|^2 = \sum_i \lambda_i^2(V),$$

where the norm is the Frobenius norm.

Any pair of primal–dual search directions ΔX and ΔZ must satisfy

$$\begin{aligned} \mathrm{Tr}(A_i \Delta X) &= 0, \quad i = 1, \ldots, m \\ \Delta Z &\in \mathrm{span}\{A_1, \ldots, A_m\}. \end{aligned}$$

One can scale ΔX and ΔZ in same way as X and Z to obtain:

$$D_X := L^{-1} \Delta X L^{-T}$$

and

$$D_Z := L^T \Delta Z L.$$

The scaled directions D_X and D_Z are orthogonal by the orthogonality of ΔX and ΔZ, i.e. $\mathrm{Tr}(D_X D_Z) = 0$.

The scaled Newton step is defined by $D_V := D_X + D_Z$. After a feasible primal–dual step ΔX, ΔZ the duality gap becomes

$$\begin{aligned} \mathrm{Tr}\left((X + \Delta X)(Z + \Delta Z)\right) &= \mathrm{Tr}\left((V + D_X)(V + D_Z)\right) \\ &= \mathrm{Tr}(V^2 + V D_V). \end{aligned}$$

The Newton equations in the V-space may be derived by requiring:

$$\frac{1}{2}\left[(V + D_X)(V + D_Z) + ((V + D_X)(V + D_Z))^T\right] = \mu I,$$

which is linearized by neglecting the cross terms $D_X D_Z$ and $D_Z D_X$ to obtain

$$\frac{1}{2}\left((D_X + D_Z)V + V(D_X + D_Z)\right) = \mu I - V^2. \tag{1.3}$$

Equation (1.3) (called a Sylvester equation) has a unique symmetric solution [17], given by

$$D_V = \mu V^{-1} - V.$$

Pre and postmultiplying with L and L^T respectively yields the *Nesterov-Todd* (or NT) equations:

$$D\Delta Z D + \Delta X = \mu Z^{-1} - X$$

subject to

$$\begin{aligned} \mathrm{Tr}(A_i \Delta X) &= 0, \quad i = 1, \ldots, m \\ \Delta Z &\in \mathrm{span}\{A_1, \ldots, A_m\}. \end{aligned}$$

1.3. A MEASURE OF CENTRALITY

In a recent paper Jiang [7] uses (up to the constant $\frac{1}{2}$) the measure

$$\delta(X, Z, \mu) = \frac{1}{2}\frac{1}{\sqrt{\mu}}\|D_V\| = \frac{1}{2}\left\|\sqrt{\mu}V^{-1} - \frac{1}{\sqrt{\mu}}V\right\|.$$

This measure generalizes the LP measure of Jansen et al [6] to semidefinite programming, and will be used in this paper. It is shown in [7] that $\delta(X, Z, \mu)$ is the absolute value of the directional derivative of the primal-dual barrier (1.1) along the NT direction. In this sense it is a natural centrality measure associated with the NT direction.

1.4. THE TARGET FOLLOWING APPROACH

For a given value of μ, the point μI is regarded as a target point with associated target duality gap $n\mu$. In other words if the unique pair $(X(\mu), Z(\mu))$ is computed then the duality gap is equal to $n\mu$.

Target following algorithms iteratively compute $(X(\mu), Z(\mu))$ approximately, followed by a decrease in the value of μ. The algorithms presented here all fit in the following framework:

Generic Primal-Dual Logarithmic Barrier Algorithm

Input:
a strictly feasible pair (X_0, Z_0);
parameters $\tau < 1$ and $\mu_0 > 0$ such that $\delta(X_0, Z_0, \mu_0) \leq \tau$;
an accuracy parameter $\epsilon > 0$.

begin

$X := X_0$; $Z := Z_0$; $\mu = \mu_0$

while $\mathrm{Tr}(XZ) > \epsilon$ **do**

begin

compute the NT directions ΔX, ΔZ with respect to μ;

$X := X + \Delta X$;

$Z := Z + \Delta Z$;

compute an update parameter θ (default $\theta := \frac{1}{2\sqrt{n}}$);

$\mu := \mu(1 - \theta)$;

end

end

2. Feasibility of the Newton step

One can now prove the following two results which are analogous to the LP case: If $\delta < 1$ then the Newton step is feasible, and the duality gap after the step attains its target value.

To this end we need the following three results from De Klerk et al. [9].

LEMMA 2.1 (General condition for a feasible primal-dual step). *Let*

$$X(\alpha) := X + \alpha\Delta X, \quad Z(\alpha) := Z + \alpha\Delta Z.$$

If one has

$$\det\left(X(\alpha)Z(\alpha)\right) > 0 \;\; \forall\, 0 \leq \alpha \leq \bar{\alpha},$$

then $X(\bar{\alpha}) \succ 0$ *and* $Z(\bar{\alpha}) \succ 0$. □

LEMMA 2.2 *The spectral norm of* $D_{XZ} := \frac{1}{2}(D_X D_Z + D_Z D_X)$ *is bounded by*

$$\|D_{XZ}\|_2 \le \frac{1}{4}\|D_X + D_Z\|^2.$$

□

LEMMA 2.3 *Let* Q *be an* $n \times n$ *real symmetric matrix,* S *an* $n \times n$ *real skew–symmetric matrix. One has* $\det(Q+S) > 0$ *if* $Q \succ 0$. □

We now prove the feasibility of the Newton step in the following lemma. The condition $\delta(X, Z, \mu) < 1$ in the lemma is a significant improvement over the corresponding condition $\delta(X, Z, \mu) < \frac{1}{2\sqrt{2}}$ derived by Jiang [7].

LEMMA 2.4 *If* $\delta := \delta(X, Z, \mu) < 1$ *then the full Newton step is strictly feasible.*

Proof. We show that the determinant of $X(\alpha)Z(\alpha)$ remains positive for all $\alpha \le 1$. One then has $X(1), Z(1) \succ 0$ by Lemma 2.1.

To this end note that

$$\begin{gathered} X(\alpha)Z(\alpha) \sim (V + \alpha D_X)(V + \alpha D_Z) = \\ V^2 + \alpha D_X V + \alpha V D_Z + \alpha^2 D_X D_Z = V^2 + \alpha(\mu I - V^2) + \alpha^2 (D_{XZ}) + \\ \left[\tfrac{1}{2}\alpha^2 (D_X D_Z - D_Z D_X) + \tfrac{1}{2}\alpha(D_X V + V D_Z - V D_X - D_Z V)\right], \end{gathered}$$

using the Newton equation (1.3). The matrix in square brackets in the last equation is skew–symmetric. Lemma 2.3 therefore implies that the determinant of $[X(\alpha)Z(\alpha)]$ will be positive if the matrix

$$M(\alpha) := V^2 + \alpha(\mu I - V^2) + \alpha^2 D_{XZ}$$

is positive definite. Since we can rewrite the expression for $M(\alpha)$ as

$$M(\alpha) = (1-\alpha)V^2 + \alpha\mu\left[I + \frac{\alpha}{\mu} D_{XZ}\right],$$

one will have $M(\alpha) \succ 0$ if $\alpha \le 1$ and $\|(D_{XZ}/\mu)\|_2 < 1$. The last condition is easily shown to hold by using Lemma 2.2 and $\delta < 1$ successively:

$$\|D_{XZ}/\mu\|_2 = \frac{1}{\mu}\|D_{XZ}\|_2 \le \frac{1}{4\mu}\|D_V\|^2 = \delta^2 < 1.$$

This completes the proof. □

The following result is analogous to the LP-case, and is useful in constructing target updating schemes.

COROLLARY 2.5 *The target duality gap is attained after one full Newton step.*

Proof. Since

$$X(1)Z(1) \sim \mu I + D_{XZ} + \left[\frac{1}{2}(D_X D_Z - D_Z D_X) + \frac{1}{2}(D_X V + V D_Z - V D_X - D_Z V)\right] \tag{2.4}$$

one has

$$\mathrm{Tr}\,(X(1)Z(1)) = \mathrm{Tr}(\mu I)$$

by using the orthogonality of D_X and D_Z and the skew symmetry of the matrix in square brackets. □

NOTATION: In what follows we denote the skew-symmetric matrix in (2.4) by S. As we will only work with full Newton steps, i.e. $\alpha = 1$, it will also be convenient to write $X^+ := X(1)$, $Z^+ := Z(1)$, etc.

3. Quadratic convergence to the target

We proceed to prove quadratic convergence to the target μI. To this end we need three technical results which give information concerning the spectrum of X^+Z^+. We denote the symmetrical transformation of X^+Z^+ by $(V^+)^2$.

LEMMA 3.1 *One has*

$$\lambda_{\min}\left((V^+)^2\right) \geq \mu(1-\delta^2)\,,$$

where $\lambda_{\min}$ denotes the smallest eigenvalue.

Proof. From (2.4) it follows that has

$$\lambda_{\min}\left((V^+)^2\right) = \lambda_{\min}\left(\mu I + D_{XZ} + S\right)\,.$$

The skew-symmetry of S implies

$$\lambda_{\min}\left((V^+)^2\right) \geq \lambda_{\min}\left(\mu I + D_{XZ}\right) \geq \mu - \|D_{XZ}\|_2\,.$$

Substitution of the bound for $\|D_{XZ}\|_2$ from Lemma 2.2 now yields:

$$\lambda_{\min}\left((V^+)^2\right) \geq \mu - \frac{1}{4}\|D_V\|^2 = \mu\left(1-\delta^2\right),$$

which completes the proof. □

Lemma 2.2 gave a bound on the spectral norm of D_{XZ}. We now derive a similar bound on its Frobenius norm:

LEMMA 3.2 *One has*

$$\|D_{XZ}\|^2 \leq \frac{1}{8}\|D_V\|^4.$$

Proof. It is trivial to verify that

$$D_X D_Z + D_Z D_X = \frac{1}{2}\left[(D_X + D_Z)^2 - (D_X - D_Z)^2\right]$$

Since D_X and D_Z are orthogonal the matrices $D_V = D_X + D_Z$ and $Q_V := D_X - D_Z$ have the same norm. Consequently

$$\begin{aligned}
\|D_{XZ}\|^2 &= \left\|\frac{1}{4}\left(D_V^2 - Q_V^2\right)\right\|^2 \\
&= \frac{1}{16}\mathrm{Tr}\left(D_V^4 + Q_V^4 - 2D_V^2 Q_V^2\right) \\
&\leq \frac{1}{16}\left(\left\|D_V^2\right\|^2 + \left\|Q_V^2\right\|^2\right) \\
&\leq \frac{1}{16}\left(\|D_V\|^4 + \|Q_V\|^4\right) = \frac{1}{8}\|D_V\|^4.
\end{aligned}$$

□

The quadratic convergence result may now be proved.

LEMMA 3.3 *After a feasible Newton step the distance measure satisfies*

$$\left(\delta^+\right)^2 := \left(\delta(X^+, Z^+, \mu)\right)^2 \leq \frac{\delta^4}{2(1-\delta^2)}.$$

Proof. The distance measure after the full Newton step is given by

$$\begin{aligned}
\left(\delta^+\right)^2 &= \frac{1}{4\mu}\left\|\mu\left(V^+\right)^{-1} - V^+\right\|^2 \\
&= \frac{1}{4\mu}\left\|\left(V^+\right)^{-1}\left(\mu I - \left(V^+\right)^2\right)\right\|^2 \\
&\leq \frac{1}{4\mu}\lambda_{\max}^2\left(\left(V^+\right)^{-1}\right)\left\|\mu I - \left(V^+\right)^2\right\|^2.
\end{aligned}$$

We now substitute the bound from Lemma 3.1 to obtain

$$\left(\delta^+\right)^2 \leq \frac{1}{4\mu^2(1-\delta^2)}\left\|\mu I - \left(V^+\right)^2\right\|^2.$$

To complete the proof we show that:

$$\left\|\mu I - \left(V^+\right)^2\right\|^2 \leq \|D_{XZ}\|^2. \tag{3.5}$$

In order to prove (3.5), note that

$$\begin{aligned}\left\|\mu I-(V^{+})^{2}\right\|^{2} &= \sum_{i=1}^{n}\left(\lambda_i\left(\mu I+D_{XZ}+S\right)-\lambda_i\left(\mu I\right)\right)^{2}\\ &= \sum_{i=1}^{n}\left(\lambda_i\left(D_{XZ}+S\right)\right)^{2}\\ &= \mathrm{Tr}\left(D_{XZ}+S\right)^{2}.\end{aligned}$$

Using the skew-symmetry of S one obtains

$$\begin{aligned}\left\|\mu I-(V^{+})^{2}\right\|^{2} &= \mathrm{Tr}\left((D_{XZ})^{2}-SS^{T}\right)\\ &\leq \mathrm{Tr}(D_{XZ})^{2}=\|D_{XZ}\|^{2}.\end{aligned}$$

The final result now follows from Lemma 3.2. □

The quadratic convergence result may be stated concisely as:

COROLLARY 3.4 *If* $\delta<\frac{1}{\sqrt{2}}$ *then* $\delta^{+}<\delta^{2}$, *i.e. quadratic convergence is obtained. The weaker condition* $\delta<\sqrt{\frac{2}{3}}$ *implies* $\delta^{+}<\delta$ *and is therefore sufficient for convergence.*

4. Updating the target parameter μ

If the current iterates are sufficiently close to the target, say $\delta\leq\frac{1}{2}$, then the target parameter is updated via

$$\mu^{+}=(1-\theta)\mu$$

where $0<\theta<1$. We show that a default value of $\theta=\frac{1}{2\sqrt{n}}$ ensures that $\delta(X,Z,\mu^{+})\leq\frac{1}{\sqrt{2}}$. The next Newton step then again yields a feasible pair (X^{+},Z^{+}) with $\delta(X^{+},Z^{+},\mu^{+})\leq\frac{1}{2}$ due to the quadratic convergence property (Corollary 3.4).

We first prove a lemma which relates the distance measure after the Newton step with the measure before the step.

LEMMA 4.1 *Let* $\delta:=\delta(X,Z,\mu)$ *and let* $\mathrm{Tr}(XZ)=\mu$. *If* $\mu^{+}=(1-\theta)\mu$ *one has*

$$(\delta(X,Z,\mu^{+}))^{2}=\frac{n\theta^{2}}{4(1-\theta)}+(1-\theta)\delta^{2}.$$

Proof. To simplify notation we introduce

$$U := \frac{1}{\sqrt{\mu}} V.$$

In terms of this notation one has

$$\begin{aligned} 4\left(\delta(X, Z, \mu^+)\right)^2 &= \left\| \sqrt{(1-\theta)} U^{-1} - \frac{1}{\sqrt{(1-\theta)}} U \right\|^2 \\ &= \left\| \frac{\theta U}{\sqrt{1-\theta}} + \sqrt{1-\theta}\left(U^{-1} - U\right) \right\|^2 . \end{aligned}$$

Note that $\|U\|^2 = \mathrm{Tr}(U^2) = \frac{1}{\mu}\mathrm{Tr}(V^2) = n$. This implies that U is orthogonal to $U^{-1} - U$:

$$\mathrm{Tr}\left(U\left(U^{-1} - U\right)\right) = n - \|U\|^2 = 0 .$$

Consequently

$$4\left(\delta(X, Z, \mu^+)\right)^2 = (1-\theta)\left\|U^{-1} - U\right\|^2 + \frac{\theta^2 \|U\|^2}{1-\theta} .$$

The required result now follows from the observation $\|U^{-1} - U\| = 2\delta$ together with $\|U\|^2 = n$. □

An immediate corrollary of the lemma is the following: If one has a primal-dual pair (X, Z) and target parameter μ such that $\delta(X, Z, \mu) \leq \frac{1}{2}$, and μ is updated via $\mu^+ := (1 - \frac{1}{2\sqrt{n}})\mu$, then one has $\delta(X, Z, \mu^+) \leq \frac{1}{\sqrt{2}}$.[3]

As discussed above, the next Newton step now yields a pair (X^+, Z^+) satisfying $\delta(X^+, Z^+, \mu^+) \leq \frac{1}{2}$.

The algorithm therefore generates a sequence of iterates which always satisfy $\delta \leq \frac{1}{2}$. Moreover, the duality gap is reduced by a factor $(1 - \frac{1}{2\sqrt{n}})$ at each iteration, since the duality gap after the Newton step equals the target duality gap.

These observations imply the following result which ascertains the polynomial convergence of the algorithm. The proof is identical to that in the LP case and is omitted here.

[3] This result can be improved to the following: If $\delta \leq \frac{1}{2}$ and μ is updated to $\mu^+ = (1 - 1/\sqrt{n+1})\mu$, then after a full Newton step with respect to μ^+ one has $\delta(X^+, Z^+, \mu^+) \leq \frac{1}{2}$. We omit the proof since the improvement is only marginal — one still has $\theta = O(1/\sqrt{n})$.

THEOREM 4.2 *If* $\tau = \frac{1}{\sqrt{2}}$ *and* $\theta = \frac{1}{2\sqrt{n}}$ *then the Primal-Dual Logarithmic barrier Algorithm with full Newton steps requires at most*

$$\left\lceil 2\sqrt{n}\ln\frac{n\mu^0}{\epsilon}\right\rceil$$

iterations. The output is a primal-dual pair (X, Z) *satisfying* $\mathrm{Tr}(XZ) \leq \epsilon$.

5. Adaptive target updates

The short step method presented in the previous section suffers from the usual drawback that the number of iterations needed for convergence will be close to the upper bound given by Theorem 4.2. This is due to the small, fixed μ-updates. In an implementation it is desirable to make the largest possible updates at each iteration, albeit at the cost of extra computation.

This is the topic of this section: Given a pair (X, Z) and parameter μ such that $\delta(X, Z, \mu) \leq \frac{1}{2}$ we show how to obtain a larger θ such that the next Newton step with respect to $\mu^+ := (1-\theta)\mu$ again yields a pair (X^+, Z^+) satisfying $\delta(X^+, Z^+, \mu^+) \leq \frac{1}{2}$. This allows much larger updates than with the strategy presented in the previous section. It is known that dynamic updating strategies can reduce the iteration count considerably in the LP case [16].

5.1. A CONDITION FOR ADAPTIVE UPDATES

As before, let (X^+, Z^+) arise from a Newton step from (X, Z) with respect to μ. From the proofs of Lemma 3.1 and Lemma 3.3 it follows that

$$(\delta^+)^2 := \delta(X^+, Z^+, \mu)^2 \leq \frac{\|D_{XZ}/\mu\|^2}{4\left(1 - \|D_{XZ}/\mu\|_2\right)}.$$

One will therefore have $\delta^+ \leq \tau$ for a given tolerance $\tau > 0$ if

$$\frac{\|D_{XZ}/\mu\|^2}{\left(1 - \|D_{XZ}/\mu\|_2\right)} \leq 4\tau^2$$

or

$$\|D_{XZ}/\mu\|^2 \leq 4\tau^2\left(1 - \|D_{XZ}/\mu\|_2\right). \tag{5.6}$$

Note that this condition guarantees $\|D_{XZ}/\mu\|_2 \leq 1$ which in turn is sufficient to guarantee a feasible step by the proof of Lemma 2.4.

The updating condition (5.6) is independent of the primal-dual scaling — for any given factorization of the scaling matrix, say $D = LL^T$, one has

$$\begin{aligned} D_{XZ} &:= \frac{1}{2}L^{-1}\Delta X \Delta Z L + \frac{1}{2}L^T \Delta Z \Delta X L^{-T} \\ &\sim \frac{1}{2}\Delta X \Delta Z + \frac{1}{2}D\Delta Z \Delta X D^{-1}. \end{aligned}$$

The eigenvalues of D_{XZ} are therefore independent of L, and consequently so are $\|D_{XZ}\|$ and $\|D_{XZ}\|_2$ in (5.6).

5.2. AN ADAPTIVE UPDATING STRATEGY

The matrices (D_X, D_Z) depend on the parameter μ, as they correspond to the solution of

$$\frac{1}{2}\left[V(D_X + D_Z) + (D_X + D_Z)V\right] = \mu I - V^2. \tag{5.7}$$

In what follows D_X, D_Z and $D_V = D_X + D_Z$ correspond to a fixed parameter μ, and D_X^+ and D_Z^+ correspond to an unknown parameter $\mu^+ = (1-\theta)\mu$, which is dependent on a variable θ. The idea is that (D_X^+, D_Z^+) can be expressed in terms of (D_X, D_Z) as a function of θ.

In terms of this notation the updating condition is to choose the largest value of θ such that

$$\left\|D_{XZ}^+/\mu^+\right\|^2 \leq 4\tau^2\left(1 - \left\|D_{XZ}^+/\mu^+\right\|_2\right). \tag{5.8}$$

In what follows we show how to approximate the largest value of θ that satisfies (5.8).

The first step is to express D_{XZ}^+ in terms of D_X and D_Z. To this end, note that the solution of the Newton equation (5.7) can been seen as the sum of solutions of the linear systems with right hand sides μI and V^2 respectively. The solution of

$$\frac{1}{2\sqrt{\mu}}\left[V(D_X + D_Z) + (D_X + D_Z)V\right] = \mu I,$$

will be denoted by D_X^c, D_Z^c (the so-called *centering component*), and the solution of

$$\frac{1}{2}\left[V(D_X + D_Z) + (D_X + D_Z)V\right] = -V^2, \tag{5.9}$$

will be denoted by D_X^a, D_Z^a (the so-called *affine component*). It is trivial to show that

$$\begin{aligned} D_X^+ &= (1-\theta)D_X^c + D_X^a, \\ D_Z^+ &= (1-\theta)D_Z^c + D_Z^a. \end{aligned}$$

Using that $D_X = D_X^c + D_X^a$ and $D_Z = D_Z^c + D_Z^a$ one has

$$\begin{aligned} D_X^+ &= (1-\theta)D_X + \theta D_X^a, \\ D_Z^+ &= (1-\theta)D_Z + \theta D_Z^a. \end{aligned}$$

It follows that

$$D_{XZ}^+/\mu^+ := \frac{\frac{1}{2}(D_X^+ D_Z^+ + D_Z^+ D_X^+)}{(1-\theta)\mu}$$

can be rewritten as

$$\begin{aligned} \frac{D_{XZ}^+}{\mu^+} = \frac{1-\theta}{\mu} D_{XZ} + \frac{1}{2\mu}\theta\left(D_X D_Z^a + D_Z D_X^a + D_X^a D_Z + D_Z^a D_X\right) + \\ \frac{1}{2}\frac{\theta^2}{\mu(1-\theta)}\left(D_X^a D_Z^a + D_Z^a D_X^a\right) . \end{aligned} \tag{5.10}$$

Note that the entries of $(1-\theta)D_{XZ}^+/\mu^+$ are quadratic functions of θ. The updating condition (5.8) can be multiplied by $(1-\theta)$ and rewritten as

$$4\tau^2(1-\theta)^2 - \left\|(1-\theta)D_{XZ}^+/\mu^+\right\|^2 \geq 4\tau^2(1-\theta)\left\|(1-\theta)D_{XZ}^+/\mu^+\right\|_2 . \tag{5.11}$$

Note that the left hand side is a fourth order polynomial in θ, while the right hand side can be a more general nonlinear function of θ. Condition (5.11) is guaranteed to hold if we replace the right hand side expression by a larger value. To this end, we use the triangle inequality

$$\begin{aligned} \left\|(1-\theta)D_{XZ}^+/\mu^+\right\|_2 \leq &\; \frac{1-\theta}{\mu}\left\|D_{XZ}\right\|_2 \\ &+ \frac{\theta(1-\theta)}{2\mu}\left\|D_X D_Z^a + D_Z D_X^a + D_X^a D_Z + D_Z^a D_X\right\|_2 \\ &+ \frac{1}{2}\frac{\theta^2}{\mu}\left\|D_X^a D_Z^a + D_Z^a D_X^a\right\|_2 . \end{aligned}$$

If we replace the right hand side of (5.11) by this expression, the updating condition reduces to finding the largest root of a fourth order polynomial — an analytically solvable problem.

Remark: If the current iterates are perfectly centered with respect to μ, then $D_X = D_Z = 0$, $D_X^a = -D_X^c$, and $D_Z^a = -D_Z^c$. It follows that equation (5.10) simplifies to

$$D_{XZ}^+/\mu^+ = \frac{\theta^2}{2\mu(1-\theta)}\left(D_X^a D_Z^a + D_Z^a D_X^a\right).$$

In this case the original updating condition (5.8) can be solved, since $(1-\theta)\left\|D^{+}_{XZ}/\mu^{+}\right\|_2$ is now a quadratic function of θ. This observation is of interest as the starting point can be assumed perfectly centered by embedding the original primal-dual problem pair into a larger problem, where a feasible starting point on the central path is known [8]. The first μ-update can thefore be done by solving (5.8). In subsequent iterations, centering steps will be needed to approximate this feature. The effect of centering steps on the updating strategy will be illustrated in the next section.

5.3. Numerical examples

We investigate a small example ($n = 3, m = 3$) where different updating strategies are used. First the fixed defualt value of $\theta = 1/(2\sqrt{n})$ is used. The plot in Figure 1 shows the smallest and largest eigenvalues of the iterates XZ, and the solid line indicates the central path (where the smallest and largest eigenvalues are equal).

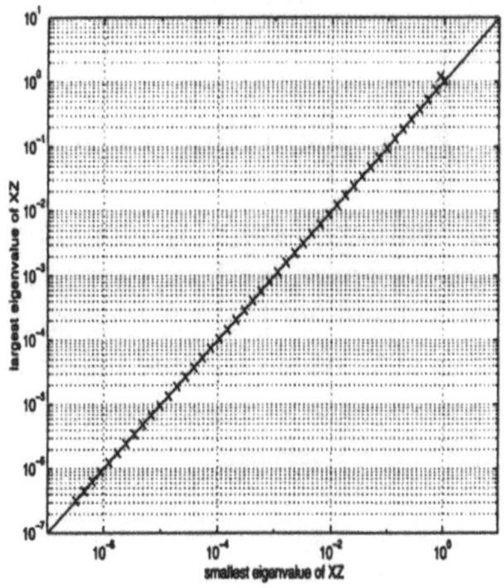

Figure 1. Iterates for a fixed updating scheme. The solid line corresponds to the central path.

In this example 45 iterations are required to meet the convergence condition $\text{Tr}(XZ) \leq 10^{-6}$. This is close to the upper bound of 52 iterations given by Theorem 4.2 for the example.

If the adaptive updating strategy is employed, the iterates are as shown in Figure 2.

Only eight iterations are needed for convergence now, and the values of θ at each iteration are as shown in Table I. Note that the proximity after step is typically well below $\frac{1}{2}$, which shows that the update made using the updating condition is not the largest possible.

Finally, we investigate the effect of an extra centering step at each iteration. In other words, a step $(\Delta X, \Delta Z)$ satisfying:

$$D\Delta Z D + \Delta X = \mu Z^{-1}$$

subject to

$$\begin{aligned} \mathrm{Tr}(A_i \Delta X) &= 0, \quad i = 1, \ldots, m \\ \Delta Z &\in \mathrm{span}\{A_1, \ldots, A_m\} \end{aligned}$$

is made before a new μ update is done using the adaptive updating strategy.

The plot in Figure 3 shows that the result is fast asymptotic convergence. Unfortunately there is no theoretical justification for superlinear convergence properties at this time.

Only four μ-updates are needed for convergence. The values of θ in subsequent updates, as well as values of the proximity δ before and after centering are shown in Table II.

6. Concluding remarks

The algorithm with centering steps in the previous section can be viewed as a predictor-corrector method, though not in the traditional sense — in the Mizuno-Todd-Ye [14] type predictor corrector methods the predictor direction is fixed as the primal-dual affine scaling direction. The predictor step length is then chosen to be the maximal step which yields a pair (X, Z) satisfying

$$\left\| I - \frac{n}{\mathrm{Tr}(X^{\frac{1}{2}} Z X^{\frac{1}{2}})} X^{\frac{1}{2}} Z X^{\frac{1}{2}} \right\| \leq \frac{1}{2}$$

(see e.g. [17]). This steplength can be computed analytically as the solution of a fourth order polynomial. The underlying concept of the predictor step in the previous section is different: the predictor direction (and not the step length) is determined using the updating condition, and only full Newton steps are taken.

The updating condition seems less effective than in the LP case in predicting the proximity value δ^+ after the Newton step. In the LP case superlinear convergence is observed (though unproven) in experiments [16]. For the small example presented here (and for similar ones), this effect is not observed unless centering steps are done.

This observation seems consistent with numerical examples by Kojima [4] which show that superlinear convergence cannot be achieved using the NT direction in a Mizuno-Todd-Ye predictor-corrector scheme — it is necessary to use more than one centering step. [5]

With the updating strategy presented here it may therefore also turn out that centering steps are essential.

[4] Talk presented at the IFORS 96 conference in Vancouver, July 8–12, 1996.

[5] Superlinear convergence has been proved for the Mizuno-Todd-Ye predictor-corrector scheme under assumptions of a strictly complementary solution and increasingly centered iterates [12]. The requirement for 'increasing centrality' necessitates extra centering steps in the formulation of a superlinearly convergent algorithm, while extra centering steps are never done in practice.

Table I. Sequence of updating parameters θ for the adaptive updating scheme. The centrality measure δ and duality gap after each step is also shown.

iteration	θ	δ	gap
0	–	0.1163	0.3101
1	0.8045	0.4024	6.061×10^{-1}
2	0.8606	0.3115	8.447×10^{-2}
3	0.9380	0.2066	5.229×10^{-3}
4	0.8804	0.2372	6.252×10^{-4}
5	0.8594	0.2536	8.790×10^{-5}
6	0.8512	0.2582	1.307×10^{-5}
7	0.8478	0.2598	1.989×10^{-6}

Table II. Iterations of the predictor-corrector type method. Centering steps are indicated by 'c' and μ-update steps by 'p'. The values of the proximity δ and duality gap after each step are shown, as well as the update parameter θ at each μ-update.

Iteration	δ	θ	gap
0	0.1163	–	3.101
1c	0.0050	–	3.101
1p	0.4157	0.8235	0.5474
2c	0.0075	–	0.5474
2p	0.3381	0.9468	2.910×10^{-2}
3c	0.0012	–	2.910×10^{-2}
4p	0.3378	0.9957	1.265×10^{-4}
4c	0.0012	–	1.265×10^{-4}
5p	0.0269	0.9987	1.647×10^{-7}

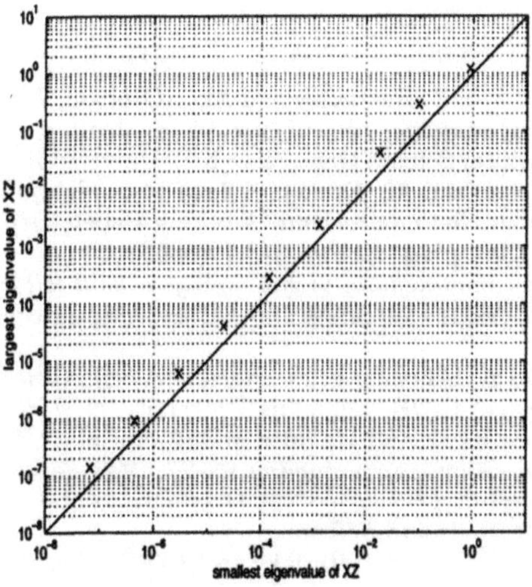

Figure 2. Iterates for the adaptive updating scheme. The solid line corresponds to the central path.

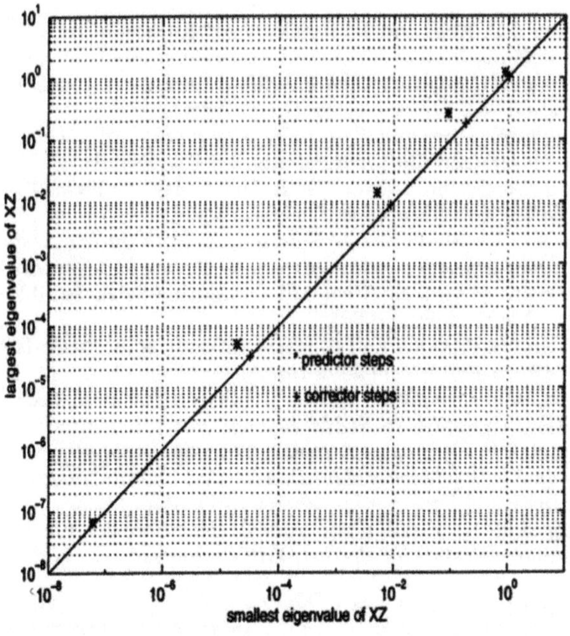

Figure 3. Iterates using adaptive updates followed by centering steps. Both centering and updating (predictor) steps are shown. The solid line corresponds to the central path.

References

1. F. Alizadeh and J.-P.A. Haeberley and M.L. Overton, *Primal-dual methods for semidefinite programming: convergence rates, stability and numerical results,* 721, NYU Computer Science Dept, New York University, New York, NY (1996).
2. K. M. Anstreicher and M. Fampa, *A long-step path following algorithm for semidefinite programming problems,* Working Paper, Department of Management Sciences, University of Iowa, USA, (1996).
3. L. Faybusovich, *Semi-definite programming: a path-following algorithm for a linear-quadratic functional,* SIAM Journal on Optimization, 6(1996), pp. 1007–1024.
4. D. Goldfarb and K. Scheinberg, *Interior point trajectories in semidefinite programming,* Dept. of IEOR, Columbia University, New York, NY, (1996).
5. B. He and E. de Klerk and C. Roos and T. Terlaky, *Method of approximate centers for semi–definite programming,* 96–27, 1996, Faculty of Technical Mathematics and Computer Science, Delft University of Technology, (To appear in *Optimization Methods and Software.*)
6. B. Jansen and C. Roos and T. Terlaky and J.–Ph. Vial, *Primal–dual algorithms for linear programming based on the logarithmic barrier method,* Journal of Optimization Theory and Applications, 83(1994), pp. 1–26.
7. J. Jiang, *A long step primal–dual path following method for semidefinite programming,* Dept of Applied Mathematics, Tsinghua University, Beijing, China (1996).
8. E. de Klerk and C. Roos and T. Terlaky, *Initialization in semidefinite programming via a self–dual, skew–symmetric embedding,* 96–10, Faculty of Technical Mathematics and Computer Science, Delft University of Technology, Delft, The Netherlands, (1996). (To appear in *OR Letters.*)
9. E. de Klerk and C. Roos and T. Terlaky, *Polynomial primal–dual affine scaling algorithms in semidefinite programming,* 96–42, Faculty of Technical Mathematics and Computer Science, Delft University of Technology, Delft, The Netherlands, (1996). (To appear *J. Comb. Opt.*)
10. M. Kojima and M. Shida and S. Shindoh, *A Note on the Nesterov-Todd and the Kojima-Shindoh-Hara search directions in Semidefinite Programming,* B-313, Department of Mathematical and Computing Sciences, Tokyo Institute of Technology, Tokyo, Japan (1996).
11. M. Kojima and S. Shindoh and S. Hara, *Interior point methods for the monotone semidefinite linear complementarity problems,* No. 282, Department of Mathematical and Computing Sciences, Tokyo Institute of Technology, Tokyo, Japan (1994). (To appear in *SIAM Journal on Optimization*)
12. Z.-Q. Luo and J. F. Sturm and S. Zhang, *Superlinear convergence oa a Symmetric primal–dual path following algorithm for semidefinite programming,* 9607/A, Tinbergen Institute, Erasmus University Rotterdam (1996).
13. I. J. Lustig and R. E. Marsten and D. F. Shanno, *Interior point methods: Computational state of the art,* ORSA Journal on Computing, 6(1994), pp. 1–15.
14. S. Mizuno and M. J. Todd and Y. Ye, *On adaptive step primal–dual interior–point algorithms for linear programming,* Mathematics of Operations Research, 18(1993) pp. 964–981.
15. Y. Nesterov and M. J. Todd, *Self–scaled barriers and interior–point methods for convex programming,* 1091, School of OR and IE, Cornell University, Ithaca, New York, USA, (1994). (To appear in Mathematics of Operations Research)
16. C. Roos and T. Terlaky and J.-Ph. Vial, *Theory and Algorithms for Linear Optimization: An interior point approach,* To appear December 1996, John Wiley & Sons, New York.

17. J. F. Sturm and S. Zhang, *Symmetric primal–dual path following algorithms for semidefinite programming*, 9554/A, Tinbergen Institute, Erasmus University Rotterdam, (1995).
18. M. J. Todd and K. C. Toh and R. H. Tütüncü, *On the Nesterov-Todd direction in semidefinite programming,* School of OR and Industrial Engineering, Cornell University, Ithaca, New York 14853–3801, (1996).

A Piecewise Linear Dual Procedure in Mixed Integer Programming

István Maros
Imperial College, London
and Computer and Automation Institute, Budapest

Abstract. Most solution methods for Mixed Integer Programming (MIP) problems require repeated solution of continuous linear programming (LP) problems. It is typical that subsequent LP problems differ just slightly. In most cases it is practically the right-hand-side that changes. For such a situation the dual simplex algorithm (DSA) appears to be the best solution method.

The LP relaxation of a MIP problem contains many bounded variables. This necessitates such an implementation of the DSA where variables of arbitrary type are allowed. The paper presents an algorithm called BSD for the efficient handling of bounded variables. This leads to a "mini" nonlinear optimization in each step of the DSA. Interestingly, this technique enables several cheap iterations per selection making the whole algorithm very attractive. The implementational implications and some computational experiences on large scale MIP problems are also reported. BSD is included in FortMP optimization system of Brunel University.

1. Introduction

The dual simplex algorithm (DSA) developed by Lemke [6] has long been known as a better alternative to Dantzig's primal simplex [2] for solving certain types of linear programming (LP) problems. Its importance has been recognized in many areas, most importantly in case of the Branch and Bound (BB) method for Mixed Integer Programming (MIP). BB requires the repeated solution of continuous LP problems (c.f. Nemhauser and Wolsey [9]). It is typical that subsequent problems differ just slightly. In most cases it is practically the right-hand-side that changes. For the required "reoptimizations" the dual simplex method appears to be the best solution algorithm.

The LP relaxation of a MIP problem contains many bounded variables. This necessitates such an implementation of the DSA where variables of arbitrary type are allowed and especially, the bounded primal variables are treated efficiently.

2. Problem statement

2.1. The primal problem

We consider the following primal linear programming (LP) problem:

F. Giannessi et al. (eds.), New Trends in Mathematical Programming, 159–170.

Table I. Types of variables

Feasibility range	Type	Reference
$x_j = 0$	0	Fixed variable
$0 \leq x_j \leq u_j < +\infty$	1	Bounded variable
$0 \leq x_j \leq +\infty$	2	Non-negative variable
$-\infty \leq x_j \leq +\infty$	3	Free variable

$$\begin{array}{ll} \text{minimize} & c^T x, \\ \text{subject to} & Ax = b, \\ & l \leq x \leq u, \end{array} \tag{1}$$

where A is an $m \times n$ matrix, c, x, l, and u are n vectors, and b is an m vector. Some or all components of l and u can be $-\infty$ or $+\infty$, respectively. A itself contains a unit matrix I, that is, $A = [I, \bar{A}]$, so it is of full row rank. Variables which multiply columns of I transform every constraint to an equation and are often referred to as *logical variables.* Variables which multiply columns of $\bar{A}$ are called *structural variables.*

After some elementary translations it can be achieved that the variables (whether logical or structural) fall into four categories as shown in Table I (for further details, see Orchard-Hays [10]).

2.2. The dual problem

For notational convenience, we restate the primal problem to contain bounded variables only. However, the algorithm to be presented will work with all types of variables.

$$\begin{array}{lll} (P1) & \text{minimize} & c^T x, \\ & \text{subject to} & Ax = b, \\ & & 0 \leq x \leq u \end{array} \tag{2}$$

A basis to (P1) is denoted by B and A is partitioned as $A = [B, N]$. The components of x and c are partitioned accordingly. Column j of A is denoted by a_j. A basic solution is

$$x_B = B^{-1}(b - \sum_{j \in U} u_j a_j), \tag{3}$$

where U is the index set of nonbasic variables at upper bound. The ith basic variable is denoted by x_{Bi}. The d_j reduced cost of variable j is defined as $d_j = c_j - \pi^T a_j = c_j - c_B^T B^{-1} a_j$ which is further equal to $c_j - c_B^T \alpha_j$ if the notation $\alpha_j = B^{-1} a_j$ is used.

The dual of (P1) is:

$$(D1) \quad \begin{array}{ll} \text{maximize} & b^T y - u^T w, \\ \text{subject to} & A^T y - w \le c, \\ & w \ge 0 \end{array} \tag{4}$$

In case of *Zero-One Mixed Integer Programming*, some or all of the variables may be *0/1 variables*. It means that in the LP relaxation there are many variables with upper bound of 1.0.

The Branch and Bound type algorithms require the solution of many closely related LP problems (c.f. [9]). Since the optimal basis usually remains dual feasible for a subsequent problem, the dual algorithm can be used efficiently for "reoptimization" starting from this basis.

In practice, dual algorithms work on the primal problem using the computational tools of the sparse primal simplex method but perform basis changes according to the rules of the dual.

The upper bounded version of the DSA was first described by Orchard-Hays [10] and later by Chvátal [1]. One of the computationally most expensive steps of the algorithm is the creation of the updated pivot row p, i.e., the computation of α_{pj} for all nonbasic j (c.f. [8]). None of the quoted descriptions discusses the multiple use of the updated pivot row which becomes possible in case of the presence of upper bounded variables. In our BSD (for Bound Swapping Dual) algorithm to be presented in section 4 we concentrate on this issue and show how it may be possible to make — occasionally many — iterations with one updated pivot row. BSD is included in the FortMP optimization system of Brunel University [3].

The underlying idea comes from the FEWPHI algorithm [7] of the author that can make several primal iterations with one updated column. Similar ideas for the primal are also discussed by Wolfe [11] and Greenberg [5]. BSD is based on the maximization of a piecewise linear function that describes the gain in the dual objective achievable with a selected outgoing variable. In this respect, its closest counterpart is the excellent theoretical exposition by Fourer [4].

3. Upper bounded dual

If there are only type-2 variables in the problem then w is not present and dual feasibility is expressed by

$$B^T y = c_B \tag{5}$$

$$N^T y \le c_N, \text{ or } d_N = c_N - N^T y \ge 0. \tag{6}$$

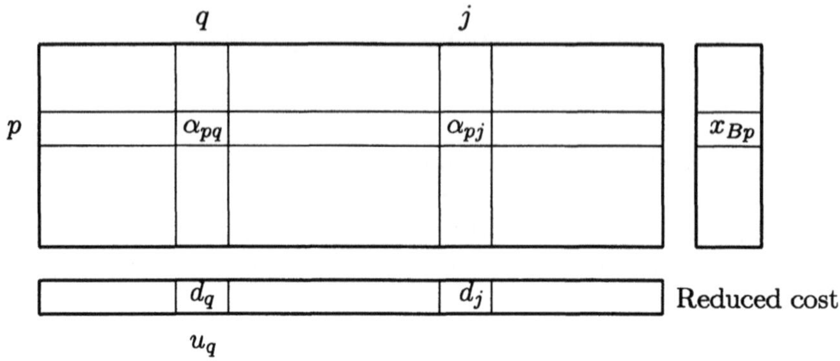

Figure 1. The key elements of the dual pivot step

The idea of a dual step is based on the relaxation of the pth dual equation from $a_p^T y = c_p$ to $a_p^T y \leq c_p$. If the change in the pth equation is parameterized by $t > 0$ then we have $B^T y(t) + te_p = B^T y$, from which the dual solution is $y(t) = y - t\rho_p$, if ρ_p denotes the pth row of B^{-1}. The corresponding dual objective value is:

$$b^T y(t) = b^T y - tb^T \rho_p = b^T y - tx_{Bp} \tag{7}$$

The rate of change in the dual objective is $-x_{Bp}$ which leads to an improvement over $b^T y$ if $x_{Bp} < 0$. This suggests that $x_{Bp} < 0$ (an infeasible primal basic variable) is to be selected to leave the basis at a feasible level (usually 0). The entering variable is determined by the dual ratio test to ensure that dual feasibility ($\bar{d}_N \geq 0$) is maintained after basis change.

3.1. Dual steps with type-1 variables present

If there are upper bounded (type-1) variables in the LP problem then dual feasibility for such a variable means $d_j \geq 0$ if x_j is nonbasic at zero and $d_j \leq 0$ if x_j is nonbasic at upper bound.

The steps of the upper bounded dual algorithm (following Chvátal [1, p. 157]) can be described as:

Dual-1:

Step 1. If the corresponding primal solution is feasible, that is, $0 \leq x_B \leq u_B$ then it is optimal, otherwise there is at least one primal infeasible basic variable. It can either be $x_{Bi} < 0$ or $x_{Bi} > u_i$. Any of them can be selected as a variable to be the outgoing variable. Its basic position is denoted by p and the corresponding row is referred to as pivot row.

Step 2. Compute the nonbasic components of the updated pivot row, that is, α_{pj} for $j \notin B$ by $\alpha_{pj} = \rho_p^T a_j$, where ρ_p^T is the pth row of the basis inverse: $\rho_p^T = e_p^T B^{-1}$ with e_p being the pth unit vector.

Step 3. If $x_{Bp} < 0$ then let J be the set of those nonbasic variables x_j for which $\alpha_{pj} < 0$ and $x_j = 0$, or $\alpha_{pj} > 0$ and $x_j = u_j$. If $x_{Bp} > u_{Bp}$ then let J be the set of those nonbasic variables x_j for which $\alpha_{pj} > 0$ and $x_j = 0$, or $\alpha_{pj} < 0$ and $x_j = u_j$. If J is empty then stop: the problem is infeasible. Otherwise, find the x_q, $q \in J$ such that

$$\theta = \left|\frac{d_q}{\alpha_{pq}}\right| = \min_{j \in J} \left|\frac{d_j}{\alpha_{pj}}\right| \tag{8}$$

and let x_q be the entering variable (θ is the selected dual ratio).

Step 4. Update the entering column: $\alpha_q = B^{-1} a_q$

Step 5. Assume: the original index of the outgoing variable (x_{Bp}) in A is r. Set $\theta_p = x_{Bp}/\alpha_{pq}$ and $\bar{x}_r = 0$ in case $x_{Bp} < 0$ and $\theta_p = (x_{Bp} - u_{Bp})/\alpha_{pq}$ and $\bar{x}_r = u_r$ in case $x_{Bp} > u_{Bi}$. Update the basic solution: $\bar{x}_{Bi} = x_{Bi} - \theta_p \alpha_{iq}$ for $1 \leq i \leq m$, $i \neq p$ and $\bar{x}_{Bp} = x_q + \theta_p$. Update reduced cost of nonbasic variables: $\bar{d}_r = -d_q/\alpha_{pq}$ and $\bar{d}_j = d_j + \bar{d}_r \alpha_{pj}$, $j \neq r$.

3.2. Dual steps with all types of variables

Type-0 variables do not play any role in the dual method since any d_j is feasible if x_j is a fixed variable. Such nonbasic variables do not take part in the dual ratio test and, consequently, never enter the basis. This is completely in line with the selection principles of the primal simplex algorithm.

Type-3 (free) variables can also be present in an LP problem. Dual feasibility requires $d_j = 0$ if x_j is nonbasic and of type-3.

It is easy to see that *Dual-1* algorithm requires very few changes to take care of all types of variables, namely, type-0 nonbasic positions are left out of J but type-3 variables take part in the (8) dual ratio test if the corresponding $\alpha_{pj} \neq 0$. It clearly shows that free variables will rush into the basis because they produce a zero ratio. We refer to this modified algorithm as *Dual-2*.

4. Bound swap in dual

There is one important thing about *Dual-1* and *Dual-2* that must be investigated in more detail. At the beginning of an iteration the incoming

x_q variable is nonbasic at a feasible level. It becomes basic in position p with a value of $x_q + \theta_p$ which will be infeasible if it is a bounded variable and

$$|\theta_p| > u_q. \tag{9}$$

In such a case, Orchard-Hays [10] sets $\bar{x}_q = 0$ or u_q, depending on the status of x_q. At the same time, however, the corresponding d_q becomes infeasible (no basis change, no updating of d_js but bound swap for x_q). We propose a rectification of this situation in the following way.

First, we assume that p was selected because $x_{Bp} < 0$. In this case the pth dual constraint is released negatively and the change of the objective function is described by (7). If, after applying the (8) ratio test, for the resulting θ_p (9) holds then a bound swap of x_q is triggered and the x_B basic solution is updated by: $\bar{x}_B = x_B \pm u_q \alpha_q$, more specifically, $\bar{x}_{Bp} = x_{Bp} \pm u_q \alpha_{pq}$. Here, '$-$' applies if $x_q = 0$, and '$+$' if $x_q = u_q$ before the iteration. Taking into account the definition of J (see Step 3 of Dual-1) the two cases can be expressed by the single equation:

$$\bar{x}_{Bp} = x_{Bp} + |\alpha_{pq}| u_q. \tag{10}$$

Here, $\bar{x}_{Bp} < 0$ holds because of $|x_{Bp}/\alpha_{pq}| > u_q$, meaning that the originally selected infeasible basic variable remains infeasible. It implies that the dual objective function keeps improving if t is increased beyond θ, however, the rate of improvement decreases by $|\alpha_{pq}|u_q$ from $-x_{Bp}$ to $-x_{Bp} - |\alpha_{pq}|u_q$. Now, we leave out index q from J, go to the (8) ratio test, and repeat the above procedure. This can easily be done since the updated pth row remains the same. If $|\theta_p| \leq u_q$ is encountered then the corresponding x_q is the incoming variable and a basis change is performed. If $\bar{x}_{Bp} < 0$ holds and J becomes empty then the primal problem is infeasible.

Figure 2 shows the change of the dual objective as a function of relaxing the pth dual constraint. This is a piecewise linear function of $\Delta y_p \equiv t$. The first break point corresponds to the smallest ratio the second to the second smallest, and so on. At the maximum of the function the sign of the slope changes. This is determined by $|\theta_p| \leq u_q$, that is, when a basis change is performed.

Next, we consider the case when p was selected because $x_{Bp} > u_{Bp}$. With a little linear algebra on (4) and taking (3) into account, it can be seen that if the pth dual constraint is relaxed positively (which is again parameterized by $t \geq 0$) then the change in the dual objective is $\Delta_{d\_obj} = t(x_{Bp} - u_{Bp})$, that is, the rate of change is $x_{Bp} - u_{Bp} > 0$.

Similarly to the first case, (8) is performed and a bound swap of x_q is triggered if (9) holds and the updated value of the pth basic variable is $\bar{x}_{Bp} = x_{Bp} \pm u_q \alpha_{pq}$. Here '$+$' applies if $x_q = 0$, and '$-$' if $x_q = u_q$

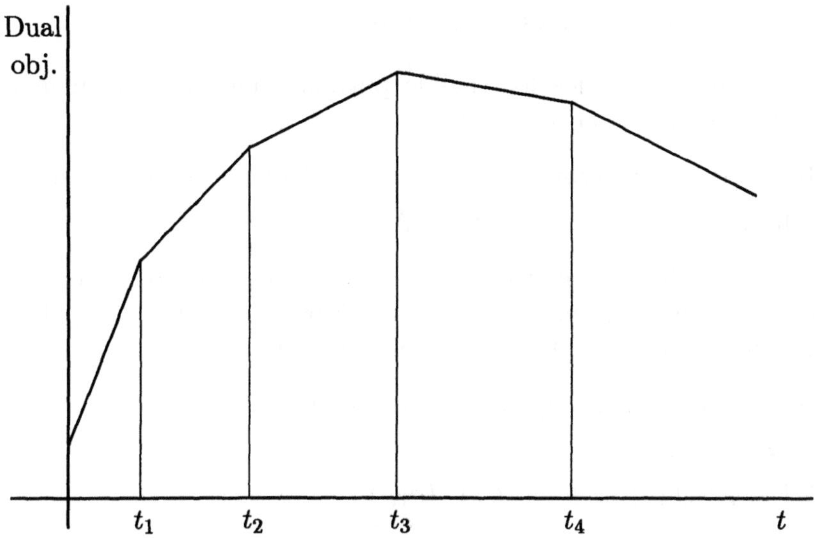

Figure 2. The dual objective as the function of releasing a dual basic equation

before the iteration, but in this case (see definition of J in Dual-1) the single updating equation is

$$\bar{x}_{Bp} = x_{Bp} - |\alpha_{pq}|u_q. \tag{11}$$

Now, $\bar{x}_{Bp} > u_{Bp}$ holds because of $|(x_{Bp}-u_{Bp})/\alpha_{pq}| > u_q$, meaning that the originally selected infeasible basic variable remains infeasible. It implies that the dual objective function keeps improving if t is increased beyond θ, however, the rate of improvement decreases again by $|\alpha_{pq}|u_q$. The rest of the arguments of the first case applies here unchanged.

The important common feature of the two cases is that the slope of the corresponding piecewise linear function (the dual objective) decreases at every bound swap (performed with the same outgoing candidate) by $|\alpha_{pq}|u_q$.

In other words, the initial slope of the dual objective is

$$s_p = \begin{cases} -x_{Bp}, & \text{if } x_{Bp} < 0 \\ x_{Bp} - u_{Bp}, & \text{if } x_{Bp} > u_{Bp}. \end{cases} \tag{12}$$

This slope changes at every bound swap to $\bar{s}_p = s_p - |\alpha_{pq}|u_q$ until a basis change is encountered.

This observation suggests the following dual iteration step, assuming that B is a dual feasible basis for (D1) and the upper bounds for type-2 and type-3 variables are $u_j = +\infty$.

Bound Swapping Dual (BSD):

1. Select a primal infeasible basic variable and denote its basic position by p (pivot row).

   If none exists: solution is optimal, stop.

2. Compute dual ratios for eligible positions and store them in a sorted order: $0 \leq t_1 \leq \ldots \leq t_Q$.

3. If $Q = 0$ the problem is dual unbounded (primal infeasible), stop; otherwise set $k = 1$, $T^+ = T^- = \emptyset$.

4. Let q denote the subscript of the variable defining t_k. If for the corresponding θ_p (as defined in Step 5 of Dual-1) $|\theta_p| \leq u_q$ then entering variable (x_q) is found, go to Step 6.

5. In accordance with the direction of the bound swap, set

   $x_q = u_q$ or $x_q = 0$,

   $T^+ = T^+ \cup \{q\}$ or $T^- = T^- \cup \{q\}$,

   and

   increment k,

   go to Step 4.

6. Update solution:

   a) Take care of bound swaps:

   $$\bar{x}_B = x_B - \sum_{j \in T^+} u_j \alpha_j + \sum_{j \in T^-} u_j \alpha_j, \tag{13}$$

   where the sum is defined to be 0 if the corresponding index set is empty.

   b) Take care of basis change:
   Replace $\bar{x}_B$ by $F\bar{x}_B$, F denoting the elementary transformation matrix created from α_q.

Verbally this algorithm can be interpreted in the following way: If not the smallest ratio is selected for pivoting (bound swap in BSD) then the d_js of the bypassed small ratios change sign and this is exactly what is needed for them to stay feasible after bound swap and basis change.

4.1. Work per iteration

It can easily be seen that the extra work required for BSD is generally small.

1. Ratio test: the same as in Dual-2.

2. The break points of the piecewise linear dual objective function have to be stored and sorted. This requires extra memory for the storage, and extra work for the sorting. However, the t_k values have to be sorted only up to the point when a basis change is defined. Therefore, if an appropriate priority queue is set up for these values then the extra work can be reduced sharply.

3. Taking care of bound swap according to (13) would require multiple update (FTRAN) of the corresponding columns. However, the same can be achieved by only one extra FTRAN because

$$\begin{aligned}\bar{x}_B &= x_B - \sum_{j\in T^+} u_j \alpha_j + \sum_{j\in T^-} u_j \alpha_j \\ &= x_B - B^{-1}\left(\sum_{j\in T^+} u_j a_j - \sum_{j\in T^-} u_j a_j\right) \\ &= x_B - B^{-1}\tilde{a}\end{aligned}$$

with the obvious interpretation of $\tilde{a}$.

4.2. Implementation

For the efficient implementation of BSD sophisticated data structure (priority queue) is needed to store and (partially) sort the break points. Additionally, since row and column operations are performed on A, it is important to have a data structure that supports row and columnwise access of A.

4.3. Advantages

The main advantage of the BSD algorithm that it can make many iterations with one updated pivot row and these iterations cost hardly more than one iteration. Such situations frequently occur when LP relaxations of MIP problems are solved.

On the other hand, BSD incorporates the usual dual algorithm and makes steps according to it when bound swap cannot be made.

Consequently, it can be said that BSD is an efficient generalization of the traditional dual simplex algorithm.

5. Experiences

We implemented BSD in our FortMP system as a replacement for the Dual-2 algorithm. Now BSD is the standard dual procedure in FortMP.

Table II. The test problems

Problem	# of rows	# of columns	# of binary variables	# of nonzeros in A
aa03	826	8627	8627	79433
tuss21	140	7785	7509	67339

Table III. Solution statistics of aa03 for the first LP relaxation

Algorithm	Iterations	Solution time (sec)
Primal	17064	2795.49
Dual-2	10859	1220.27
BSD	4841	647.42

However, we have kept the previous version (based on Dual-2) so that it could be used for comparison.

Our experiences are mostly based on solving LP relaxation of MIP problems. To compare the effect of using BSD we demonstrate our findings with two problems: `aa03` and `tuss21`. Their main characteristics are shown in Table II. In the tests we did not use presolve to reduce problem size.

In case of `aa03` we used the dual algorithm to solve the first LP relaxation starting from scratch with the unit matrix. The Dantzig criterion (most violating variable) was used to select the pivot row. For comparison, we also included solution statistics with the primal simplex method of FortMP. Table III shows the number of iterations and the solution time (in seconds) achieved on a 90 MHz Pentium PC. The results show that BSD could make many bound swap operations with one selected pivot row giving a more than 50% reduction in the number of iterations, while the time per iteration increased less than 20%. The overall gain is obvious.

We have made a different set of tests with `tuss21`. In this case dual was used in a "natural environment" of MIP for reoptimization. All runs started from the optimal basis of the first LP relaxation. The results are shown in Table IV. Again, for comparison, we have included the solution characteristics of primal. As expected, primal is heavily outperformed by the dual algorithms. Among the duals, the excellent performance of BSD is remarkable which justifies its inclusion in FortMP.

Table IV. Solution statistics of **tuss21** reaching the first integer solution starting from the first LP relaxation.

Algorithm	Number of reoptimizations	Iterations	Solution time (sec)
Primal	18	49307	1047.91
Dual-2	16	6612	142.09
BSD	15	3305	54.78

Table V. Typical iteration counts per reoptimization for 20 real life industrial MIP problems

Algorithm	Itns/reopt
Primal	500 – 5000
Dual-2	50 – 500
BSD	20 – 200

Finally, we give a summary table of our experiences on a larger set of real-life MIP problems encountered in the collaborative work with our business partners at Brunel. Table V shows typical numbers for iterations per reoptimization in case of 20 problems.

6. Summary

We have presented a generalization of the dual ratio test that handles bounded primal variables efficiently. It is based on the piecewise linear nature of the dual objective if defined as a function of releasing one basic equation. The main advantage is that a number of very cheap iterations can be made with one updated pivot row. In this respect it is a counterpart of the FEWPHI algorithm in the primal simplex which can make progress of several iterations with one updated column.

We showed that it can be implemented efficiently and the computational testing verified the theoretical advantages in practice. Though our tests were limited to LP relaxation of MIP problems, it is expected that the favorable characteristics of BSD will materialize in case of general LP problems with upper bounded primal variables.

7. Acknowledgements

The work presented in this paper benefited a lot from fruitful discussions with Tamás Terlaky in 1992. The encouragements from Gautam Mitra are also warmly acknowledged.

References

1. Chvátal, V., *Linear Programming*, Freeman and Co., 1983.
2. Dantzig, G. B., *Linear Programming and Extensions*, Princeton University Press, Princeton, N.J., 1963.
3. Ellison, E. F. D., Hajian, M., Levkovitz, R., Maros, I., Mitra, G., Sayers, D., "FortMP Manual", Brunel, The University of West London, May 1994.
4. Fourer, R., "Notes on the Dual Simplex Method", Unpublished, March, 1994.
5. Greenberg, H. J., "Pivot selection tactics", in Greenberg, H. J. (ed.), *Design and Implementation of Optimization Software*, Sijthoff and Nordhoff, 1978.
6. Lemke, C. E., "The Dual Method of Solving the Linear Programming Problem", *Naval Research Logistics Quarterly*, 1, 1954, p. 36–47.
7. Maros, I., "A general Phase-I method in linear programming", *European Journal of Operational Research*, 23(1986), p. 64–77.
8. Maros, I., Mitra, G., "Simplex Algorithms", Chapter 1 in Beasley J. (ed.) *Advances in Linear and Integer Programming*, Oxford University Press 1996.
9. Nemhauser, G. L., Wolsey, L. A., *Integer and Combinatorial Optimization*, John Wiley, 1988.
10. Orchard-Hays, W., *Advanced Linear-Programming Computing Techniques*, McGraw-Hill, 1968.
11. Wolfe, P., "The composite simplex algorithm", *SIAM Review*, 7 (1), 1965, p. 42–54.

Geometrical solution of weighted Fermat problem about triangles

Mario Martelli
Università di Pisa, Dipartimento di Matematica Applicata U. Dini,
Via Diotisalvi 2, 56126 Pisa, Italia

Abstract. The paper considers weighted generalization of a classic Fermat problem about triangles: given A, B, C vertices of a triangle and three real positive numbers $\bar{a}, \bar{b}, \bar{c}$, find in plane a point P minimizing the sum of its distances to A, B, C, multiplied by $\bar{a}, \bar{b}, \bar{c}$, respectively. It is shown by simple geometrical methods that, if $\bar{a}, \bar{b}, \bar{c}$ satisfy the triangle inequalities and further conditions also involving the angles of the triangle ABC, then there is one and only one minimum interior point of the triangle and a construction is supplied which enables us to find the minimum point by "ruler and compasses". In all other cases, one identified vertex of the triangle is the minimum point.

1. Introduction

The following Fermat problem is one of the first examples of non-linear optimization: Given in a plane three points A, B, C, find a point P minimizing the sum of its distances to A, B, C.
Torricelli [5], [8] proved, for a given triangle with no angle exceeding 120°, that there is only one interior point which minimizes this sum and showed a construction which enables us to find the minimum point by "ruler and compasses". Until now, several problems [3], [4], [9], [10] derived from this Fermat problem, have been proposed and studied. In particular the Weber problem [3], [9], known under an abundance of other names, asks for a point minimizing the sum of the weighted Euclidean distances from itself to n given points $A_1, \ldots, A_n$ in a plane, each associated with a given positive weight. The Weber problem is optimally solvable via a mechanical analogue device named the Varignon frame [9].

In this paper the following weighted Fermat problem about triangles, is studied. Let A, B, C be vertices of a triangle and let $\bar{a}, \bar{b}, \bar{c}$ be three real positive numbers; find in the plane a point P minimizing the sum of its distances to A, B, C, multiplied by $\bar{a}, \bar{b}, \bar{c}$, respectively, i.e. a point minimizing the function

$$f(P) := \bar{a}|PA| + \bar{b}|PB| + \bar{c}|PC|,$$

where $| \ |$ denotes the Euclidean distance.

F. Giannessi et al. (eds.), New Trends in Mathematical Programming, 171–180.

The problem, which is a particular case of Weber problem, is solved in this paper only by geometrical methods that allow us both to claim that there is only one solution of the problem, and to find it by "ruler and compasses". The unweighted case $\bar{a} = \bar{b} = \bar{c} = 1$ is a particular case.

It is right to expect that other problems [3] can be solved by geometrical tecniques.

Let us introduce some notations. If P, Q are two points, an arc or a segment whose edges are P, Q is denoted by PQ; if it does not contain its edges P, Q, is denoted by $\widetilde{PQ}$.

Let A, B, C be vertices of a triangle. The angles opposite the sides BC, AC, AB of the triangle ABC are denoted by α, β, γ, respectively. If α is an angle $\mu(\alpha)$ denotes its measure by radians.

When $\bar{a}, \bar{b}, \bar{c}$ satisfy the triangle inequalities $-\bar{a}+\bar{b}+\bar{c} > 0$, $\bar{a}-\bar{b}+\bar{c} > 0$, $\bar{a}+\bar{b}-\bar{c} > 0$, then obviously can be viewed as the lengths of the sides of a triangle; the angles opposite these sides are denoted by $\bar{\alpha}, \bar{\beta}, \bar{\gamma}$, respectively.

We shall use the following two theorems. The first one is attributed to Ptolemy, Alexandrine astronomer geodesist and mathematician of 1st–2nd century A.D. [6].

PTOLEMY'S THEOREM: Let $ABCD$ be a convex quadrilateral inscribed in a circumference. Then the product of the lengths of the diagonals equals the sum of the products of the lengths of the opposite sides: $|AC|\,|BD| = |AD|\,|BC| + |AB|\,|CD|$.

PTOLEMY'S COMPLETE THEOREM [1]. Let $ABCD$ be a convex quadrilateral. Then the product of the lengths of the diagonals is equal to or less than the sum of the products of the lengths of the opposite sides: $|AC|\,|BD| \leq |AD|\,|BC| + |AB|\,|CD|$. Equality holds if and only if A, B, C and D are on a circumference.

2. Solution of weighted Fermat problem

We now give some immediate properties.

PROPOSITION 1. *Let A, C, D be the vertices of a triangle, χ its circumscribing circumference and P any point of the arc AC of χ which does not contain D; then:*

$$\frac{|CD|}{|AC|}|PA| + \frac{|AD|}{|AC|}|PC| = |PD|.$$

Proof. The cases $P \equiv A$ and $P \equiv C$ are trivial. If $P \neq A$ and $P \neq C$ the thesis follows from Ptolemy's Theorem.

Remark 2. Given a triangle ABC whose angles are α, β, γ, let $\bar{\alpha}, \bar{\beta}, \bar{\gamma}$, be the angles of another triangle. Then only one of the following three cases, at the most, holds:
i) $\mu(\alpha) + \mu(\bar{\alpha}) \geq \pi$,
ii) $\mu(\beta) + \mu(\bar{\beta}) \geq \pi$,
iii) $\mu(\gamma) + \mu(\bar{\gamma}) \geq \pi$.

Remark 3. If a point M minimizes $\bar{a}|PA| + \bar{b}|PB| + \bar{c}|PC|$, then M minimizes $k\bar{a}|PA| + k\bar{b}|PB| + k\bar{c}|PC|$, where $k \in \mathbb{R},\ k > 0$.

The solution of weighted Fermat problem about triangles is given by a theorem that, if suitable conditions are verified, suggests a construction by "ruler and compasses".

CONSTRUCTION BY "RULER AND COMPASSES". Given a triangle ABC, let us consider the triangle ACD, where vertex D is opposite B with respect to the line AC and such that the angles $\bar{\alpha}, \bar{\gamma}$ are opposite the sides CD, AD respectively and let us consider its circumscribing circumference; then the minimum point, interior to triangle ABC, is on the intersection between the circumscribing circumference and the segment DB.

Described construction and other equivalent constructions that use suitable pairs of circumferences and straight lines, can be drawn from fig. 1.

The following theorem proves the existence and uniqueness of the minimum point of the problem only by geometrical methods and therefore does not exploit properties of the convex functions.

THEOREM 4. *Let A, B, C be the vertices of a triangle, $\bar{a}, \bar{b}, \bar{c}$ be three real positive numbers and $f(P) := \bar{a}|PA| + \bar{b}|PB| + \bar{c}|PC|$. Then:*

i) There are no minimum points outside the triangle.

ii) If $\bar{a}, \bar{b}, \bar{c}$ satisfy the triangle inequalities and all inequalities $\mu(\alpha) + \mu(\bar{\alpha}) < \pi$, $\mu(\beta) + \mu(\bar{\beta}) < \pi$, $\mu(\gamma) + \mu(\bar{\gamma}) < \pi$ are verified, then there is only one point minimizing f; this point is interior to the triangle ABC and can be found by the above described construction.

iii) If $\bar{a}, \bar{b}, \bar{c}$ satisfy the triangle inequalities and satisfy one of the inequalities $\mu(\alpha) + \mu(\bar{\alpha}) \geq \pi$, $\mu(\beta) + \mu(\bar{\beta}) \geq \pi$, $\mu(\gamma) + \mu(\bar{\gamma}) \geq \pi$, then there is only one point minimizing f: vertex A if $\mu(\alpha) + \mu(\bar{\alpha}) \geq \pi$, vertex B if $\mu(\beta) + \mu(\bar{\beta}) \geq \pi$, or vertex C if $\mu(\gamma) + \mu(\bar{\gamma}) \geq \pi$.

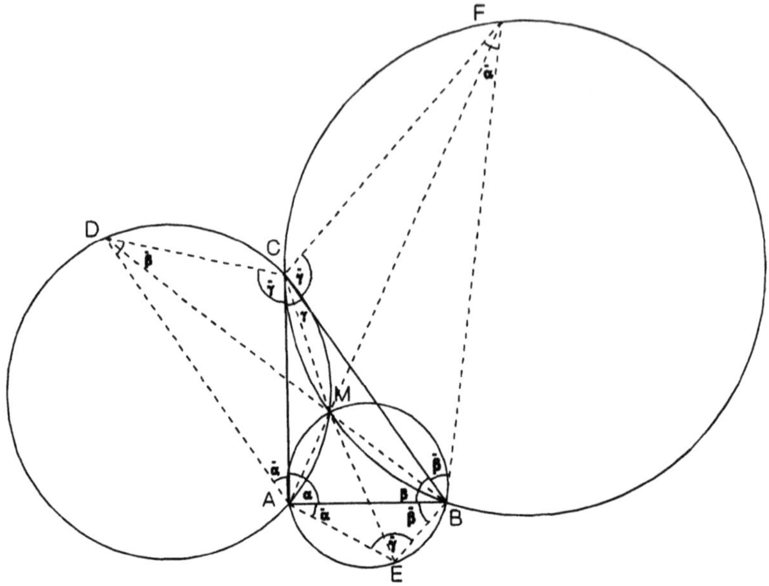

Figure 1.

iv) If $\bar{a}, \bar{b}, \bar{c}$ do not satisfy the triangle inequalities, there is only one point minimizing f: vertex A if $\bar{a} > \bar{b} + \bar{c}$, vertex B if $\bar{b} > \bar{a} + \bar{c}$ or vertex C if $\bar{c} > \bar{a} + \bar{b}$.

Proof. i) Let ρ' and ρ'' the half–planes defined by line AC such that $B \in \rho''$ and line AC be contained neither in ρ' nor in ρ''. For each point P' in ρ', let us consider the point P'' orthogonally symmetric with respect to the line AC. Obviously it is

$$|P'A| = |P''A|, \ |P'B| > |P''B|, \ |P'C| = |P''C|;$$

then $f(P') > f(P'')$.

Considering also pairs of points orthogonally symmetric with respect to the lines AB and BC, i) follows.

ii) Let us consider (fig. 1) the triangle ACD with vertex D belonging to ρ' and with the angles $\bar{\alpha}, \bar{\beta}$ opposite the sides CD, AC, respectively. Let χ_b be the circumscribing circumference of ACD. From the Proposition 1 we have

$$f_b(P) := \frac{\bar{a}}{\bar{b}}|PA| + \frac{\bar{c}}{\bar{b}}|PC| = \frac{|CD|}{|AC|}|PA| + \frac{|AD|}{|AC|}|PC|,$$

for every point P belonging to arc AC of χ_b which does not contain point D.

Since $\mu(\alpha)+\mu(\bar{\alpha}) < \pi$, $\mu(\gamma)+\mu(\bar{\gamma}) < \pi$, then the line DB intersects circumference χ_b at point D and at another point M of the arc $\widetilde{AC}$ of χ_b which does not contain point D; since also $\mu(\beta) + \mu(\bar{\beta}) < \pi$, M lies between B and D and is interior to the open triangle ABC.

Point M minimizes $|PD| + |PB|$ and, by Ptolemy's Theorem,

$$f_b(M) + |MB| = |DM| + |MB|.$$

Every other point P^* of the arc AC of χ_b which does not contain point D, does not minimize $f_b(P)+|PB|$. In fact, let us consider the triangle BDP^*; by the Proposition 1,

$$f_b(P^*) + |P^*B| = |P^*D| + |P^*B| > |DM| + |MB|.$$

Every other point P^* of the half–plane ρ'' and belonging to closed triangle ABC, does not minimize $f_b(P)+|PB|$. In fact, let us consider the quadrilateral $ADCP^*$; by the hypothesis and Ptolemy's Complete Theorem, $f_b(P^*) > |P^*D|$ and therefore

$$f_b(P^*) + |P^*B| > |P^*D| + |P^*B| \geq |DM| + |MB|.$$

We can conclude, for the moment, that point M, by Remark 3, minimizes f at least if we consider only points not belonging to segment $\widetilde{AC}$.

Let $f_c(P) := \frac{\bar{a}}{\bar{c}}|PA| + \frac{\bar{b}}{\bar{c}}|PB|$. Let us consider the triangle ABE, where E and C are opposite with respect to the line AB, and where the angles $\bar{\alpha}, \bar{\beta}$ are opposite the sides BE, AE, respectively. Let χ_c be the circumscribing circumference of ABE. By Remark 3, point M minimizes $f_c(P)+|PC|$, at least if we consider only points not belonging to segment $\widetilde{AC}$. By Ptolemy's Complete Theorem and according to what has been seen above, M is also a point of χ_c. Analogously, by the hypothesis, we can conclude that no point P^*, different from A, of the segment AC minimizes $f_c(P) + |PC|$ and therefore does not minimize f. Hence M is the only point minimizing f and ii) follows.

iii) Let $\mu(\alpha) + \mu(\bar{\alpha}) = \pi$. Vertex A of triangle ABC (fig. 2) minimizes $|PD| + |PB|$ and, by the Proposition 1, $f_b(A) = |AD|$; it follows $f_b(A) + |AB| = |AD| + |AB|$.

Every other point P^*, different from A, of the arc AC of χ_b which does not contain point D, does not minimize $f_b(P) + |PB|$. In fact, let us consider the triangle ABP^*; from the Proposition 1,

$$f_b(P^*) + |P^*B| = |P^*D| + |P^*B| > |AD| + |AB|.$$

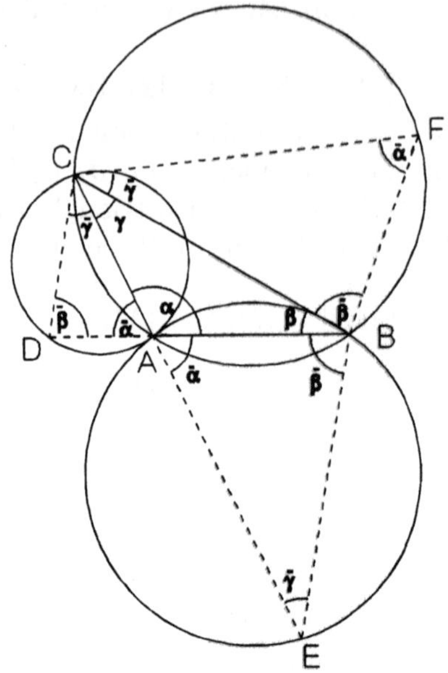

Figure 2.

Every other point P^*, of the half–plane ρ'', belonging to closed triangle ABC and not belonging to segment AB, does not minimize $f_b(P)+|PB|$. In fact, by Ptolemy's Complete Theorem, $f_b(P^*) > |P^*D|$ and therefore

$$f_b(P^*) + |P^*B| > |P^*D| + |P^*B| > |AD| + |AB|.$$

We can conclude, for the moment, that point A, by Remark 3, minimizes f at least if we consider only points different from B and not belonging to the segments $\widetilde{AC}$, $\widetilde{AB}$.

Vertex A, by Remark 3, minimizes $f_c(P) + |PC|$, at least if we consider only the above mentioned points, and $f_c(A)+|AC| = |AE|+|AC|$. The point B does not minimize $f_c(P)+|PC|$. In fact, from the Proposition 1, $f_c(B)+|BC| = |BE|+|BC|$ and, by triangle inequality involving the sides of triangle BCE, $|BE| + |BC| > |AE| + |AC|$. It follows, by Remark 3, that B does not minimize f.

Let $f_a(P) := \frac{\bar{b}}{\bar{a}}|PB| + \frac{\bar{c}}{\bar{a}}|PC|$; by Remark 3, vertex A minimizes $f_a(P) + |PA|$, at least if we consider only points not belonging to segments $\widetilde{AC}$, $\widetilde{AB}$.

Let us now consider the triangle BCF, where F and A are opposite with respect to the line BC, and where the angles $\bar{\alpha}, \bar{\beta}$ are opposite the sides BC, CF, respectively. Let χ_a be the circumscribing circumference of BCF. By Ptolemy's Complete Theorem, $A \in \chi_a$. Any point P^* belonging to segments $\widetilde{AB}$, $\widetilde{AC}$, by the hypothesis and Remark 2, neither belongs to the line BF nor to the line CF. Hence, analogously, it is proved that P^* does not minimize $f_a(P)+|PA|$ and therefore, from

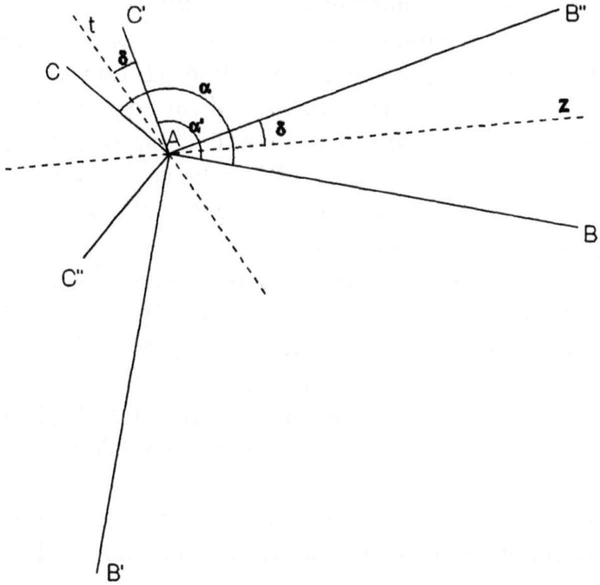

Figure 3.

Remark 3, it minimizes neither $f_b(P) + |PB|$, nor $f(P)$. In conclusion A is the only point minimizing f.
Analogously it is proved that if $\mu(\beta) + \mu(\bar{\beta}) = \pi$, $\mu(\gamma) + \mu(\bar{\gamma}) = \pi$, vertex B or vertex C is the only one point minimizing f, respectively.

Let $\mu(\alpha) + \mu(\bar{\alpha}) > \pi$. Let (fig. 3) be the triangle where $|AC| = |AC'|$, where C, C' are points of the same half–plane defined by line AB and where α' is the angle opposite side BC', such that $\mu(\alpha') + \mu(\bar{\alpha}) = \pi$; $\alpha > \alpha' > 0$. Let $\delta := \frac{\alpha - \alpha'}{2}$, $f(P; ABC) := \bar{a}|PA| + \bar{b}|PB| + \bar{c}|PC|$.

The line t, orthogonal axis of segment CC', contains A. Let τ, τ' be the half–planes defined by t, such that $C \in \tau$, $C' \in \tau'$ and line t be contained neither in τ nor in τ'. Any point P^* of half–plane τ' is such that $|P^*C| > |P^*C'|$; since $\bar{c} > 0$, $\bar{c}|P^*C| > \bar{c}|P^*C'|$; hence $f(P^*; ABC) > f(P^*; ABC')$. By previous considerations, A minimizes $f(P; ABC')$ and $f(A; ABC') = \bar{b}|AB| + \bar{c}|AC'|$; therefore

$$f(P^*; ABC) > \bar{b}|AB| + \bar{c}|AC'| \,.$$

As $f(A; ABC) = \bar{b}|AB| + \bar{c}|AC|$, from the hypothesis, we can conclude that $f(P^*; ABC) > f(A; ABC)$ for every point P^* of τ'.

Let $AB'C$ be the symmetrical triangle of ABC' with respect to the orthogonal axis t. $|AB'| = |AB|$ and line t is obviously the orthogonal axis of segment BB'. Analogously we obtain $f(P^*; ABC) > f(A; ABC)$ for every point P^* of half–plane τ. Summarizing,

$$f(P^*; ABC) > f(A; ABC)$$

holds for every point P^* not belonging to line t.

Let $AB''C$ be the triangle where $|AB''| = |AB|$, where C, B'' are points of the same half–plane defined by line AB and where α'', angle in A, is equal to α'. The orthogonal axis z of the segment BB'', forms the angle δ with line AB and contains A. Let ζ, ζ' be the half–planes defined by z, such that $B \in \zeta$, $B'' \in \zeta'$ and line z be contained neither in ζ nor in ζ'. Working analogously as above, we get $f(P^*; ABC) > f(A; ABC)$ for every point P^* of the half–plane ζ'.

Finally, let ABC'' be the symmetrical triangle of $AB''C$ with respect to the orthogonal axis z; we obtain $f(P^*; ABC) > f(A; ABC)$ for every point P^* of the half–plane ζ. Summarizing, $f(P^*; ABC) > f(A; ABC)$ holds for every point P^* not belonging to line z.

Lines t, z are different and both contain point A. Hence the proof follows in the case $\mu(\alpha) + \mu(\bar{\alpha}) > \pi$. We work analogously in the other two cases.

iv) If the numbers $\bar{a}, \bar{b}, \bar{c}$ do not satisfy the triangle inequalities, then one of these numbers is necessarily equal to or larger than the sum of the other two.

Let $\bar{a} \geq \bar{b} + \bar{c}$. Then there is one positive number $\bar{a}^*$, smaller than $\bar{a}$, such that numbers $\bar{a}^*, \bar{b}, \bar{c}$ are the lengths of the sides of a triangle and such that $\mu(\alpha) + \mu(\bar{\alpha}^*) = \pi$. Let $f^*(P) := \bar{a}^*|PA| + \bar{b}|PB| + \bar{c}|PC|$; we have $f^*(P) \leq f(P)$ for every point P in the plane; by iii), A is the only point which minimizes f^* and $f^*(A) = f(A)$; hence A is the only point minimizing f. We proceed analogously in the other two cases.

Notes about the Theorem 4. Torricelli's construction (fig. 4) and Simpson's subsequent construction [5], [7] (fig. 5) are particular cases of those illustrated in fig. 1. iii) generalizes Torricelli's observation proved by Heinen [2], [5].

Note. The degenerate case, A, B, C collinear, is simple to treat by easy calculations. For example, when A and B coincide and $\bar{a} + \bar{b} = \bar{c}$, we have at once that any point on the segment AC is here an optimal solution, and minimum of f is $|AC|\bar{c}$.

Remark 5. Let $\bar{a}, \bar{b}, \bar{c}$ be three real positive numbers and let M be a point of the plane. Therefore it is possible to construct every triangle ABC such that M minimizes $\bar{a}|PA| + \bar{b}|PB| + \bar{c}|PC|$.

Let $\bar{a}, \bar{b}, \bar{c}$ not satisfy the triangle inequalities. Moreover, for instance, let $\bar{a} \geq \bar{b} + \bar{c}$. In order to obtain every triangle as required, we put, by iv) of Theorem 4, vertex A in point M.

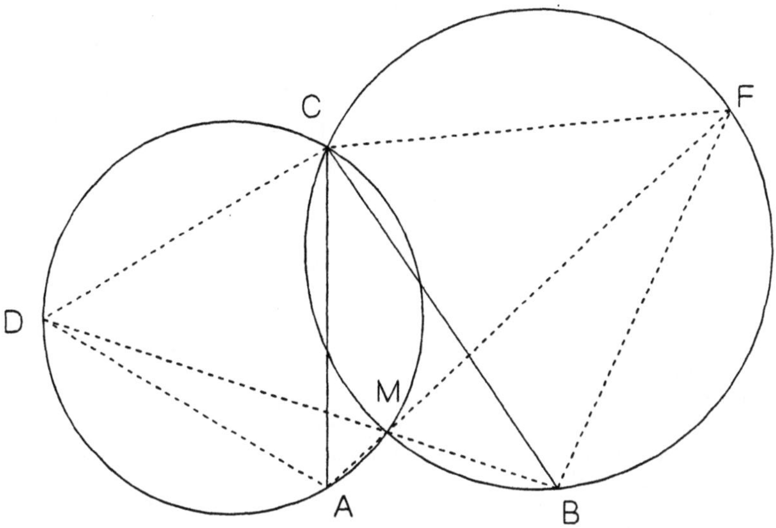

Figure 4.

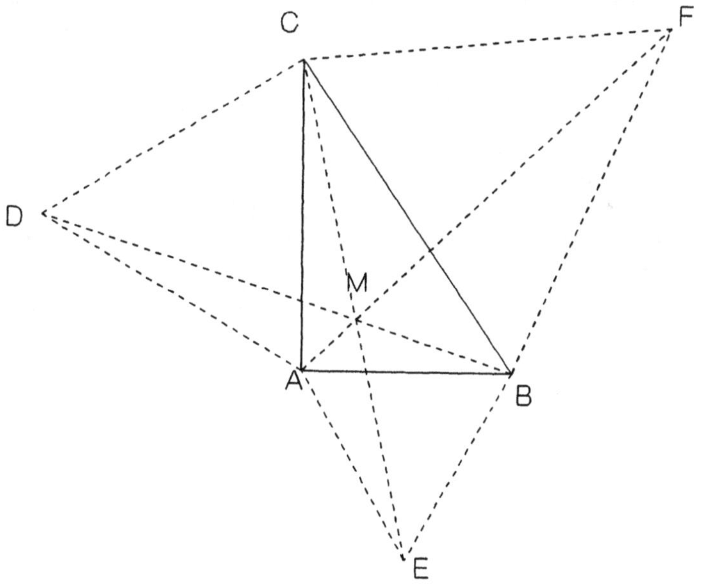

Figure 5.

Let $\bar{a}, \bar{b}, \bar{c}$ satisfy the triangle inequalities. In order to obtain every triangle as required, we proceed in accordance with one of the following two ways:

– By iii) of Theorem 4, it is possible to choose, for instance, the angle α such that $\mu(\alpha) + \mu(\bar{\alpha}) \geq \pi$ and then to put vertex A in point M.

– It is possible to choose angles α, β, γ such that $\mu(\alpha) + \mu(\bar{\alpha}) < \pi$, $\mu(\beta)+\mu(\bar{\beta}) < \pi$, $\mu(\gamma)+\mu(\bar{\gamma}) < \pi$. Let us observe in the above construction that point M which minimizes f, is the only one belonging to closed triangle ABC such that $\mu(\widehat{AMC}) = \pi - \mu(\bar{\beta})$, $\mu(\widehat{AMB}) = \pi - \mu(\bar{\gamma})$, $\mu(\widehat{BMC}) = \pi - \mu(\bar{\alpha})$. Hence, let us choose three half–lines a^*, b^*, c^* with origin M, and such that $\mu(\widehat{a^*b^*}) = \pi - \mu(\bar{\gamma})$, $\mu(\widehat{a^*c^*}) = \pi - \mu(\bar{\beta})$, $\mu(\widehat{b^*c^*}) = \pi - \mu(\bar{\alpha})$. Then we choose three points A, B, C, different from M, on each of the three half–lines a^*, b^*, c^*.

References

1. O. Bottema, R. Z. Djordjević, R. R. Janić, D. S. Mitrinović, P. M. Vasić, "Geometric Inequalities", Wolters–Noordhoff Publishing Groningen, pp. 128–129, 1968.
2. F. Heinen, "Progr. Cleve", pp. 18–19, 1834.
3. H. Idrissi, O. Lefebvre, C. Michelot, "A primal dual algorithm for a constrained Fermat–Weber problem involving mixed norms", Operations Research 22, pp. 313–330, 1988.
4. H. W. Kuhn, "A note on Fermat's problem", Mathematical Programming 4, pp. 98–107, 1973.
5. J. Krarup, S. Vajda, "On Torricelli geometrical solution to a problem of Fermat", to appear in Special Edition of Journal of Mathematics Applied in Business and Industry: Duality: In celebration of work of S. Vajda; S. Powell and H. P. Williams eds., 1996.
6. G. Loria, "Storia delle Matematiche", vol. 1, U. Hoepli, pp. 159–163, Milano, 1929.
7. T. Simpson, "The Doctrine and Application of Fluxions", London, 1750.
8. E. Torricelli, "De Maximis et Minimis", Opere di Evangelista Torricelli, G. Loria and G. Vassura eds., par. 22, Faenza, Italy, 1919.
9. G. O. Wesolowsky, "The Weber problem: history and perspectives", Location Science 1, pp. 5–23, 1993.
10. D. J. White, "An analogue derivation of the dual of the general Fermat problem", Management science 23, pp. 92–94, 1976.

Separation and regularity in the image space

Giandomenico Mastroeni and Massimo Pappalardo
Department of Mathematics,
Università di Pisa
Via Buonarroti 2, 56100 Pisa

Abstract. The optimality for a constrained extremum problem can be studied proving the disjunction between suitable subsets of the image space associated to the problem. This disjunction can be stated showing that these sets are linear or conically separated [1,2,3]. Under the "image regularity condition" (in short IRC) in [2], it has been proved the existence of generalized Karush–Kuhn–Tucker multipliers associated to a conic separation. Moreover it has been proved that IRC is equivalent to the calmness introduced by Clarke and the existence of an exact penalty function. Here we study the possibility of relating the local image regularity condition with a generic nonlinear separation not necessarily conic or linear and we provide refinements of some results obtained in [2].

Key words: Separation theorems, generalized Karush–Kuhn–Tucker multipliers, image regularity conditions, image space.

1. Introduction

The image space analysis [1] has been widely used to treat several aspects of the theory of constrained extremum problems: optimality conditions, regularity, duality, penalty methods. One of the main tools for developing this analysis is the separation in the image space. In [1], for example, it has been shown that the separation in the image space is equivalent to the existence of a saddle point of a suitable generalized Lagrangean function; in [2,3] the regularity for conic and linear separation has been studied. The aim of this work is to clarify some aspects concerning separation and regularity and to generalize some results obtained in [2]; there it has been proved that the "image regularity conditions" (in short, IRC) is one of the most general regularity conditions and it guarantees the regular conic separation. This is important because it has been shown that regular conic separation guarantees the existence of the generalized Karush–Kuhn–Tucker (in short, KKT) multipliers. In this paper, among other things, we analyse the possibility of obtaining a local version of IRC (in short, LIRC), through a general nonlinear separation not necessarily of conic type.

Let us introduce some notations which will be used in the sequel. If $M \subseteq \mathbb{R}^n$, $a \geq_M b$ shall mean $a - b \in M$; $\operatorname{cl} M$, $\operatorname{conv} M$ and M^C will denote, the closure, the convex hull and the complementarity set of M,

F. Giannessi et al. (eds.), New Trends in Mathematical Programming, 181–190.

respectively. *cone* M is the cone generated by M at x; while

$$C^* := \{z \in \mathbb{R}^n : \langle y, z \rangle \geq 0, \quad \forall y \in C\}$$

is the positive polar of a cone $C \subset \mathbb{R}^n$. Moreover, if $x \in \operatorname{cl} M$, we shall denote by $T(M, x)$ the Bouligand tangent cone to M at x, and by

$$\rho(b; M) := \inf\{\|b - b'\| : b' \in M\}$$

the distance between $b \in \mathbb{R}^n$ and M.

We say [2] that A and B admit a conic separation iff there exists a cone K such that $B \subset \operatorname{cl} K$ and $A \subset \operatorname{cl} K^C$; if, in addition, $B \subset \operatorname{int} K$ then A and B are said regularly cone separated. The sets

$$\operatorname{epi} f := \{(x, u) \in \mathbb{R}^{m+1} \,|u \geq f(x)\}$$

$$\operatorname{lev}_{\geq \alpha} f := \{x \in X | f(x) \geq \alpha\}$$

are the epigraph and the level set of the function $f : \mathbb{R}^m \to \mathbb{R}$, respectively. Finally, $f^T(x^*; z)$ will denote the lower Dini–Hadamard derivative at the point x^* in the direction z.

We recall the main definitions concerning the image space. Consider the following constrained extremum problem:

$$\min \phi(x) \quad \text{s.t.} \quad x \in R := \{x \in X : g(x) \underset{C}{\geq} 0\}, \qquad \text{(P)}$$

where $C \subseteq \mathbb{R}^m$ is a closed convex cone with apex in 0, $X \subseteq \mathbb{R}^n$ is a nonempty set, $\phi : X \to \mathbb{R}$, $g : X \to \mathbb{R}^m$ and $g(x) := (g_1(x), \ldots, g_m(x))$. Suppose that $x^* \in R$ and $\mathcal{V}$ is a basis of neighbourhoods of x^* with $V \in \mathcal{V}$ and define $f(x) := \phi(x^*) - \phi(x)$ and $F(x) := (f(x), g(x))$. Consider the sets

$$K := \{(u, v) \in \mathbb{R} \times \mathbb{R}^m : u = f(x), \; v = g(x), \; x \in X\} = F(X)$$

$$\mathcal{H} := \{(u, v) \in \mathbb{R} \times \mathbb{R}^m : u > 0, \; v \geq 0\}; \; \mathcal{E}(x^*) := K(x^*) - \operatorname{cl} \mathcal{H};$$

$$\mathcal{H}_u := \{(u, v) \in \mathbb{R} \times \mathbb{R}^m : u > 0, \; v = 0\}; \; \mathcal{E}_V := F(X \cap V) - \operatorname{cl} \mathcal{H}$$

$K(x^*)$ and $\mathcal{E}(x^*)$ are called, respectively, the image and the extended image of (P), while the space $\mathbb{R} \times \mathbb{R}^m$ is the image space. To simplify the notation we will understand the dependence of the sets $\mathcal{E}$ and K on x^*. It is easy to show that $x^* \in R$ is optimal for (P) iff the generalized system $F(x) \in \mathcal{H}, \; x \in X$ has no solutions or iff $K \cap \mathcal{H} = \emptyset$ or iff

$\mathcal{E} \cap \mathcal{H}_u = \emptyset$.

The reason that leads to the introduction of the set $\mathcal{E}$ lies in the fact that the set $\mathcal{E}$ preserves the optimality conditions and has, often, better properties than the set K. For example, if ϕ and $-g$ are C–convex then $\mathcal{E}$ is convex, ever if K is not. Recall that a function ϕ is C–convex iff

$$\phi(\lambda x + (1-\lambda)y) - \lambda\phi(x) - (1-\lambda)\phi(y) \in C, \quad \forall x, y, \quad \forall \lambda \in (0,1).$$

2. Nonlinear separation in the image space

In the image space analysis the problem of stating whether or not the sets $\mathcal{E}$ and $\mathcal{H}$ have empty intersection plays a crucial role. This can be done showing that $\mathcal{E}$ and $\mathcal{H}$ lie in two disjoint level sets of a suitable separating functional.

DEFINITION 1. Given a set of parameters $\Lambda \subseteq \mathbb{R}^P$, $\mathcal{E}$ and $\mathcal{H}$ are (nonlinearly) separated iff there exist $\vartheta \in \mathbb{R}$ and a function $\gamma_\lambda : \mathbb{R}^m \to \mathbb{R}$, depending on the parameter $\lambda \in \Lambda$, such that

$$\begin{array}{lll} a) & \theta u + \gamma_\lambda(v) \geq 0 & \forall (u,v) \in \mathcal{H} \\ b) & \theta u + \gamma_\lambda(v) \leq 0 & \forall (u,v) \in \mathcal{E}. \end{array}$$

The components of the vector (θ, λ) will be called the (nonlinear) "multipliers" of the separation.

Remark 2. Observe that, for example, $\mathrm{lev}_{\geq 0}\gamma_\lambda \supseteq C$, $\forall \lambda \in \Lambda$, implies condition a) of Definition 2.1.

In [3] it has been shown that, when γ is linear and the problem is differentiable, the elements (θ, λ) are the KKT multipliers, if $\theta \neq 0$.

In [1] it has been proved that, under suitable assumptions on the family of functions γ_λ, $\lambda \in \Lambda$, the existence of a separating function is equivalent to the existence of a saddle point of the generalized Lagrangean function defined by

$$L(x; \theta, \lambda) := \theta\phi(x) - \gamma_\lambda(g(x)).$$

In particular the following assumptions have been considered

a) $\bigcap_{\lambda \in \Lambda} \mathrm{lev}_{\geq 0}\gamma_\lambda(v) = \mathbb{R}^m_+$;

b) $\gamma_{\lambda^*}(v^*) > 0 \Rightarrow \exists \lambda' \in \Lambda$ such that $\gamma_{\lambda'}(\mathrm{v}^*) < \gamma_{\lambda^*}(\mathrm{v}^*)$.

(a) and (b) allow to prove that

$$x^* \in R, \quad \gamma_{\lambda^*}(g(x^*)) = 0, \tag{2.1}$$

when L fulfils the saddle point condition. It is interesting to note that any assumptions on the family γ_λ, $\lambda \in \Lambda$, which ensure that (2.1) holds in presence of a saddle point on L, is sufficient to prove the equivalence between a saddle point for L and the existence of a separating function.

Now, let us consider the relationships between the existence of a nonlinear separating function and that of an optimal solution of (P).

THEOREM 3. *i) If $x^* \in R$ is a global solution for (P), then $\mathcal{E}$ and $\mathcal{H}$ are (nonlinearly) separated. ii) If $\mathcal{E}$ and $\mathcal{H}$ admit a nonlinear separation with $\theta > 0$, then $x^* \in R$ is a global solution for (P).*

Proof. i) The function $\gamma_\lambda(v) := -\lambda\rho(v; C)$ gives a separating function.
ii) Since $\theta > 0$ then $\theta u + \gamma_\lambda(v) > 0$, $\forall (u, v) \in \mathcal{H}$. Moreover, from Definition 2.1 we have $\theta u + \gamma_\lambda(v) \le 0$, $\forall (u, v) \in K$; this implies $K \cap \mathcal{H} = \emptyset$.

□

One of the most interesting cases of nonlinear separation functions is the function $u - \lambda\rho(v; C)$ that we will relate to the conic separation.

THEOREM 4. *$\mathcal{E}$ and $\mathcal{H}$ admit a regular conic separation iff there exist $\theta, \lambda > 0$ such that $u - \lambda\rho(v; C)$ is a separating function.*

Proof. In [2] it has been proved that $\mathcal{E}$ and $\mathcal{H}$ admit a regular conic separation iff there exists $\lambda > 0$ such that

$$f(x) \le \lambda\rho(g(x); C), \quad \forall x \in X. \tag{2.2}$$

Recalling that $u = f(x)$ and $v = g(x)$ we obtain the equivalence between (2.2) and $w(u, v) \le 0$, $\forall (u, v) \in F(X)$, with $\theta = 1$. Now we prove that this is equivalent to the thesis. The following relations hold:

$$u - h_1 \le u \le \lambda\rho(v; C) \le \lambda\rho(v - h_2; C) \quad \forall (u, v) \in F(X), (h_1, h_2) \in \operatorname{cl}\mathcal{H}. \tag{2.3}$$

The first two inequalities are immediate; to prove the last observe that

$$\rho(v - h_2; C) := \inf_{x \in C} \|v - h_2 - x\| \ge \inf_{x \in C} \|v - x\| = \rho(v; C)$$

where the inequality follows from the fact that C is a convex cone and therefore $h_2 + x \in C$, $\forall x$. (2.3) show that to $u + \gamma_\lambda(v) \le 0$, $\forall (u, v) \in \mathcal{E}$.

Recall [2] that, if x^* is a global solution to (P), then $\mathcal{E}$ and $\mathcal{H}$ are conically separated, and that, if $\mathcal{E}$ and $\mathcal{H}$ are regularly conically separated, then x^* is a global solution to (P). Moreover, $\mathcal{E}$ and $\mathcal{H}$ are regularly conically separated iff

$$\text{cl cone } \mathcal{E} \cap \mathcal{H}_u = \emptyset. \tag{2.4}$$

Condition (2.4) is the IRC; it has been shown [2] that it is a necessary and sufficient condition to obtain regular conic separation and therefore generalized KKT multipliers. We say that the LIRC holds at x^* if there exists a neighbourhood I of x^* such that

$$\text{cl cone } \mathcal{E}_I \cap \mathcal{H}_u = \emptyset,$$

where $\mathcal{E}_I = F(X \cap I) - \text{cl}\,\mathcal{H}$. Let $\mathcal{V}$ be a basis of neighbourhoods of x^*; recall [2] that LIRC holds iff

$$\bigcap_{V \in \mathcal{V}} \text{cl cone } \mathcal{E}_V \cap \mathcal{H}_u = \emptyset.$$

In the next sections we will investigate some relationships between a nonlinear separation and the LIRC.

3. Regularity and nonlinear image separation

In this section we will show that, under suitable assumptions, the IRC (2.4) guarantees the existence of a separating function with $\theta = 1$. Let $f : \mathbb{R}^n \to \mathbb{R}$. Observe that f is lower semicontinuous (in short, lsc) at x_0 iff the set $S : \{(x_0, y) \in \mathbb{R}^{n+1} : y < f(x_0)\} \cap \text{cl}(\text{epi } f) = \emptyset$, and that f is upper semicontinuous (in short, usc) at x_0 if the set $S' := \{(x_0, y) \in \mathbb{R}^{n+1} : y > f(x_0)\} \subseteq \text{int}(\text{epi}\, f)$. See [4] for details on these properties. Moreover, recall that

$$\text{int epi } f^T(x^*; z) = \text{int}\, T(\text{epi}\, f, (x^*, f(x^*)));$$

$$\text{cl epi } f^T(x^*; z) = \text{cl}\, T(\text{epi}\, f, (x^*, f(x^*))).$$

The following result holds:

THEOREM 5. *Let $\gamma_\lambda(\cdot), \phi$ and g be continuous, $\gamma_\lambda^T(0, \cdot)$ lsc at $0 \in \mathbb{R}^m$ and $g(x^*) = 0$. If $u + \gamma_\lambda(v)$ is a separating function for $\mathcal{E}_I$ and $\mathcal{H}$, then the LIRC holds for problem (P).*

Proof. The hypotheses imply

$$u \le -\gamma_\lambda(v) \quad \forall (u,v) \in \mathcal{E}_I.$$

Therefore

$$\mathcal{E}_I \subseteq \text{hypo}(-\gamma_\lambda(v)). \tag{3.1}$$

$\gamma_\lambda^T(0,\cdot)$ is lsc at 0 iff

$$\text{cl}(\text{epi}\,\gamma_\lambda^T(0,\cdot)) \cap S = \emptyset,$$

where $S := \{(0, u_\in \mathbf{R}^{m+1} : u < \gamma_\lambda^T(0,0) = 0\}$, or, equivalently

$$T(\text{epi}\,\gamma, 0) \cap S = \emptyset. \tag{3.2}$$

Consider the transformation $A(u,v) := (-u,v)$. Since A is an isometry, (3.2) is equivalent to

$$A[T(\text{epi}\,\gamma, 0)] \cap A(S) = \emptyset. \tag{3.3}$$

We observe that $\gamma_\lambda^T(0,0) = 0$ and therefore (3.3) is equivalent to

$$T(\text{hypo}(-\gamma), 0) \cap \mathcal{H}_u = \emptyset.$$

From (3.1) we obtain that

$$T(\mathcal{E}_I, 0) \subseteq T(\text{hypo}(-\gamma_\lambda(v)), 0)$$

and therefore

$$T(\mathcal{E}_I, 0) \cap \mathcal{H}_u = \emptyset,$$

which, in the considered hypotheses, is equivalent to LIRC [see 2]. □

The previous theorem shows that LIRC can be a necessary condition for the existence of a separating function with $\theta = 1$; in the next result we will see that it can be sufficient if we specialize the class of a separating functions we deal with.

THEOREM 6. *Suppose that* $intC \neq \emptyset$. *Let* $\{\gamma_{\lambda_i}(v)\}_{i=1}^{\infty}$, *be a sequence of functions that fulfill the following conditions:*

$$\begin{array}{ll} i) & \bigcap_{i \in \mathcal{N}} \text{lev}_{\ge 0}(u + \gamma_{\lambda_i}(v)) = \text{cl}\,\mathcal{H}; \\ ii) & \text{epi}(u + \gamma_{\lambda_i}(v)) \subset \text{epi}\,(u + \gamma_{\lambda_{i+1}}(v)); \\ iii) & \gamma_{\lambda_i}(v) = 0 \;\; \forall v \ge 0 \;\; \forall i. \end{array}$$

If LIRC holds then there exists i^* *such that* $u + \gamma_{\lambda_{i^*}}(v)$ *is a separating function for* $\mathcal{E}_I$ *and* $\mathcal{H}$.

Proof. From ii) it follows that

$$\text{lev}_{\geq 0}(u+\gamma_{\lambda_i}(v)) \supset \text{lev}_{\geq 0}(u+\gamma_{\lambda_{i+1}}(v)).$$

Since cl cone $\mathcal{E}_I \cap \mathcal{H}_u = \emptyset$, and because of i), there exists λ_{i^*} such that

$$\text{lev}_{\geq 0}(u+\gamma_{\lambda_{i^*}}(v)) \supseteq \text{cl}[(\text{cl cone}\,\mathcal{E}_I)^C].$$

The previous relation implies that

$$\text{lev}_{\leq 0}(u+\gamma_{\lambda_{i^*}}(v)) \supseteq \text{cl}\{\text{cl}[(\text{cl cone}\,\mathcal{E}_I)^C]\}^C =$$
$$= \text{cl}[\text{int cl cone}\,\mathcal{E}_I] \subseteq \text{cl}(\text{cl cone}\,\mathcal{E}_I) \subseteq \text{cl cone}\,\mathcal{E}_I.$$

Therefore $u+\gamma_{\lambda_i}(v) \leq 0$, $\forall (u,v) \in \mathcal{E}_I$, and the thesis follows. □

4. Image and extended image

To deepen the analysis of the concept of image and extended image of a constrained extremum problem we will show that it is possible to replace the extended image $\mathcal{E}$ with the image K.

LEMMA 7. $\text{cl cone}\,\mathcal{E} - \mathcal{H} \subseteq \text{cl cone}\,\mathcal{E}$.

Proof. Let $y \in \text{cl cone}\,\mathcal{E}$ and $h \in \mathcal{H}$. Therefore $\exists \{t_k\}_{k\in\mathcal{N}} \subseteq \mathbb{R}_+ \setminus \{0\}$ and $\exists \{y_k\}_{k\in\mathcal{N}} \subseteq \mathcal{E}$ such that $t_k y_k \to x$. Let $h_k = \frac{1}{t_k} h \in \mathcal{H}$, then $y_k - h_k \in \mathcal{E}$, since $\mathcal{H} + \text{cl}\,\mathcal{H} = \mathcal{H}$, and $t_k y_k - h = t_k(x_k - h_k) \in \text{cone}\,\mathcal{E}$. Since $t_k y_k \mapsto y$ we have

$$y - h \in \text{cl cone}\,\mathcal{E}.$$

THEOREM 8. *cl cone* $K \cap \mathcal{H} = \emptyset \Leftrightarrow \text{cl cone}\,\mathcal{E} \cap \mathcal{H}_u = \emptyset$.

Proof. $\Leftarrow$) Since cl cone $K \subseteq$ cl cone $\mathcal{E}$ it is sufficient to prove that cl cone $\mathcal{E} \cap \mathcal{H}_u = \emptyset$ implies cl cone $\mathcal{E} \cap \mathcal{H} = \emptyset$. Let us suppose, ab absurdo, that $\exists (\bar{u},\bar{v}) \in$ cl cone $\mathcal{E} \cap \mathcal{H} = \emptyset$ with $\bar{v} \neq 0$. This means that $I = \{i : \bar{v}_i > 0\} \neq \emptyset$. By applying Lemma 7 we obtain that the point $(\bar{u},\bar{v}) - (\frac{\bar{u}}{2},\bar{v}) =\in$ cl cone $\mathcal{E}$ and this contradicts the hypotheses.
$\Rightarrow$) Ab absurdo let cl cone $\mathcal{E} \cap \mathcal{H}_u \neq \emptyset$. This implies $\mathcal{H}_u \subseteq$ cl cone $\mathcal{E}$, and therefore $\exists (u^*,0), u^* > 0$, with $(u^*,0) \in$ cl cone $\mathcal{E}$.
Therefore $\exists \{(u_h,v_h)\}_{i=1}^{\infty} \subset$ cone $\mathcal{E}$ such that $(u_h,v_h) \to (u^*,0)$. Since $(u_h,v_h) \in$ cone $\mathcal{E}$ we have $u_h = \lambda_h \bar{u}_h$, $v_h = \lambda_h \bar{v}_h$ with $(\bar{u}_h,\bar{v}_h) \leq_{\text{cl}\,H} (f(x_h),g(x_h)), x_h \in X$, $(\bar{u}_h,\bar{v}_h) \in \mathcal{E}$, $\forall h$. Since $\lambda_h > 0$, $(u_h,v_h) = \lambda_h(\bar{u}_h,\bar{v}_h) \leq_{\text{cl}\,H} \lambda_h(f(x_h),g(x_h))$, $\forall h$.
From the last inequality, since $(u_h,v_h) \to (u^*,0)$ there exists a subsequence $\{x_{h_k}\}_1^{\infty} \subseteq \{x_h\}$ such that $\lambda_{h_k}(f(x_{h_k}),g(x_{h_k})) \to (u^{**},v^{**}) \geq_{\text{cl}\,H} (u^*,0)$. Since $u^{**} \geq u^* > 0$ and $v^{**} \in C$, we have that $(u^{**},v^{**}) \in \mathcal{H}$. Moreover $(u^{**},v^{**}) \in$ cl cone K and this contradicts the hypotheses. □

The previous proposition allows to prove that, even if $\mathcal{E}$ and $\mathcal{H}$ are linearly separated and the LIRC holds, the linear separation may not be regular as shown by the following counterexample:

EXAMPLE 9. Let $\mathcal{K} := \{(u,v) \in \mathbb{R}^3 \mid (0,0,0), (-3,1,-1), (8,-2,2)\}$. It is easy to see that $v_1 + v_2 = 0$ is a separating hyperplane for K and $\mathcal{H}$ but that a regular linear separation cannot exist. However, the condition cl cone $K \cap \mathcal{H} = \emptyset$ holds and therefore, from Theorem 4.1, (2.5) is fulfilled.

Under the assumption that the function F be continuous at x^* and all the constraints be binding at x^* it is possible to show that the LIRC can be obtained taking in consideration only the image of the problem (P_I).

LEMMA 10. *Let x^* a local minimum point for (P), let $\{V_k\}_{k\in\mathcal{N}}$ a decreasing basis of neighbourhoods of x^*. Then*

$$\bigcap_{k\in\mathcal{N}} \text{cl cone}\, F(X \cap V_k) \cap \mathcal{H} = \emptyset \quad \Leftrightarrow \quad \bigcap_{k\in\mathcal{N}} \text{cl cone}\, \mathcal{E}_{V_k} \cap \mathcal{H}_u = \emptyset.$$

Proof. $\Rightarrow$) Ab absurdo, suppose that $\bigcap_{k\in\mathcal{N}} \text{cl cone}\, \mathcal{E}_{V_k} \cap \mathcal{H}_u \neq \emptyset$. Therefore, $\exists u^* > 0$ such that $(u^*, 0) \in \bigcap_{k\in\mathcal{N}} \text{cl cone}\, \mathcal{E}_{V_k}$, and

$$(u^*, 0) \in \text{cl cone}\, \mathcal{E}_V, \quad \forall k \in \mathcal{N}. \tag{4.1}$$

Because of Theorem 4.1, (4.1) is equivalent to

$$\text{cl cone}\, F(X \cap V_k) \cap \mathcal{H} \neq \emptyset, \quad \forall k \in \mathcal{N}.$$

Since cl cone $F(X \cap V_k) \supset$ cl cone $F(X \cap V_{k+1})$, we have that

$$\bigcap_{k\in\mathcal{N}} \text{cl cone}\, F(X \cap V_k) \cap \mathcal{H} \neq \emptyset,$$

and this contradicts the hypotheses.
$\Leftarrow$) Ab absurdo, suppose that $\bigcap_{k\in\mathcal{N}} \text{cl cone}\, F(X \cap V_k) \cap \mathcal{H} \neq \emptyset$. Therefore, $\exists (u^*, v^*) \in \mathcal{H}$ such that

$$(u^*, v^*) \in \text{cl cone}\, F(X \cap V_k),$$

that is

$$(u^*, v^*) \in \text{cl cone}\, F(X \cap V_k) \quad \forall k \in \mathcal{N}. \tag{4.2}$$

Because of Theorem 4.1, (4.1) is equivalent to

$$\text{cl cone}\, \mathcal{E}_{V_k} \cap \mathcal{H}_u \neq \emptyset, \quad \forall k \in \mathcal{N}.$$

Since cl cone $\mathcal{E}_{V_k} \supset$ cl cone $\mathcal{E}_{V_{k+1}}$, we have that

$$\bigcap_{k \in \mathcal{N}} \text{cl cone}\, \mathcal{E}_{V_k} \cap \mathcal{H}_u \neq \emptyset,$$

and this contradicts the hypotheses. □

THEOREM 11. *Let F be continuous at x^* and $g(x^*) = 0$. (P) satisfies LIRC iff there exists a neighbourhood I of x^* such that:*

$$T(K_I; 0) \cap \mathcal{H} = \emptyset.$$

Proof. "Only if" Suppose that LIRC holds. By definition there exists a neighbourhood I of x^* such that cl cone $\mathcal{E}_I \cap \mathcal{H}_u = \emptyset$ which is equivalent to cl cone$K_I \cap \mathcal{H} = \emptyset$. Since $K_I \subseteq \mathcal{E}_I$ we obtain:

$$T(K_I; 0) \subseteq \text{cl cone}\, K_I$$

and the thesis follows.
"If" Let $T(K_I; 0) \cap \mathcal{H} = \emptyset$. Let $\{I_k\}_{k \in \mathcal{N}}$ a base of neighbourhoods of 0. It is known that

$$T(K_I; 0) = \bigcap_{k \in \mathcal{N}} \text{cl cone}\, K_I \cap I_k; \tag{4.3}$$

let $\{V_k\}_{k \in \mathcal{N}}$ a basis of neighbourhoods of x^*, with $V_k \subseteq I$. From Theorem 7 it follows that LIRC holds iff

$$\bigcap_{k \in \mathcal{N}} \text{cl cone}\, \mathcal{E}_{V_k} \cap \mathcal{H}_u = \emptyset. \tag{4.4}$$

Because of Lemma 10 (4.4) is equivalent to

$$\bigcap_{k \in \mathcal{N}} \text{cl cone}\, F(X \cap V_k) \cap \mathcal{H} = \emptyset.$$

Since $F(x)$ is continuous at x^*

$$\forall I_k \exists V_{i(k)} : F(X \cap V_{i(k)} \subseteq I_k.$$

Moreover, as $F(X \cap V_{i(k)}) \subseteq K_I$ we have that $F(X \cap V_{i(k)}) \subseteq K_I \cap I_k$. Therefore

$$\bigcap_{k \in \mathcal{N}} \text{cl cone}\, F(X \cap V_k) \subseteq \bigcap_{k \in \mathcal{N}} \text{cl cone}\, F(X \cap V_{i(k)}) \subseteq \bigcap_{k \in \mathcal{N}} \text{cl cone}\, K_I \cap I_k,$$

from which the thesis follows. □

Remark 12. We observe that, when we replace the extended image $\mathcal{E}$ with the image K, at the same time we have to replace the set $\mathcal{H}_u$ with the set $\mathcal{H}$.

References

1. F. Giannessi, "Theorems of the Alternative and Optimality Conditions", J.O.T.A., Vol. 42, N. 3, 1984, p. 331–365.
2. P. H. Dien, G. Mastroeni, M. Pappalardo, P. H. Quang, "Regularity conditions for constrained exteremum problems via image space: the nonlinear case; J.O.T.A., Vol. 80, N. 1, 1994, p. 19–37.
3. P. H. Dien, G. Mastroeni, M. Pappalardo, P. H. Quang, "Regularity conditions for constrained extremum problems via image space: the linear case" Springer Lectures Notes in Economics, Vol. 405, Komlosi–Rapcsak–Schaible (eds), 1994, pp. 145–153.
4. G. Mastroeni, "Regularity Properties of the Marginal Function Via Image Space", Report N. 3.203(765), Dept. of Mathematics, University of Pisa, 1993.

Dynamic models and generalized equilibrium problems

Antonino Maugeri
Dipartimento di Matematica, Università di Catania
Viale A. Doria, 6, 95125 Catania

Abstract. The aim of this paper is to consider equilibrium problems with data depending explicitly on the time and to study how the equilibrium conditions change and under which assumptions the existence of a solution time depending can be ensured. We consider the special case of equilibrium flows in a network, because this model represents a valid test to know the problems arising from this new formulation suggested by F. Giannessi (see [1]) in order to achieve a better definition of equilibrium, whose nature is essentially dynamic.

1. The dynamic model. An existence theorem.

Let us recall the model of equilibrium flows in a network in the static case: $L = \{P_1, P_2, \dots, P_q\}$ denotes the set of nodes,
$N = \{a_1, a_2, \dots, a_n\}$ the set of links,
$W = \{w_1, w_2, \dots, w_\ell\}$ the set of Origin-Destination pairs
$R_r\ r = 1, \dots, m$ the paths connecting the O/D pairs
$\mathcal{R}_j = \{R_r : R_r \text{ connects the O/D pair } w_j\}\quad j = 1 \dots \ell$
$\mathcal{R} = \bigcup_{j=1}^{\ell} \mathcal{R}_j$
$F' = (F_1, F_2, \dots, F_m)$ the path–flow vector (the apex denotes the transposition)
$C'(F) = (C_1(F), C_2(F), \dots, C_m(F))$ the path–cost vector;
$\rho' = (\rho_1, \rho_2, \dots, \rho_\ell)$ the vector travel demand between the O/D pairs w_j;

$$\varphi = \{\varphi_{jr}\}_{\substack{j=1,\dots,\ell\\ r=1,\dots,m}} \qquad \varphi_{jr} = \begin{cases} 1 & \text{if } R_r \in \mathcal{R}_j \\ 0 & \text{otherwise} \end{cases}$$

(φ is the pair–path incidence matrix),

$$\Delta = \{\varphi_{jr}\}_{\substack{j=1,\dots,n\\ r=2,\dots,m}} \qquad \varphi_{jr} = \begin{cases} 1 & \text{if } a_i \in R_r \\ 0 & \text{otherwise} \end{cases}$$

(Δ is the link–flow incidence matrix).
If

$$c(f) = (c_1(f), \dots, c_n(f)) \quad \mathcal{F} = (f_1, \dots, f_n)$$

are the link–flow cost vector and the link–flow vector, it results $f = \Delta F$, $C(F) = \Delta' C(\Delta F)$.

F. Giannessi et al. (eds.), New Trends in Mathematical Programming, 191–202.

If we denote by K the set of feasible path–flow vectors

$$\mathbb{K} = \{F \in \mathbb{R}^m : F_r \geq 0 \;\; r = 1, \ldots, m \;, \; \sum_{r=1}^{m} \varphi_{jr} F_r = \rho_j \;\; j = 1 \ldots \ell\}$$

the well known user's equilibrium principle due to Wardrop is the following:
"*A vector $H \in K$ will be an equilibrium vector in the static case if $\forall \mathcal{R}_j$, $\forall R_r$, $R_s \in \mathcal{R}_j$ if it results*

$$C_r(H) > C_s(H) \tag{1.1}$$

then

$$H_r = 0$$

and in the year 1979 it was proved by M.S. Smith, that
"*H is an equilibrium solution to the traffic equilibrium problem in the Wardrop sense if and only if*

$$H \in \mathbb{K} \;\; \textit{is such that} \;\; \langle C(H), \; F - H \rangle \geq 0 \quad \forall F \in \mathbb{K} \text{ ''} \tag{1.2}$$

(see [2] for more informations).

Now let us suppose that the travel demands ρ_j, $j = 1 \ldots \ell$, evolve in the time

$$\rho_j : [0, T] \to R^+ \quad j = 1, \ldots, \ell$$

and, hence, the flows and the costs in turn become functions of the time,

$$F(t), \; C(t, F(t)) \qquad F : [0, T] \to R^m_+ \qquad C : [0, T] \times \mathbb{R}^m_+ \to \mathbb{R}^m_+ \; .$$

Moreover, we consider a model without capacity constraints and assume that the optimal paths joining the OD pairs w_j are the same for each t. In other words, the optimal paths for each t are those chosen at the moment $t = 0$.

In this new situation the equilibrium conditions could be transformed in the following one.
"*The (vectorial) function $H(t) \in \mathbb{K}$, where*

$$\mathbb{K} = \{F(t) \in R^m, \; F_r(t) \geq 0 \quad r = 1, \ldots, m \quad \forall t \in [0, T],$$
$$\sum_{r=1}^{m} \varphi_{jr} F_r(t) = \rho_j(t) \quad \forall t \in [0, T], \quad j = 1, \ldots \ell\},$$

is an equilibrium solution if $\forall \mathcal{R}_j$, $\forall R_r$, $R_s \in \mathcal{R}_j$ if it results

$$C_r(t, H(t)) > C_s(t, H(t)) \quad \forall t \in [0, T] \tag{1.3}$$

then

$$H_r(t) = 0 \quad \forall t \in [0,T]. \text{''} \tag{1.4}$$

But the conditions (1.3) and (1.4) that require that in each moment the user adopts the optimal choice are unreasonable. Hence it is more reasonable to consider a formulation such that the conditions (1.3) and (1.4) hold almost everywhere in $[0,T]$:
If

$$C_r(t, H(t)) > C_s(t, H(t)) \quad \text{a.e.} \quad t \in [0,T] \tag{1.3'}$$

then

$$H_r(t) = 0 \quad \text{a.e.} \quad t \in [0,T] \tag{1.4'}$$

and also in the definition of $\mathbb{K}$ we may consider relations that hold almost everywhere and with not much regularity in t for example:

$$\mathbb{K} = \{F(t) \in L^2(0,T,\mathbb{R}^m) : F_r(t) \geq 0 \quad \text{a.e} \quad [0,T] \tag{1.5}$$

$$r = 1, \ldots, m; \; \sum_{r=1}^{m} \varphi_{jr} F_r(t) = \rho_j(t) \quad \text{a.e.} \quad [0,T] \quad j = 1, \ldots, \ell\}.$$

Taking into account that the optimal paths are the same for each $t \in [0,T]$, conditions (1.3') and (1.4') become equivalent to the next ones:
If there exists a subset $S \subset [0,T]$ of positive measure such that

$$C_r(t, H(t)) > C_s(t, H(t)) \quad \text{for each} \quad t \in S, \tag{1.3''}$$

then

$$H_r(t) = 0 \quad \text{a.e.} \quad t \in [0,T]. \tag{1.4''}$$

In these settings, our "environment" is $L^2(0,T,\mathbb{R}^m)$ and in order to give a sense to our function we make the following assumptions:
i) *The vector function*

$$C(t,F) : [0,T] \times \mathbb{R}^m \to \mathbb{R}^m_+$$

is measurable in t, continuous in F and such that

$$\|c(t,F)\| = A(t)\|F\| + B(t) \tag{1.6}$$

with $B(t) \in L^2(0,T)$ and $A(t) \in L^\infty(0,T)$.
ii) *The travel demands $\rho_j(t)$ belong to $L^2(0,T)$.*
Taking into account the properties of the convergence in L^2 it is easy to show that

PROPOSITION 1. $\mathbb{K}$ is a bounded, convex, closed nonempty subset of $L^2(0,T,\mathbb{R}^m)$.

Now the following result holds:

THEOREM 2. *If $H(t) \in \mathbb{K}$ is an equilibrium solution in the sense of (1.3'), (1.4') and (1.5) of the Dynamic Traffic Equilibrium Problem, then it results*

$$\int_0^T \langle C(t, H(t)),\ F(t) - H(t)\rangle dt \geq 0 \quad \forall F(t) \in \mathbb{K} \tag{1.7}$$

and viceversa if $H \in \mathbb{K}$ verifies (1.6) it is an equilibrium solution according with (1.3'), (1.4').

Proof. Let $H(t) \in \mathbb{K}$ an equilibrium solution in accordance with (1.3') and (1.4'); then as a consequence setting for t a.e. in $[0, T]$:

$$C^j(t, H(t)) = \min\{C_r(t, H(t)) : R_r \in \mathcal{R}_j\}$$

it will be:

$$C_r(t, H(t)) \quad \begin{cases} = C^j(t, H(t)) & \text{if } H_r(t) > 0 \text{ a.e. } [0, T] \\ \geq C^j(t, H(t)) & \text{if } H_r(t) = 0 \text{ a.e. } [0, T] \end{cases}$$

and, if we set

$$\begin{aligned} A_j &= \{r : R_r \in \mathcal{R}_j\} = \{r : \varphi_{jr} = 1\} \quad j = 1 \dots \ell \\ B_j &= \{r \in A_j \text{ and } H_r(t) > 0 \text{ a.e. } [0, T]\} \\ C_j &= A_j \backslash B_j = \{r \in A_j : H_r(t) = 0 \text{ a.e. } [0, T]\}, \end{aligned}$$

we obtain:

$$\begin{aligned}
&\textstyle\int_0^T \langle C(t, H(t)),\ F(t) - H(t)\rangle dt = \\
&= \sum_{j=1}^{\ell} \sum_{r \in A_j} \textstyle\int_0^T C_r(t, H(t))(F_r(t) - H_r(t))dt = \\
&= \sum_{j=1}^{\ell} \left\{ \sum_{r \in B_j} \textstyle\int_0^T C_r(t, H(t))(F_r(t) - H_r(t))dt + \right. \\
&\left. \sum_{r \in C_j} \textstyle\int_0^T C_r(t, H(t)) F_r(t) dt \right\} \geq \\
&\geq \sum_{j=1}^{\ell} \left\{ \sum_{r \in B_j} \textstyle\int_0^T C^j(t, H(t))(F_r(t) - H_r(t))dt + \right. \\
&\left. \sum_{r \in C_j} \textstyle\int_0^T C^j(t, H(t))(F_r(t) - H_r(t))dt \right\} = \\
&= \sum_{j=1}^{\ell} \sum_{r \in A_j} \textstyle\int_0^T C^j(t, H(t))(F_r(t) - H_r(t))dt = \\
&= \sum_{j=1}^{\ell} \textstyle\int_0^T C^j(t, H(t))(\sum_{r=1}^{m} \varphi_{jr} F_r(t) - \sum_{r=1}^{m} \varphi_{jr} H_r(t))dt = 0
\end{aligned}$$

and hence the assert.

Viceversa let H be a solution of V.I. (1.6) and let us suppose that H does not verify the conditions (1.3'), (1.4'). Then there exist R_r, $R_s \in \mathcal{R}_j$ such that

$$C_r(t, H(t)) > C_s(t, H(t)) \quad \text{a.e.} \quad [0, T]$$

and

$$H_r(t) > 0 \quad \text{a.e.} \quad [0, T];$$

now we can consider a vector $F(t)$ with components

$$F_\ell(t) = \begin{cases} H_\ell(t) & \ell \neq r, s \\ H_r(t) + H_s(t) & \ell = s \\ 0 & \ell = r \end{cases}$$

that belongs to $\mathbb{K}$ and is such that:

$$\begin{aligned} &\textstyle\int_0^T \langle C(t, H(t)),\ F(t) - H(t) \rangle dt = \int_0^T \sum_{\ell \neq r,s} C_\ell(t, H(t))(F_\ell(t) - H_\ell(t))dt + \\ &+ \textstyle\int_0^T C_s(t, H(t))(F_s(t) - H_s(t))dt + \int_0^T C_r(t, H(t))(F_r(t) - H_r(t))dt = \\ &= \textstyle\int_0^T C_s(t, H(t))H_r(t)dt - \int_0^T C_r(t, H(t))H_r(t)dt = \\ &= \textstyle\int_0^T (C_s(t, H(t)) - C_r(t, H(t))H_r(t)dt < 0 \end{aligned}$$

that is absurd.
We are able to give an existence result. Let us add the following monotonicity assumption:
iii) *For each $F_1, F_2 \in \mathbb{K}$ it results*

$$\int_0^T \langle C(t, F_2(t)) - C_2(t, F_2(t)),\ F_1(t) - F_2(t) \rangle dt \geq 0$$

and we have the following

THEOREM 3. *If the assumptions i)–iii) are fulfilled, the V.I. (1.6) admits solutions.*

Proof. We apply the well known Hartmann–Stampacchia theorem. $L^2(0, T, \mathbb{R}^m)$ is a reflexive Banach space, the assumptions i) and ii) ensure that $C(t, F(t))$ belongs to $L^2(0, T, \mathbb{R}^m)$ and that $\mathbb{K}$ is a closed, convex, bounded subset of $L^2(0, T, \mathbb{R}^m)$. Moreover the assumption i) ensures that the function

$$\lambda \in [0, 1] \to \int_0^T \langle C(t, \lambda H(t) + (1 - \lambda)F(t), H(t) - F(t) \rangle dt$$

is continuous with respect to the variable λ for each $F, H \in \mathbb{K}$, that is the operator $C(t, F(t))$ is "*hemicontinuous*". Then, taking into account the assumption iii) on monotonicity, the V.I. (1.6) admits solutions.

2. A weak formulation. Further existence theorems

It could be interesting to replace the convex $\mathbf{K}$ with the convex

$$\mathbf{K}_w = \left\{ F \in L^2(0,T,\mathbf{R}^m) : \sum_{r=1}^{m} \varphi_{jr} \int_0^T F_r(t)dt = \right.$$

$$\left. = \int_0^T \rho_j(t)dt \quad j = 1,\dots,\ell, \int_0^T F_r(t)dt \geq 0 \quad r = 1,\dots,m \right\}$$

in which the constraints are considered in average. In this case $\mathbf{K}_w$ is convex, closed but not bounded; moreover $\mathbf{K}_w$ has a non empty relative interior (see the appendix due to B. Ricceri).
Then if one wishes to achieve the existence of a solution, he must add to the assumptions i)–iii) the assumption of coercivity:
iv) *There exists a positive constant* ν *such that*

$$\int_0^T \langle C(t, H(t)),\ H(t)\rangle dt \geq \nu \int_0^T \|H(t)\|^2 dt \quad \forall H \in \mathbf{K}_w .$$

Another existence theorem can be obtained using the following recent result due to B. Ricceri (see [3]).

THEOREM 4. *Let us suppose that the operator*

$$A : \mathbf{K}_w \to L^2(0,T,\mathbf{R}^m)$$

is weakly–continuous and that* $\mathbf{K}_w$ *is a closed convex set of* $L^2(0,T,\mathbf{R}^m)$ *whose relative interior is non–empty. Then if there exist two subsets of* $\mathbf{K}_w$, $\mathbf{K}_1$ *and* $\tilde{\mathbb{K}}$ *with*

$$\mathbf{K}_1 \subseteq \tilde{\mathbb{K}},$$

K_1 *finite dimensional and* $\tilde{\mathbb{K}}$ *compact, such that for each* $H \in K_w \backslash \tilde{\mathbb{K}}$

$$\sup_{F \in \mathbf{K}_1} \langle A(H),\ F - H\rangle > 0, \tag{2.1}$$

then the V.I.

$$H \in \mathbf{K}_w \quad \langle A(H),\ F - H\rangle \geq 0 \quad \forall F \in \mathbf{K}_w$$

admits solutions.

In our case the weakly* continuity means that

$$\lim_{n\to\infty} \int_0^T \langle C(t, H_n(t)),\ F(t)\rangle dt = \int_0^T \langle C(t, H(t)),\ F(t)\rangle dt \qquad (2.2)$$

for each sequence $\{H_n\}_{n\in N}$ converging to $H(t)$ in the sense of $L^2(0, T, \mathbb{R}^m)$ weak topology. By a result due to E. De Giorgi [4], R. Landes [5], we obtain (2.2) if and only if $C(t, F(t))$ takes the form [1]

$$C(t, F(t)) = A(t)F(t) + B(t)$$

where $A(t)$ is a $m \times m$ matrix belonging to $L^\infty(0, T)$ and $B(t) \in L^2(0, T, \mathbb{R}^m)$. As a finite dimensional set $\mathbb{K}_1$ we may choose the set constitued by only an element $\tilde{F} \in \mathbb{K}$, for istance the point obtained in the following way:
setting

$$A_j = \{r : 1 \le r \le m : \varphi_{jr} = 1\} \quad j = 1 \ldots \ell$$
$$r_j = \min A_j,$$

the components of $\tilde{F}$ are the numbers $\tilde{F}_r$ given by

$$\tilde{F}_r = \begin{cases} \int_0^T \rho_j(t)dt & \text{if } r = r_j \\ 0 & \text{otherwise.} \end{cases}$$

As a compact $\tilde{\mathbb{K}}$ (in the sense of the weak topology) it is enough to consider

$$\tilde{\mathbb{K}} = \{H(t) : H(t) \in \mathbb{K}_w,\ \|H(t)\|_{L^2} \le \rho\}$$

for a suitable ρ. It remains to show that the inequality

$$\int_0^T \langle A(t)H(t) + B(t),\ \tilde{F}(t) - H(t)\rangle dt > 0 \qquad (2.3)$$

for each $H \in \mathbb{K}_w - \tilde{\mathbb{K}}$. To this end let us suppose that there exists ν such that

$$\int_0^T \langle A(t)H(t), H(t)\rangle dt \ge \nu \|H\|^2_{L^2} \qquad (2.4)$$

for each H such that

$$\|H\|^2_{L^2} > \frac{1}{(\nu - 2\epsilon)4\epsilon}\{\|A\|^2_{L^\infty}\|\tilde{F}\|^2_{L^2} + \|B\|_{L^2}\} + \qquad (2.5)$$

---

[1] In effect the De Giorgi–Landes result holds if $\mathbf{K}_w = L^2(0, T, \mathbb{R}^m)$ and it has not been proved when $\mathbf{K}_w$ is an infinite–dimensional subset of $L^2(0, T, \mathbb{R}^m)$. Then the affine form of $C(t, F(t))$ could be unnecessary. Of course this problem deserves much attention.

$$\frac{1}{(\nu-2\epsilon)}\|B\|_{L^2}\|\tilde{F}\|_{L^2}=\rho^2(\epsilon,\nu)$$

for some $0<\epsilon<\frac{\nu}{2}$. Then, under (2.4)–(2.5), it results:

$$\int_0^T\langle A(t)H(t),H(t)\rangle dt\geq\nu\|H\|^2_{L^2}\geq \tag{2.6}$$

$$\geq 2\epsilon\|H\|^2_{L^2}+\frac{1}{4\epsilon}\|A\|^2_{L^\infty}\|\tilde{F}\|^2_{L^2}+\|B\|_{L^2}\|\tilde{F}\|_{L^2}+\frac{1}{4\epsilon}\|B\|^2_{L^2}\geq$$

$$\geq\|A\|_{L^\infty}\|H\|_{L^2}\|\tilde{F}\|_{L^2}+\|B\|_{L^2}\|\tilde{F}\|_{L^2}+\|B\|_{L^2}\|H\|_{L^2}\geq$$

$$\geq\int_0^T\langle A(t)H(t),\tilde{F}(t)\rangle dt+\int_0^T\langle B(t),\tilde{F}(t)\rangle dt-\int_0^T\langle B(t),H(t)\rangle dt$$

that is (2.3) for each $H\in\mathbb{K}_w-\tilde{\mathbb{K}}$ where $\tilde{\mathbb{K}}$ is defined for $\rho=\rho(\epsilon,\nu)$.

3. Separation theory and Gap Function

Following the hints by F. Giannessi (see [1]) we wish to extend to the V.I. (1.6) the Image Space approach already proposed in the finite dimensional case.
To this end, let us set

$$I^0=\{1,\dots,\ell\}\quad I^+=\{1,\dots,m\} \tag{3.1}$$

$$I=\{1,\dots,\ell,\ \ell+1,\dots,\ell+m\}$$

and let us consider the set

$$C=O_\ell\times\mathbb{R}^m_+. \tag{3.2}$$

Then, if we define

$$\psi_i(F)=\left\{\sum_{r=1}^m\varphi_{jr}F_r(t)-\rho_j(t):\ \text{a.e. }[0,T]\right\}\quad i\in I^0 \tag{3.3}$$

$$\psi_i(F)=\{F_i(t):\ \text{a.e. }[0,T]\}\ \ i\in I^+ \tag{3.4}$$

we can rewrite $\mathbb{K}$ as

$$\mathbb{K}=\{F\in L^2(0,T,\mathbb{R}^m):\psi(F)\subseteq C\}. \tag{3.5}$$

Now let us consider the multifunction

$$M(H,F)=\{(u,v)\in R\times R^{m+\ell}:u= \tag{3.6}$$

$$\int_0^T \langle C(t, H(t)), H(t) - F(t)\rangle dt, \quad v = \psi(F)\}$$

and let us observe that $H \in L^2(0, T, \mathbb{R}^m)$ is a solution of (1.6) if and only if the system in the unknown F

$$\begin{cases} \int_0^T \langle C(t, H(t)),\ H(t) - F(t)\rangle dt > 0 \\ \psi(F) \subseteq C,\ \ F \in L^2(0, T, \mathbb{R}^m) \end{cases} \tag{3.7}$$

is impossible.

Now in order to use the image space approach that essentially requires the finite dimensionality of the image, we must overcome the infinite dimensionality of $L^2(0, T, \mathbb{R}^m)$. To this end let us consider the functions

$$\omega_i : L^2(0, T, \mathbb{R}^m) \times [0, T] \times L^2(0, T, \mathbb{R}^m) \to R \quad i \in I,$$

and let us denote by $\Omega(H, F)$ the set of the vectors

$$\omega(H, F) = \{\omega_i \quad i \in I\}$$

whose elements are not identically zero on $[0, T]$ and such that $\omega_i \geq 0$ $i \in I^+$. Consider the vector function

$$\Phi : L^2(0, T, \mathbb{R}^m) \times \mathbb{R}^{1+\ell+m} \times \Omega(H, F) \to R^{1+\ell+m}$$

defined $\forall F \in L^2(0, T, \mathbb{R}^m)$ by

$$\Phi(H, M(H, F), \omega(H, F)) =$$

$$\left(\int_0^T \langle C(t, H(t)), H(t) - F(t)\rangle dt,\ \int_0^T \omega_i(H(t), t, F(t))\psi_i(F(t))dt \ \ i \in I\right)$$

and set

$$\mathcal{H} = \{(u, u_1, \ldots, v_\ell,\ v_{\ell+1}, \ldots, v_{\ell+m}) \in \mathbb{R}^{1+\ell+m} :$$

$$u > 0,\ \ v_i = 0 \quad i \in I^0,\ \ v_{\ell+i} \geq 0 \ \ i \in I^+\}.$$

We ask Φ to be such that, $\forall H \in L^2(0, T, \mathbb{R}^m)$

$$M(H, F) \subseteq \mathcal{H} \Leftrightarrow \Phi(H, M(H, F), \omega(H, F)) \in \mathcal{H}\ \ \forall \omega(H, F) \in \Omega(H, F). \tag{3.8}$$

The meaning of (3.8) is that $\forall H$ the set

$$M(H, L^2(0, T, \mathbb{R}^m))$$

is replaced with an element

$$\Phi(H, M(H,F), \omega(H,F))$$

that plays the role of the "representative" of the set. In fact if we introduce the set

$$\mathcal{K}(H,\omega) = \{\Phi(H, M(H, L^2(0,T,\mathbb{R}^m)),\ \omega(H)),$$
$$H \in L^2(0,T,\mathbb{R}^m),\ \ \omega(H,F) \in \Omega(H,F)\}$$

H is a solution of V.I. (1.6) if and only if there exists $\omega(H)$ such that

$$\mathcal{H} \cap \mathcal{K}(H,\omega) = \emptyset$$

as it follows from (3.8). $\omega(H,F)$ can be called a "*Selection Multiplier*", whereas Φ a "*Generalized Selection Function*".

Finally by means of the separation theory we can introduce the following function

$$\Psi(H) = \min_{\lambda \in C^*} \max_{F \in L^2(0,T,\mathbb{R}^m)} \left[\int_0^T \langle C(t,H(t)), H(t) - F(t) \rangle dt + \right. \tag{3.9}$$
$$\left. + \sum_{i \in I} \lambda_i \int_0^T \omega_i(H(t),t,F(t)) \Psi_i(F(t)) dt \right]$$

where C^* is the positive polar of C:

$$C^* \{\lambda \in R^{\ell+m} : \lambda_{\ell+i} \geq 0 \quad i = 1,\dots,m\}.$$

$\Psi(H)$ is a Gap function, that is

1) $\Psi(H) \geq 0 \ \ \forall H \in L^2(0,T.\mathbb{R}^m)$

2) H is a solution of V.I. (1.6) if and only if $\Psi(H) = 0$.

(see P. Daniele [6] for an analogous problem).

4. Appendix by B. Ricceri

Let us show the following

THEOREM 5. *Let E be a real topological vector space and let $H_1, \dots, H_n$ be n closed half-spaces of E such that*

$$\bigcap_{i=1}^{n} H_i \neq \emptyset.$$

Then the relative interior of $\bigcap_{i=1}^{n} H_i$ is non-empty.

Proof. The assert is clearly true for $n = 1$. Then let us suppose that it is true when $n \leq k$ and that is the assert holds for $n = k+1$. So let $H_1, \ldots, H_{k+1}$ be $k+1$ closed–half–spaces of E such that

$$\bigcap_{i=1}^{k+1} H_i \neq \emptyset.$$

Let $f_1, \ldots, f_{k+1} \in E^*\backslash\{0\}$ and $a_1, \ldots, a_{k+1}$ satisfy

$$H_i = \{x \in E = f_i(x) \leq a_i\} \quad i = 1, \ldots, k+1.$$

Denote by A the affine hull of $\bigcap_{i=1}^{k} H_i$ and consider two different cases.

I) Let us suppose that

$$\bigcap_{i=1}^{k} H_i \cap \operatorname{int}(H_{k+1}) \neq \emptyset.$$

Since, by assumption, the relative interior of $\bigcap_{i=1}^{k} H_i$ is non–empty, $\bigcap_{i=1}^{k} H_i$ is a subset of the closure of the relative interior of $\bigcap_{i=1}^{n} H_i$.
Consequently, in our case, the intersection of the relative interior of $\bigcap_{i=1}^{k} H_i$ with $\operatorname{int}(H_{k+1})$ is a non–empty and open set in A. Hence the relative interior of $\bigcap_{i=1}^{k+1} H_i$ is non–empty.

II) Let us suppose that $\bigcap_{i=1}^{k} H_i \cap \operatorname{int}(H_{k+1}) = \emptyset$. Consequently, it results

$$\bigcap_{i=1}^{k+1} H_i = \bigcap_{i=1}^{k} H_i \cap f_{k+1}^{-1}(a_{k+1}).$$

Let y be such that $y \in \bigcap_{i=1}^{k+1} H_i$ and let us set

$$V_i = \{z \in E : f_i(z) \leq a_i - f_i(y)\} \quad i = 1, \ldots, k$$

and

$$V = f_{k+1}^{-1}(a_{k+1}) - y.$$

It is easy to show that

$$\bigcap_{i=1}^{k+1} H_i = y + \left(\bigcap_{i=1}^{k} V_i \cap V\right).$$

Now, being V a vector subspace of E and taking into account that the sets $V_i \cap V$ different from V are closed–half–spaces of V, it results, by assumption, that the relative interior of $\bigcap_{i=1}^{k} V_i \cap V$ is non–empty and therefore the relative interior of $y + \left(\bigcap_{i=1}^{k} \cap V \right)$ is non–empty.

Analogously we may show that the next theorem holds:

THEOREM 6. *Let E be a real topological vector space, let A be a closed affine manifold of E and let $H_1, \dots, H_n$ be n closed–half–spaces of E such that*

$$A \cap \bigcap_{i=1}^{n} H_i \neq \emptyset.$$

Then the relative interior of $A \cap \bigcap_{i=1}^{n} H_i$ is non–empty.

References

1. F. Giannessi, A remark on infinite dimensional variational inequalities, Le Matematiche, 49 (1994), 243–247.
2. A. Maugeri, Variational and Quasi–Variational Inequalities in Network flow models, Recent developments in theory and algorithm, In "Variational Inequalities and Network Equilibrium Problems, F. Giannessi and A. Maugeri (Ed.s), Plenum, New York 1995, 195–211.
3. B. Ricceri, Basic existence theorems for generalized Variational Inequalities, in "Variational Inequalities and Network Equilibrium Problems", F. Giannessi and A. Maugeri (Ed.s), Plenum, New York 1995, 251–256.
4. E. De Giorgi, Teoremi si semicontinuità nel Calcolo delle Variazioni, Ist. Naz. di Alta Matematica (1968).
5. R. Landes, On a necessary condition in the Calculus of Variations, Rendiconti del Circolo matematico di Palermo, 61 (1992), 369–387.
6. P. Daniele, Remark on a dynamic model of a Quasi–Variational Inequality, (to appear).

Ordering Heuristics in Interior Point LP Methods

Csaba Mészáros
Computer and Automation Research Institute,
Hungarian Academy of Sciences,
H-1518 Budapest, P.O. Box 63. Hungary

Abstract. The success of the implementation of the normal equations approach of interior point methods (IPM) for linear programming depends on the quality of its *analysis* phase, i.e. reordering for sparsity. The goal of this analysis is to find a permutation matrix P such that the Cholesky factor of $PAD^2A^TP^T$ is the sparsest possible. In practice, heuristics are used to solve this problem because finding an optimal permutation is an NP-complete problem. Two such heuristics, namely the *minimum degree* and the *minimum local fill–in* orderings are particularly useful in the context of IPM implementations. In this paper a parametric set of symbolic orderings is presented, which connects these two major approaches. It will be shown that in the "neighborhood" of the minimum degree ordering a practically efficient method exist. Implementation details will be discussed as well, and on a demonstrative set of linear programming test problems the performance of the new method will be compared with Sparspak's GENQMD subroutine which was for a long time public accesible from NETLIB, and with the minimum local fill–in ordering implementation of CPLEX version 4.0.

1. Introduction

During the past years issues around efficient implementations of interior point methods (IPM) for linear programming has gained extraordinary interest. This subject is so complex that it allows improvements at many places: a significant amount of effort was put in to decrease the number of necessary iterations [8,12,16], to make the numerical computations faster by exploiting the hardware environment better [3,18,11,17], or to recognize and handle special structures [15,21].

In this paper the following linear programming problem is considered

$$\begin{array}{rl} \min & c^T x, \\ \text{subject to} & Ax = b, \\ & x \geq 0, \end{array} \tag{1}$$

where A is an $m \times n$ matrix, c and x are n vectors, and b is an m vector. Without loss of generality we may assume that $\text{rank}(A) = m$. Some recent papers [13,9,1] give a survey of the developments achieved with the different algorithmic variants of interior point methods of linear programming. In all these IPMs the most costly operation is to

F. Giannessi et al. (eds.), New Trends in Mathematical Programming, 203–221.

determine the search direction. For most problems the most efficient way is to solve the so–called *system of normal equations*

$$AD^2A^T\Delta y = h, \tag{2}$$

where D is an $n \times n$ diagonal matrix with positive diagonal values, and h is an m vector. In every iteration the diagonal matrix D is updated, and system (2) is to be solved once or twice with different right–hand sides to compute the next iterate. The big advantage of the normal equation approach is that it deals with a system of linear equations with a positive definite coefficient matrix. As such, the Cholesky factorization can be used with a great success. Since theoretically all pivots are positive, one can pay full attention to the sparsity issues in the symbolic factorization phase. The symbolic factorization is made once in advance, i.e., the pivot agenda is the same during iterations. The pivot order of the Cholesky factorization has a very strong influence on the number of fill–in and, therefore, on the storage requirement and the speed of factorizations. In the symbolic phase we try to find a good row permutation of A that minimizes the number of nonzeros in the Cholesky factors of AA^T. Since finding the "optimal" permutation is an NP complete problem [22], heuristics are used in practice.

The *minimum degree* ordering [5] is one of the most important and popular methods. The implementation of this ordering algorithm is a well–developed area, the modern implementations are extremely powerful in the sense of speed and storage requirements. The most important aspect of this ordering is the connection with graph theory, and the technique of the implicit representation of the elimination, which makes compact storage and efficient implementation possible. For further details on the above, the reader is referred to the excellent summary in [6]. Most of implementations of interior point methods are equipped with this ordering as default.

The other frequently used algorithm is the so–called *minimum local ill–in* [2] ordering. It is included in some of the implementations as an alternative to the minimum degree ordering. This method computes the number of fills produced by each pivot candidate, and chooses that one where this number is minimal for the next pivot. Generally, the minimum local fill–in algorithm produces sparser factorization but at higher initial cost to obtain the ordering, because the analysis in the pivot search process is very expensive [13].

The choice between the ordering methods depends on the ordering time and the sparsity of the ordering which influence the speed of numerical computation. During the past years by the help of the new numerical implementation technologies (e.g. using supernodes with loop unrolling) the speed of the numerical Cholesky factorization is increased. The higher order techniques decreased the number of necessary iterations dramatically and, therefore, the number of necessary numerical Cholesky factorizations. This leads to pushing the minimum local fill–in ordering into the background, because the high initial cost is recompensed only at few problems.

In the paper another efficient technique is proposed to determine the pivot order. The method first selects a set of attractive pivot candidates and, in the next step, from this smaller set chooses the pivot that generates the minimal predicted fill–in. Therefore the new method is much less expensive than the "exact" version of the minimum local fill–in ordering, but it preserves the efficiency in the sense of fills.

The paper is organized as follows: In Section 2 some terminology and notation is introduced followed by the description of the minimum degree ordering (see [6]). These help to make the implementation faster and more efficient. The pivot searching method and the investigation of the most important pivot search parameter are described in Section 3. Section 4 contains comments on the efficient implementation techniques. Numerical results are shown in Section 5 on a set of linear programming test problems from NETLIB [4]. The implementation of the new ordering method is compared with the minimum degree subroutine GENQMD [5], and with the minimum local fill–in algorithm implemented in CPLEX 4.0. Section 5 contains some remarks and conclusions.

2. Foundations

For presenting ideas we need the general framework of the symbolic factorization process on symmetric systems. In contrast to most implementations of the minimum degree ordering, our approach is based on the explicit representation of the elimination. As the first step to determine the ordering permutation, the sparsity pattern of matrix AA^T is to be computed and stored. The sparsity structure of AA^T can be represented as a boolean $m \times m$ matrix M:

$$m_{ij} = \begin{cases} true & \text{if position } (ij) \text{ is filled in } AA^T \\ false & \text{if position } (ij) \text{ is not filled in } AA^T \end{cases} . \tag{3}$$

Because M is of order m, the elimination process includes $m-1$ steps. Let I^s denote the set of row/column indices that were not pivot until the stage s, and M^s the active submatrix, i.e., the part of the matrix that has not yet been factorized. With this convention let $M^0 = M$. At each step we choose the pivot position from the diagonal of the current active submatrix, which preserve the symmetric structure. At the s-th step of the elimination we choose a pivot index $p \in I^s$. Then, $I^{s+1} = I^s - \{p\}$, and the elements of the active submatrix at stage $s+1$ can be computed as

$$m_{ij}^{s+1} = \begin{cases} m_{ij}^s & \text{if } m_{jp}^s = \mathit{false} \\ m_{ij}^s \text{ OR } m_{ip}^s & \text{if } m_{jp}^s = \mathit{true} \end{cases} \quad \text{for } (i,j) \in I^{s+1} \times I^{s+1}, \tag{4}$$

where OR is the conventional "logical or" function. The sequence of the pivot indices $\{p\}_1^m$ define the ordering permutation. To follow the graph theory conventions up to the end we denote the possible pivot indices $i \in I^s$ as *nodes*, and we call the nodes i and j *adjacent*, if $m_{ij}^s = \mathit{true}$. The matrix M^s is also called an *adjacency structure*. Because the rows/columns and nodes are associated to each other, further on we do not distinguish them. This way, under the sparsity structure of a node, we mean the structure of the corresponding row or column, and the active submatrix is referred to as active nodes.

2.1. Indistinguishable nodes

Both in the pivot searching and elimination processes the nodes (rows/-columns) play an important role which share the same nonzero pattern. They are referred to as *indistinguishable* nodes. In other words, we denote the nodes i and $j \in I^s$ indistinguishable if

$$m_{ik}^s = m_{jk}^s \quad \text{for } k \in I^s. \tag{5}$$

Clearly, the indistinguishable nodes i and j behave in the same way, i.e. if one of them is transformed by some other column, the other is also transformed in a similar a way. Therefore, their indistinguishable property remains unchanged. The other advantage of using indistinguishable nodes is in the pivot searching process. By this, only one member of the indistinguishable nodes should be checked, the transformation effect of the others do not differ from the representative.

The advantages of the indistinguishable nodes can be exploited with the so–called *supernodal* technique, i.e., we store, update and check by pivot searching only one representative of the indistinguishable nodes. This representative is called a supernode [6]. The administration of the other indistinguishable nodes is possible in the following way: We set up a linked list, which holds the indices of the indistinguishable nodes. Each list is linked to their supernode. When choosing the supernode as pivot, the nodes in the corresponding linked list are inserted into the pivot sequence as well.

By this technique we can represent the active submatrix in a more compact form, i.e., we can remove all rows/columns that are indistin-

$$\begin{pmatrix} 1 & & \otimes & & \otimes & \otimes \\ & 2 & & \otimes & & \\ \otimes & & 3 & \otimes & \otimes & \otimes \\ & \otimes & \otimes & 4 & & \\ \otimes & & \otimes & & 5 & \otimes \\ \otimes & & \otimes & & \otimes & 6 \end{pmatrix} \qquad \begin{pmatrix} 1 & & \otimes & \\ & 2 & & \otimes \\ \otimes & & 3 & \otimes \\ & \otimes & \otimes & 4 \end{pmatrix}$$

Figure 1. The full matrix structure and its compact representation

guishable from some supernode. As an example, we consider the sparsity pattern on the left–hand of Figure 1. The nodes (1), (5) and (6) are indistinguishable. The right–hand of the figure shows the compact representation of the matrix.

Later on, we will need the numbers of indistinguishable nodes represented by one node. We refer to this number as a *supernodal degree.* We will denote the number of nodes found indistinguishable from node i for all $i \in I^s$ by sd_i. We note that $sd_i \geq 1$ because each node is indistinguishable from itself. The indistinguishable nodes can be detected after building matrix M and during the elimination process. By this, if nodes i and j are found indistinguishable, the row and column j are removed from the matrix representation, j and the list representing indistinguishable nodes of j are merged in to the list of node i and sd_i is updated as $sd_i = sd_i + sd_j$.

2.2. True degree

Originally, the basic ideas of the symmetric elimination come from the unsymmetric case. For the first time, the well–known Markowitz idea [14] was applied by Tinney and Walker for symmetric factorization [20]. The *true degree* is the synonym of the row and column counts of the unsymmertic case, i.e., the true degree of a node is the number of nodes adjacent to it. We will denote the true degree of the node $i \in I^s$ with td_i. This value was the basis of the original minimum degree ordering [6]. With our notations the true degree of node $i \in I^s$ can be computed as

$$td_i = \sum_{(j \in I^s,\ m^s_{ji} = true)} sd_j \quad \text{for } i \in I^s. \tag{6}$$

Then the Markowitz count can be defined as $(td_i - 1)^2$ for all $i \in I^s$, which reflects the number of arithmetical operations when pivoting on node i, and gives an overestimation on the number of fill–in.

2.3. External degree

By the *external degree* of a node, we mean the number of nodes adjacent to it which are not indistinguishable from the given node itself. The external degree of node i, denoted by ed_i, can be computed as:

$$ed_i = td_i - sd_i \quad \text{for } i \in I^s. \tag{7}$$

In [10], Liu suggests the use of the external degree in the minimum degree algorithm instead of the true degree. The motivation comes from the fact that $(ed_i)^2$ estimates the number of fill–in than $(td_i - 1)^2$ better. In Section 4 we compare the use of the true degree and external degree in our new algorithm as well.

2.4. Mass elimination

This technique is also originated from the minimum degree algorithm [7]. After the elimination of node p in the elimination process at step s, all adjacent nodes of p can be eliminated, the true degree of which is less than the true degree of p, without generating new fill–in. In other words, the following subset $J \subseteq I^{s+1}$of the active nodes can be eliminated:

$$J = \left\{ j \in I^{s+1},\ m_{jp} = true \mid td_j < \alpha \right\}, \tag{8}$$

where α is the true degree of node p in the elimination step s. Clearly,

$$\alpha - 1 \leq \min_{j \in I^{s+1},\ m_{jp} = true} td_j \tag{9}$$

because the adjacent nodes of p inherit the sparsity structure of p. If the equation holds, then the sparsity pattern of node j was a subset of the sparsity pattern of node p, and therefore, the elimination of node j can be performed without fill–in after the elimination of node p.

3. The new pivot searching algorithm

The new ordering method is basically related to the minimum local fill–in heuristic. At the s-th stage of the elimination that heuristic selects that node as next pivot which introduces the least amount of fill–in. This criterion has been known for a long time. Actually, Markowitz suggested this strategy, but rejected to use it due to its computational cost [14]. One difficulty with the minimum local fill–in algorithm is that the process which computes the exact number of fill–in is costly. Noteworthy, it is not necessary to recompute the fill–in produced by all active nodes at each stage of the elimination, because each elimination step makes changes in the sparsity pattern to a small part of the active submatrix. Therefore, the update of the number of caused fill–in is necessary only on the nodes which were influenced by the last elimination step [2]. However, this and the existing ideas do not give enough improvement in the performance, and for this reason, the minimum local fill–in ordering was not considered, in general, as an alternative to the minimum degree ordering in the IPM implementations. In the paper the *inexact* variant of the minimum local fill–in algorithm is introduced. By this method there is no guarantee that the node with the minimum local fill–in property can be found at each pivot step, but we reach a good approximation on the minimum local fill–in with considerably less effort, which makes the method very powerful in the practice.

3.1. MOTIVATION

First, we go back to the unsymmetric Gaussian elimination. Here the Markowitz criterion is the general principle to the pivot selection. Typically, the number of pivot candidates is considerably large: all nonzero entries in the active submatrix (satisfying certain stability criteria in the case of the numerical elimination) are candidates for pivoting. Therefore, choosing one with the smallest Markowitz count is very expensive. To allow improvements in the speed of the pivot searching, another pivot choice principle is used: from among the elements which satisfy the numerical criteria, we want to choose a pivot with *small* Markowitz count [19]. A number of various techniques were developed to exploit this scheme. Zlatev proposed to search for pivot in the first few rows in increasing order of row counts [23], Suhl suggested an algorithm which searches in few rows and columns with a small row or column count [19]. Our suggested method can be regarded as the application of the above two ideas to the symmetric symbolical factorization. We will compute the caused fill–in of the first p nodes in the increasing order of the true or external degree, and we will choose one of them which introduces the least amount of fill–in.

3.2. THE ALGORITHM

In this section we present the algorithm Find_Pivot in Fortran–like pseudo–code. Similarly to the algorithm described in [19], we use linked lists to access to the nodes with the same true or external degree. Two cases are discussed. In the first case the true degree is used to set up the list, and to process the searching. In the second case the nodes are linked based on their external degree plus one. Thus the key, used for listing the nodes, is between 1 and m in both cases. To store and maintain the linked lists we need two integer working arrays of size m, **degpnt** and **deglst**. In *degpnt*(j) we hold the entry point into the sublist of nodes with true degree j (or with external degree $j-1$, respectively); the array **deglst** is used for holding the sublists represented by the successors.

As noted above, it is not necessary to recompute the number of fill–in for all pivot candidates. Therefore we introduce the use of the integer working array **fillin** of size m to control this and to make an access to the previously computed values as follows:

$$fillin(j) = \begin{cases} \text{number of fill–in produced by node } j, \\ -1, \text{ if the number of fill–in is to be recompute} \end{cases} \quad j \in I^s. \tag{10}$$

Let p denote the search parameter. The input parameters of the procedure are m, p, **degpnt**, **deglst** and **fillin**. The output parameter is *pivot*, which holds the selected pivot position, or 0 if no other pivots are available. The function Compute_Fill predicts the number of fill–in of a node.

```
Begin procedure Find_Pivot(m,p,pivot,degpnt,deglst,fillin)
 1.  fill = m^2 + 1;
 2.  k = 1;
 3.  deg = 1;
 4.  pp = 0;
 5.  pivot = 0;
 6.  j = degpnt(deg);
 7.  if j > 0 then
 8.     if fillin(j) < 0 let fillin(j) =Compute_Fill(j, ...);
 9.     if fillin(j) ≥ fill goto 13.
10.     fill = fillin(j);
11.     pivot = j;
12.    if fill = 0 exit
13.     j = deglst(j);
14.     pp = pp + 1;
15.     if pp ≤ p goto 7.
16.  endif
17.  deg = deg + 1;
18.  if deg ≤ m goto 6.
End procedure
```

The procedure **Find_Pivot** stops the search if all active nodes were checked (step 18), or the number of nodes examined exceeds the value of the parameter p (step 15), or a node is found which produces no fill–in (step 13). As a result, a node is obtained which produces the minimal amount of fill–in from the set of the first $p+1$ nodes with a small true or external degree. If more than one node exist with minimal fill–in, the procedure chooses one with the smallest true or external degree.

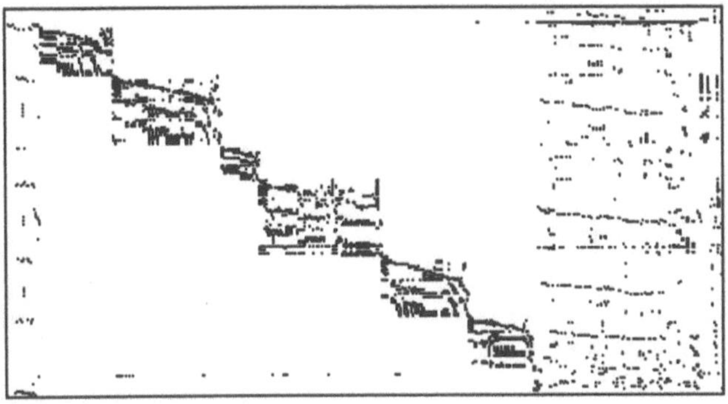

Figure 2. Nonzero pattern of problem 25fv47

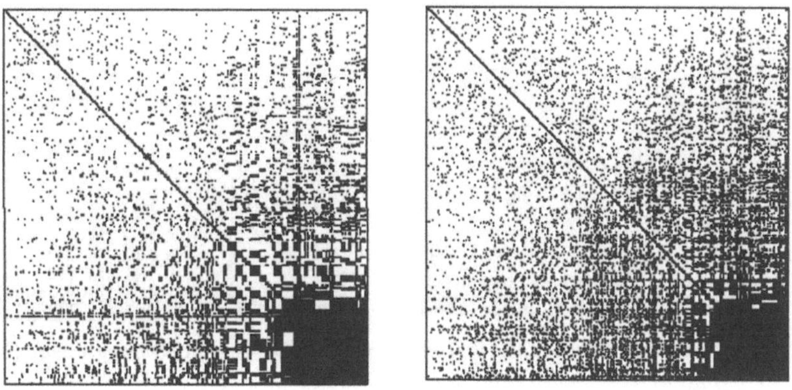

Figure 3. Sparsity pattern of the factorization via MDO and MFO on problem 25fv47

3.3. EXPERIMENTS WITH THE PIVOT SEARCH PARAMETER

The most important parameter of the above algorithm is p, which balances between the cost and the efficiency of the pivot searching. When increasing p, we search longer, and probably we gain a better pivot. For example, the setting $p = 0$ gives the minimum degree principle, and $p = m$ the exact minimum local fill–in rule. Naturally, no unique value exists for p which makes the best performance on all problems. Finding a *good* value is subject of extensive testing. Because testing like this is very sensitive, for the computing environment here a standalone IBM PC 486/DX2 50 Mhz computer was used rather than a networked multitasking Unix workstation, where the execution of a program can be influenced by many factors. In our experiments the Watcom Fortran Compiler 77^{32} version 9.5 was used with the /OX option.

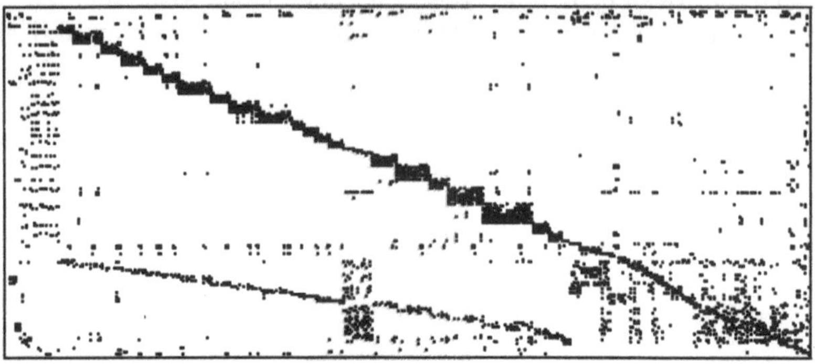

Figure 4. Nonzero pattern of problem d2q06c

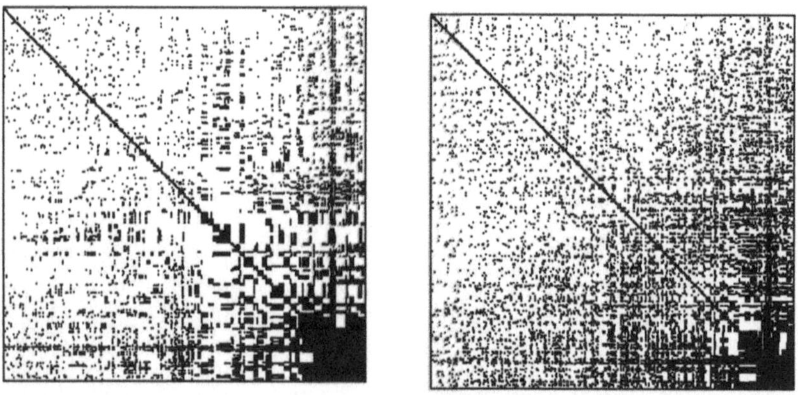

Figure 5. Sparsity pattern of the factorization via MDO and MFO on problem d2q06c

First, two test problems, 25fv47 and d2q06c from the NETLIB test set are chosen for an extensive study on the effect of the search parameter p. Problem 25fv47 behaves like most problems from our point of view, that is, the minimum local fill–in heuristic produces 20% sparser factorization than minimum degree. Figure 2 shows the nonzero pattern of this problem. The patterns of the factorization of AA^T with the minimum degree (33572 nonzeros in the lower triangular part) and minimum local fill–in ordering (27167 nonzeros in the lower triangular part) can be found in the left– and right–hand side of Figure 3, respectively.

The other test problem is d2q06c. Figure 4 shows its nonzero pattern, and Figure 5 the patterns of the factorization with the two ordering

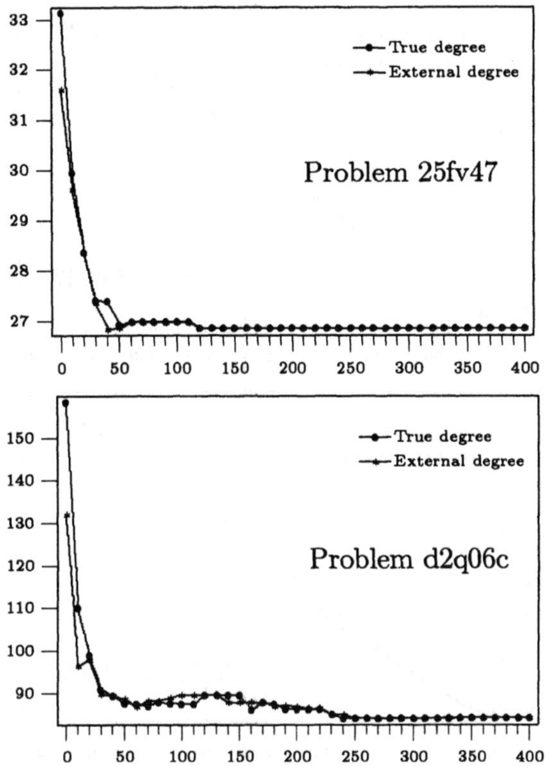

Figure 6. No. of nonzeros in the factorization (in thousand) on problems 25fv47 and d2q06c

methods, respectively. On this problem the minimum degree ordering generated 154397, the minimum local fill–in ordering 92032 nonzeros in the triangular factors of AA^T.

In the first experiment p varies between 0 and 400 with step 10. Both the true degree and external degree variant of the algorithm are run. Figure 6 shows the number of nonzeros in the factorization in thousands in the case of problems 25fv47 and d2q06c respectively, as it is plotted against the value of p. From 0 to 30 the results show significant improvement in the number of fill–in, which slows down between 30 and 50. As it can be seen, if $p > 50$, the longer search for a better pivot does not make any remarkable improvement. In both problems the time spent on obtaining the ordering (reported on Figure 7), grows nearly linearly with the value of p. These results indicates tat we should concentrate on the values $p \in [0, 50]$.

In the second experiment p is increased from 0 to 50 by step 1. Figure 8 shows the behaviour of fill–in depending on the value of p at problems

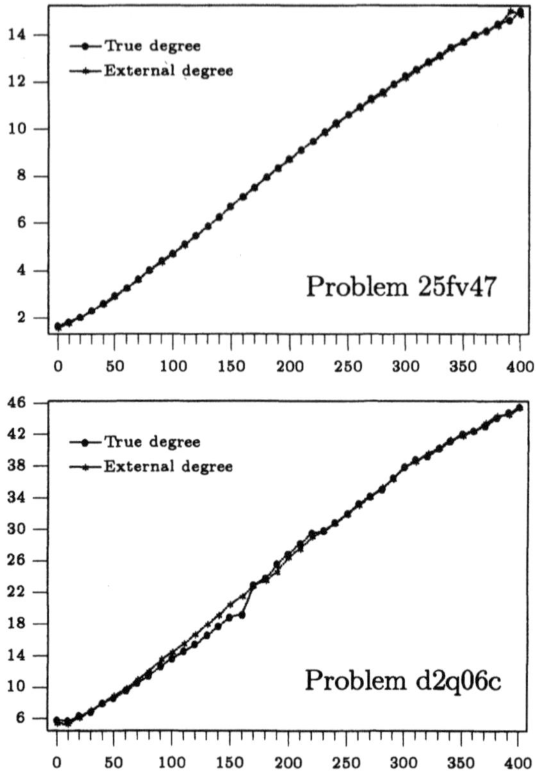

Figure 7. Ordering times (in seconds) on problems 25fv47 and d2q06c

25fv47 and d2q06c. In both problems the external degree version tends to produce better factorization, but in the later stages (after about $p = 45$ and $p = 25$ at the problems 25fv47 and d2q06c, respectively) the difference becomes insignificant. The ordering time reported on the left side of Figure 9 is in accordance with the previous experiences. A more dramatic effect is displayed in the case of problem d2q06c. Here, the ordering time is decreased in the first stage as a result of the large decrease in the number of fill–in.

The heuristical assumption, namely that the node which produces few fill–in must have small degree, is confirmed by the experiments. The other advantageous feature of our pivot searching algorithm is that it computes the predicted fill–in on nodes with small degree only. This is very important because the complexity of this process depends on the degree of the nodes.

The presented experiments show that in general the increase of p results in a nearly uniform increase in the ordering time, but the quality of the ordering remains unchanged after a certain threshold. Thus,

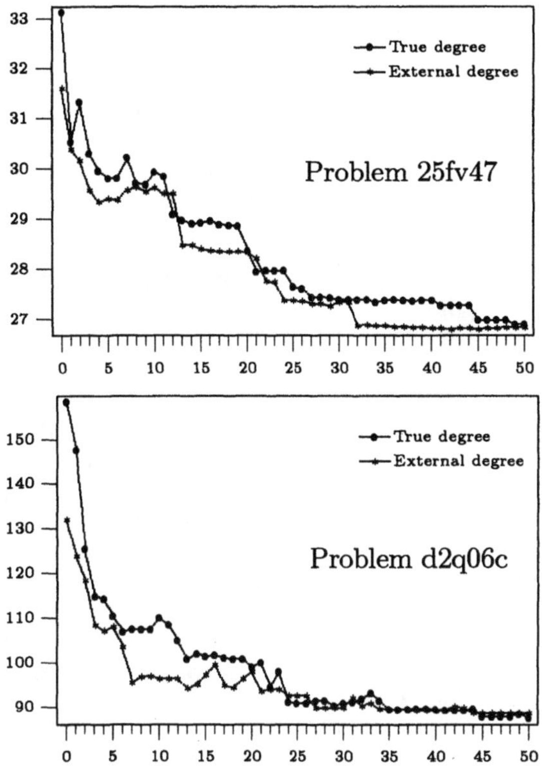

Figure 8. No. of nonzeros in the factorization (in thousand) on problems 25fv47 and d2q06c

there is no reason for using a value larger than 35. Due to the better performance, the use of the external degree version of the new algorithm is suggested. Based on solving many problems with different settings, it seems that $p = 25$ is a good value in practice. The solver BPMPD [17] is equipped with this value as default, but the possibility kept open to change it for the user.

4. Implementation

The first step of the implementation does the initialization. This includes the computation of the sparsity structure of the initial matrix $M = AA^T$, the search on it for indistinguishable nodes, furthermore the set–up of linked lists and arrays used in the elimination loop. The elimination loop begins with pivot searching. Then, update (4) is performed on the active submatrix. This process is very similar to the unsymmetric case. This part of the implementation is based on the

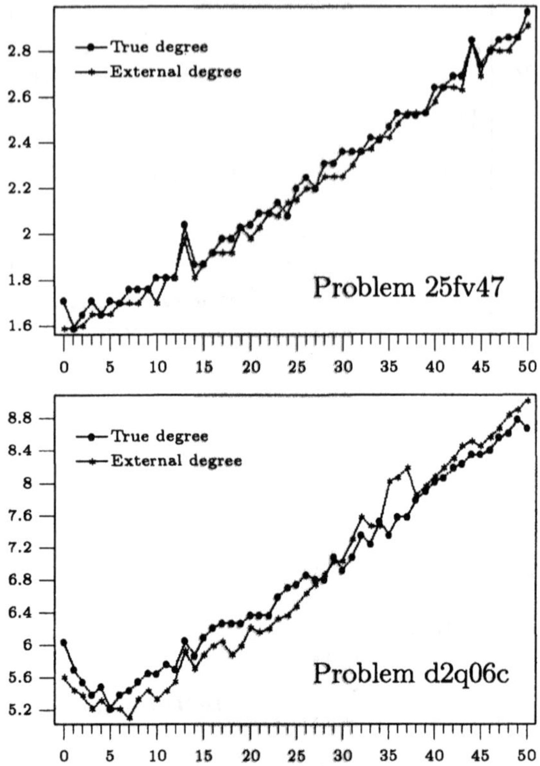

Figure 9. Ordering times (in seconds) on problems 25fv47 and d2q06c

ideas described in [19] with modifications to exploit the specialities of the symmetric and symbolic factorization. Before returning to pivot search, the implementation searches indistinguishable nodes caused by the last transformation and indicates the requirement of the recomputation of fill–in for certain nodes. The global scheme of the algorithm as implemented in BPMPD is displayed in Figure 10.

Since M, the active submatrix at each elimination step is symmetric, only its lower triangular part is stored. The off–diagonal nonzeros of the lower triangular part of the active submatrix are stored column–wise as a sequence of sparse vectors. A pointer array is used for pointing to the starting position of columns in the column file which contains the row indices. An other array holds the number of entries of each column. For efficient access, the row–wise representation is also maintained. Similarly to the column data structure, a pointer array points to the column file and an array holds the number of entries of each column. Due to the symmetric property of the active submatrix, the row–wise storage

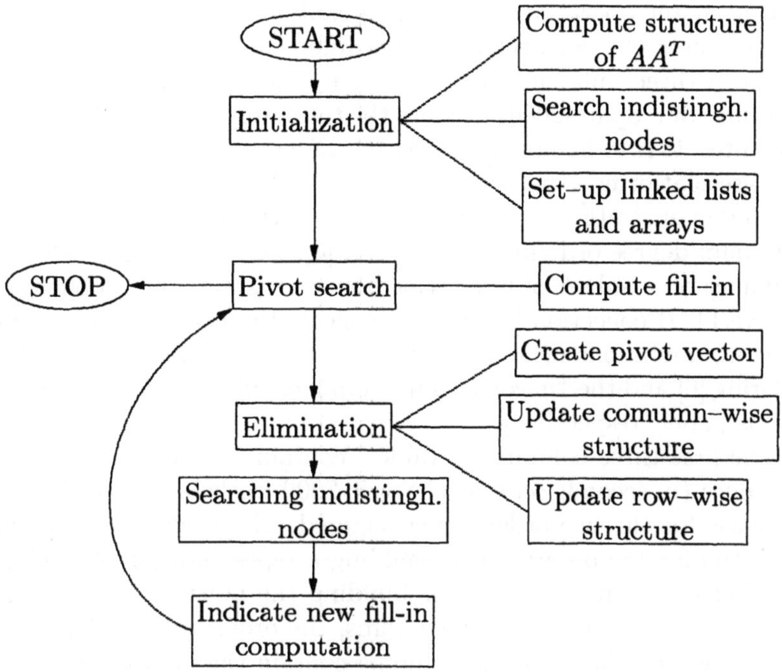

Figure 10. The outline of the implementation

scheme can be interpreted as the columns of the upper triangular part of the active submatrix.

It is to be noted that the memory requirement of the implementation is significantly larger than that of the minimum degree algorithm. But, in context of the interior point methods, it is not a decisive issue. The matrix ordering is only a preparation phase of interior point implementations, therefore all working arrays used by the main algorithm are free at this stage. Since higher order techniques and efficient numerical factorization methods require considerable amount of memory, usually no extra memory array are needed for the ordering, just the memory what is used in a later phase for other purpose is used here.

5. Numerical results

The aim of this paper is to show that the well–established minimum degree ordering can be improved by using the ideas of the minimum local fill–in heuristic. The only way of verification is computational testing. Due to the rapidly changing hardware technologies, the comparison of two algorithms is a highly complex task. In the present case, the performance of the hardware in terms of the ratio of executing integer and floating point operations plays significant role, therefore, on different architectures the conclusion can be different. In the numerical investigations three different platforms were used (IBM PC 486, IBM PC Pentium, Sun Sparc-2000) because the tests were performed at different institutes. It is to be noted that except of a speed factor, the behaviour of these platforms is nearly equal.

To demonstrate the advantages of the new ordering scheme to the minimum degree ordering described in [6], we shall compare their performance on a subset of large scale linear programming problems from the NETLIB collection [4]. Table I collects the results of this comparison. Abbreviations MDO and MFO in it denote the minimum degree ordering [6] and the "inexact" minimum local fill–in ordering, described in this paper, respectively.

The first three columns of Table I contain the problem names and the times (in seconds on a SUN Sparc-2000 workstation) of the analysis phase for the two orderings considered. The analysis time includes the setup for the ordering (i.e. building a representation of AA^T), the ordering time, and the time for building the nonzero patterns of the Cholesky factors. For both algorithms, the ordering time is the dominating factor. The following columns contain the number of nonzeros in the Cholesky factors produced by the two orderings and the average time (in seconds) to execute one factorization on a SUN Sparc-2000 workstation.

It is to be noted that the largest difference in the symbolic time occurs on problem **osa-60** where the new method is more than 100 times faster than GENQMD. The reason of it is the special behaviour of the problem, namely, by a very large size few fill–ins take place. For this problem the implicit representation of the elimination seems to be disastrous. The example of the reverse effect is the problem **dfl001** where heavy fill–in occurs. On problems like this, GENQMD works faster due to the better performance of the implicit representation. Considering the fill–in the results show the advantages of the new method.

The next experiment was to compare the barrier solver CPLEX 4.0 with its minimum local fill–in ordering option and BPMPD which uses the "inexact" minimum local fill–in ordering [1]. Both of the solvers were ran on an IBM Pentium P-90 machine with 32 Mbyte of memory with disabled presolve feature. Table II contains the results in the following order: the symbolic time, the number of nonzeros in the factorization and the total solution time to obtain a solution of a 8-digit accuracy. All

[1] This experiment was performed at the Department of Mathematics and Statistics, Brunel University, London, UK

Table I. Comparison of minimum degree (MDO) and "inexact" minimum local fill–in (MFO) orderings

Name	Analysis time		Nonzeros in L		Factorization time	
	MDO	MFO	MDO	MFO	MDO	MFO
25fv47	1.35	1.09	33572	27379	0.380	0.280
80bau3b	3.14	1.83	39821	33977	0.521	0.386
bnl2	2.47	1.67	64476	55759	1.020	0.809
cycle	2.59	2.58	51160	39796	0.550	0.363
d2q06c	5.84	3.54	154397	92492	3.654	1.481
degen3	72.43	9.89	119241	115749	2.604	2.956
dfl001	194.23	338.80	1619564	1396096	197.125	149.829
greenbea	7.93	1.75	47669	45110	0.486	0.441
nesm	0.64	1.16	20594	19720	0.225	0.220
perold	1.21	0.85	23192	19229	0.230	0.176
pilot87	83.07	74.20	464167	389356	29.214	20.934
scfxm3	0.39	0.28	12723	10257	0.082	0.070
sctap3	0.83	0.58	14769	14432	0.120	0.104
shell	0.13	0.16	3505	3301	0.020	0.020
ship12l	0.78	0.23	4920	4896	0.055	0.051
truss	1.57	1.79	64575	51593	0.942	0.634
tuff	0.91	0.17	6806	6238	0.050	0.050
woodlp	2.36	0.86	11473	11473	0.480	0.482
osa-30	1957.81	31.11	110768	110768	6.211	4.580
osa-60	11692.44	109.82	255666	255666	12.979	11.284

timing results are given in seconds, and represent the pure algorithmic time, without the input of the MPS file and the time spent on producing the output.

The results presented in Table II show that the new ordering algorithm is much faster than the "exact" minimum local fill–in ordering and competitive with it regarding the fill–in.

6. Conclusions and Recommendations

In this paper the "inexact" variant of the minimum local fill–in ordering was presented. We discussed the necessary theoretical and programming techniques which make this ordering very powerful in the practice. Computational tests proved that the new ordering method is a competitive alternative to the commonly used minimum degree and minimum local fill–in orderings.

Table II. Comparison of CPLEX and BPMPD

Problem	Symbolic time		Nonzeros in L		Solution time	
Name	CPLEX	BPMPD	CPLEX	BPMPD	CPLEX	BPMPD
25fv47	1.65	0.77	42017	38648	8.78	7.85
80bau3b	2.75	1.42	62734	57450	24.55	36.03
BNL2	8.35	1.49	93330	87019	48.88	37.41
cycle	3.46	1.21	79780	73376	19.94	14.00
d2q06c	9.11	2.41	130039	128816	47.40	44.22
degen3	27.40	6.70	156941	140496	67.83	35.60
dfl001	3265.00	204.49	1504318	1405086	7532.93	4239.92
greenbea	6.32	2.42	112249	99947	53.88	28.28
nesm	1.21	0.83	36269	32822	15.11	10.43
perold	1.21	0.66	31309	29137	10.38	9.22
pilot87	261.00	42.29	496945	457743	841.18	546.94
scfxm3	0.38	0.22	22170	19734	3.79	3.24
sctap3	1.05	0.33	28470	25096	4.23	3.68
shell	0.11	0.16	7308	6407	1.98	2.09
ship12l	0.49	0.33	29523	27307	5.27	6.31
truss	3.02	1.15	90423	79429	17.20	15.55
tuff	0.38	0.16	13530	12367	2.37	1.81
wood1p	2.58	0.83	90853	88297	18.90	16.97
woodw	4.50	1.93	87669	81994	20.98	22.41
Total	3600	270	3115877	2891171	8476	5082

The author's opinion is that a significant improvement in the inexact minimum fill–in ordering can probably be the use of another key, different from the external degree, to select the nodes for pivot searching. There is more possibility to set–up indicators which give more accurate information about the expected fill–in and hopefully will make it possible to increase the speed of pivot searching.

References

1. E. D. Andersen, J. Gondzio, Cs. Mészáros and X. Xu. Implementation of interior point methods for large scale linear programming. In T. Terlaky, editor, *Interior point methods in mathematical programming.* Kluwer Academic Publisher, 1996.
2. I. S. Duff, A. M. Erisman and J. K. Reid. *Direct methods for sparse matrices.* Oxford University Press, New York, 1986.
3. N. G. Esmond and B. W. Peyton. A supernodal Cholesky factorization algorithm for shared-memory multiprocessors. *SIAM Journal on Scientific Computing,* 2(14):761–769, 1993.
4. D. M. Gay. Electronic mail distribution of linear programming test problems. *COAL Newsletter,* 13:10–12, 1985.

5. A. George and J. W. H. Liu. *Computer Solution of Large Sparse Positive Definite Systems.* Prentice-Hall, Englewood Cliffs, NJ, 1981.
6. A. George and J. W. H. Liu. The evolution of the minimum degree ordering algorithm. *SIAM Rev.*, 31:1–19, 1989.
7. A. George and D. R. McIntyre. On the application of the minimum degree ordering to finite elements systems. *SIAM J. Numer. Anal.*, 15:90–111, 1978.
8. J. Gondzio. Multiple centrality corrections in a primal-dual method for linear programming. Technical Report 1994.20, Logilab, HEC Geneva, Section of Management Studies, University of Geneva, November 1994. Revised May 1995, to appear in Computational Optimization and Applications.
9. J. Gondzio and T. Terlaky. A computational view of interior point methods for linear programming. In J. Beasley, editor, *Advances in linear and integer programming.* Oxford University Press, Oxford, England, 1995.
10. J. W. H. Liu. Modification of the minimum-degree algorithm by multiple elimination. *ACM Trans. Math. Software*, 11(2):141–153, 1985.
11. I. J. Lustig, R. E. Marsten and D. F. Shanno. Computational experience with a primal-dual interior point method for linear programming. *Linear Algebra Appl.*, 20:191–222, 1991.
12. I. J. Lustig, R. E. Marsten and D. F. Shanno. Interior point methods for linear programming: Computational state of the art. Technical report sor 92-17, Department of Civil engineering and Operations Research, Princeton University, 1992.
13. I. J. Lustig, R. E. Marsten and D. F. Shanno. Interior point methods for linear programming: Computational state of the art. *ORSA Journal on Comput.*, 6(1):1–15, 1994.
14. H. M. Markowitz. The elimination form of the inverse and its application to linear programming. *Management Sci.*, 3:255–269, 1957.
15. I. Maros and Cs. Mészáros. The role of the augmented system in interior point methods. Technical Report TR/06/95, Brunel University, Department of Mathematics and Statistics, London, 1995.
16. S. Mehrotra. Handling free variables in interior methods. Technical Report 91-06, Department of Industrial Engineering and Managment Sciences, Northwestern University, Evanston, USA., March 1991.
17. Cs. Mészáros. Fast Cholesky factorization for interior point methods of linear programming. *Computers & Mathematics with Applications*, 31(4/5):49–54, 1996.
18. E. Rothberg and A. Gupta. Efficient sparse matrix factorization on high-performance workstations – exploiting the memory hierarchy. *ACM Trans. Math. Software*, 17(3):313–334, 1991.
19. U. H. Suhl and L. M. Suhl. Computing sparse LU factorizations for large-scale linear programming bases. *ORSA Journal on Comput.*, 2(4):325–335, 1990.
20. W. F. Tinney and J. W. Walker. Direct solution of sparse network equations by optimally ordered triangular factorization. In *Proceedings of IEEE*, volume 55, pages 1801–1809. 1967.
21. M. J. Todd. Exploiting special structure in Karmarkar's linear programming algorithm. *Mathematical Programming*, 41:81–103, 1988.
22. M. Yannakakis. Computing the minimum fill–in is NP–complete. *SIAM J. Algebraic Discrete Methods*, pages 77–79, 1981.
23. Z. Zlatev. On some pivotal strategies in Gaussian elimimination by sparse techniques. *SIAM Journal on Numerical Analysis*, 17(1):18–30, 1980.

A Tabu Based Pattern Search Method for the Distance Geometry Problem

Panos M. Pardalos *
Center for Applied Optimization,
Department of Industrial and Systems Engineering,
University of Florida, Gainesville, FL 32611
E-mail: pardalos@ufl.edu

Xin Liu
Institute of Policy and Management,
Chinese Academy of Sciences, Beijing, 100080.
E-mail: liu@deming.ise.ufl.edu

Abstract. The distance geometry problem is that of determining the coordinates of a set of points in space from a given set of pairwise distance measurements. It is a common method for formulating space conformational problems in terms of distance measurements and recently has been used in determining the structure of proteins from NMR (Nuclear Magnetic Resonance) data. In this paper we present a tabu search based approach for the distance geometry problem and report on preliminary computational experiments. The proposed algorithm can be used to solve general global optimization problems.

Key words: Distance Geometry, Molecular Conformation, Global Optimization, Pattern Search, Tabu Search.

1. Introduction

This paper describes the application of a tabu based pattern search method to the distance geometry problem. The distance geometry problem is that of determining the coordinates of a set of points in space from a given set of pairwise distance measurements. Distance geometry is a common method for formulating space conformational problems in terms of distance measurements and it has many applications in chemistry [2, 15, 16, 17, 18]. Unlike the problem of locating the global minimum energy [24], the problem of finding conformations consistent with distance constraints can be reliably solved. It is possible to analyze the NMR (Nuclear Magnetic Resonance) spectrum of a molecule to obtain pairwise atom distance information, and then to determine the shape of a molecule by solving the distance geometry problem. We can express the distance geometry problem as the problem of finding

* The work of this author was supported in part by National Science Foundation grant BIR-9505919

F. Giannessi et al. (eds.), New Trends in Mathematical Programming, 223–234.

three-dimensional coordinates $x_1, \ldots, x_n$ that satisfy:

$$\|x_i - x_j\| = d_{i,j},$$

where $i,\ j \in \{1, \ldots, n\}$ are the indices that describe n ranked atoms, and $d_{i,j}$ is the given distance between two atoms i and j. Generally, we can only know a set of pairwise distances between atoms. For any set of three atoms $\{i, j, k\}$, in which all pairwise distances are known, the following triangle inequality should be satisfied:

$$d_{i,j} \leq d_{i,k} + d_{k,j}.$$

Since the distance data available from NMR spectroscopy is necessarily imprecise, we use only upper and lower bounds on the distances $d_{i,j}$. Therefore, we can also express the distance geometry problem as the problem of finding $x_1, \ldots, x_n$ that satisfy:

$$l_{i,j} \leq \|x_i - x_j\| \leq u_{i,j},$$

where $l_{i,j}$ and $u_{i,j}$ are the lower and upper bounds of the distance, respectively. Both $l_{i,j} \leq \|x_i - x_j\| \leq u_{i,j}$ and $\|x_i - x_j\| = d_{i,j}$ are called the distance constraints.

The distance geometry problem is NP hard even in the one dimensional case [17]. Many algorithms have been investigated for the solution of the distance geometry problem [5, 15, 17, 22]. Detailed discussions are provided in [16, 17]. Most of these algorithms use some optimization formulation of the problem, usually incorporating specific insight into the structure of the molecules they are designed to analyze. The algorithm reported by [22] is among the few algorithms which are not based on any chemistry information. Each of the proposed distance geometry algorithms has its advantages and disadvantages, but most of the algorithms have been successfully applied to classes of large problems.

Many algorithms for the distance geometry problem are based on the minimization of an error function which measures the violations of the distance constraints. One such error function is defined by [16, 22]:

$$g(x) = \sum \left(\max\left(0, \frac{\|x_i - x_j\|^2}{u_{i,j}^2} - 1\right) \right)^2 + \left(\max\left(0, 1 - \frac{\|x_i - x_j\|^2}{l_{i,j}^2}\right) \right)^2,$$

where the sum $\sum$ ranges over all the given pairwise atoms whose distances are known and $x = (x_1, \ldots, x_n) \in \Re^{3n}$. Although our method applies to general problems, in our computational results we consider the function

$$f(x) = \frac{1}{2} \sum \left(\|x_i - x_j\|^2 - d_{i,j}^2 \right).$$

The distance geometry problem is defined as the following global optimization problem:

$$\min \quad f(x),$$

where the objective function $f(x)$ is the error function defined above. The objective function is always non-negative, and zero if and only if all the coordinates of atoms satisfy the distance constraints. Since the objective function is non-convex and it may have many local minimizers, the problem of finding the global minima of the distance geometry problem can be very difficult.

In recent years many techniques have been developed to solve global optimization problems [7, 20]. In particular, several heuristics have been proposed for finding good approximate solutions. Among these heuristics, simulated annealing [21], genetic algorithms [14], tabu search [3, 9, 13] and greedy randomized adaptive search procedures (GRASP) [8] have been successfully applied to many non-convex optimization problems.

In this paper we present an algorithm for the distance geometry problem that uses the idea of pattern search [19] combined with the tabu search strategy [13]. Although many traditional direction or gradient based methods are local search approaches, direct search methods [1, 4, 19, 23], a kind of stochastic search procedure developed mainly in nonlinear programming, are able to escape from local optima to find better if not optimal solutions. Because these algorithms do not compute derivatives and only use function evaluations, they are easy to implement and used to solve global optimization problems arising from practical applications.

Unlike the multistart method that randomly generates the initial solutions and uses them to conduct the local search, we are primarily concerned with designing new algorithms that implement a coarse but broad search procedure. After this search procedure, the final solution can be within a certain degree of precision or the global minimum can be "isolated" from local minima, so that the final solution can be refined to any desired accuracy by an additional search. Our work mainly focuses on the hybrid of the pattern search with the tabu search.

The paper is organized as follows. Section 2 discusses the general principles of the pattern search. Section 3 describes the tabu search, and Section 4 presents the proposed algorithm which uses the pattern search with tabu heuristics. The algorithm is finally applied to solve a distance geometry problem.

2. Pattern Search Heuristic

We consider the general global optimization problem:

$$\text{Minimize } f(x) : x \in X \subseteq \Re^n, \tag{1}$$

where f is the real-valued objective function, n is the dimension of the solution space, and X is the set of feasible solutions. We must compute a feasible solution $x^* \in X$ for which the value $f(x^*)$ is the global minimum.

The global search is conceptually hard because the number of local minima tend to increase exponentially with the problem size. Any global search must be able to escape from local minima and continue to search whatever the initial solutions are. Because it is computationally difficult to obtain a global optimum by conventional nonlinear programming techniques, which mainly depends on local search procedures, and it is practically impossible to exhaustively search the continuous solution space, a number of heuristic methods have been developed. These include the direct search methods, among which are the pattern search [19], the simplex method [23], the box method [1], and a parallel direct search algorithm [4]. Because these algorithms do not compute derivatives and use function evaluations only, they are easy to implement.

The pattern search consists of two important moves. One is the exploratory move, and the other is the pattern move. During the process of the exploratory move, it is important to find a solution with the decreasing function value in a certain step length. If such a solution cannot be found, the step length is updated and the procedure continues. The pattern move is only an accelerating technique which speeds up the search towards to a better solution. The algorithm stops if the step length satisfies a given specified precision. Fig. 1 illustrates one case of the pattern search.

To discuss the pattern search, we first introduce some mathematical notations. We denote by x_k the current solution, Δ_k is the parameter of the current iteration step length, s_k^i is the trial searching direction along the i-th coordinate axis, s_k is the direction of the current step, ϵ is the precision parameter that controls the stopping criterion, $\hat{x}_k$ is the solution during the pattern move, k denotes the parameter of the number of iterations, e_i denotes the unit coordinate vector, n is the dimension of the solution space, min is the best value of the objective function so far found, and $f(x_k)$ and $f(\hat{x}_k)$ are defined as the current iterate evaluation of the objective function.

The pattern search conducts a series of exploratory moves about the current iteration step before considering a new iteration. These moves can be viewed as sampling the function according to a predesigned

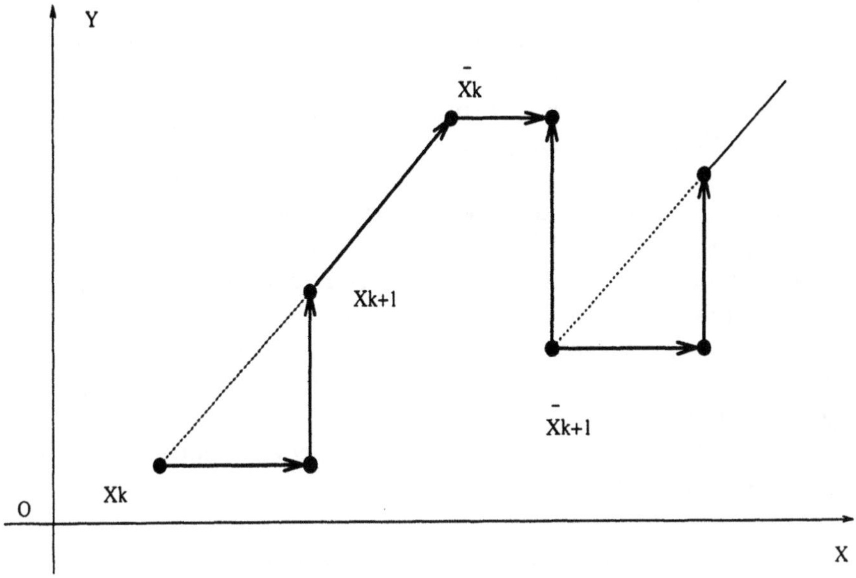

Figure 1. Example of Pattern Search in two dimensions

manner in search of a new iteration step with a lower function value. The original exploratory move of the pattern method can be described by the following pseudocode.

---

begin
  for $i = 1, \ldots, n$ **do**:
    i) $s_k^i = s_k + \Delta_k e_i$ and $x_k^i = x_k + s_k^i$. Compute $f(x_k^i)$.
    ii) if $f(x_k^i) < \min$ **then**
      $\min = f(x_k^i)$ and $s_k = s_k^i$.
    else
      i) $s_k^i = s_k - \Delta_k e_i$ and $x_k^i = x_k + s_k^i$. Compute $f(x_k^i)$.
      ii) if $f(x_k^i) < \min$, $\min = f(x_k^i)$ and $s_k = s_k^i$.
    endif
end

An iteration is called successful for two consecutive iterate steps, if $f(x_k + s_k) - f(x_k) \leq 0$; otherwise, the iteration is unsuccessful. The exploratory move is actually an alternative coordinated search along axes and an iterate step is accepted if the value of the objective function simply decreases along at least one axis.

The pattern move is an important part of Hooke and Jeeves's pattern search. Hooke and Jeeves's pattern search strategy incorporates a pattern move to accelerate the progress of the algorithm by using the previous search history. For current iterate solution x_k and the former iterate solution x_{k-1}, if $x_k \neq x_{k-1}$, the direction $x_k - \dot{x}_{k-1}$ is the direction along which the objective function decreases most rapidly. The main strategy of Hooke and Jeeves's pattern search is to investigate the possible progress in the direction of $x_k - x_{k-1}$, rather than conducting the exploratory search about the current solution x_k. The pattern move accepts a new iterate solution if the exploratory search about $x_k + (x_k - x_{k-1})$ is successful when imposing the exploratory move. The new solution is accepted even if the inequality $f(x_k + (x_k - x_{k-1})) \leq f(x_k)$ is not satisfied.

We describe the method of Hooke and Jeeves's pattern search by the following pseudocode:

---

begin
  1) $k = 1$
  2)**for** $i = 1, \ldots, n$ **do**:
    2.1) $s_k^i = s_k + \Delta_k e_i$ and $x_k^i = x_k + s_k^i$. Compute $f(x_k^i)$
    2.2) if $f(x_k^i) <$ min **then**
      min $= f(x_k^i)$ and $s_k = s_k^i$
      else
      i) $s_k^i = s_k - \Delta_k e_i$ and $x_k^i = x_k + s_k^i$. Compute $f(x_k^i)$
      ii) if $f(x_k^i) <$ min, min $= f(x_k^i)$ and $s_k = s_k^i$
      endif
  3) if $x_{k-1} \neq x_k$ **then**
      $x_k = x_{k-1}^i$ go to step 5)
      else go to step 4)
      endif
  4) if $\Delta_k \leq \epsilon$ **then**
    goto **end**
    else $\Delta_k = \alpha \Delta_k$ go to step 2)
    endif
  5) $\hat{x}_k = 2x_k - x_{k-1}$, min $= f(\hat{x}_k)$
    5.1)**for** $i = 1, \ldots, n$ **do**:
    5.2) $s_k^i = s_k + \Delta_k e_i$ and $\hat{x}_k^i = \hat{x}_k + s_k^i$. Compute $f(\hat{x}_k^i)$
    5.3) if $f(\hat{x}_k^i) <$ min **then**
      min $= f(\hat{x}_k^i)$ and $s_k = s_k^i$
      else

$i)$ $s_k^i = s_k - \Delta_k e_i$ and $\hat{x}_k^i = \hat{x}_k + s_k^i$. Compute $f(\hat{x}_k^i)$
$ii)$ if $f(\hat{x}_k^i) < \min$, $\min = f(\hat{x}_k^i)$ and $s_k = s_k^i$
endif
6) if $f(\hat{x}_k) \leq f(x_k)$ **then**
$x_{k+1} = \hat{x}_k$, $k = k + 1$, go to step 2)
else $k = k + 1$; go to step 2)
endif
end

---

3. Tabu Search

Tabu search, introduced by Glover [10, 11], has been adapted successfully to solve a wide variety of optimization problems [13]. The tabu search performs an intelligent search of the solution space; it is a kind of learning procedure that benefits from the information obtained during the search history. The approach is based on the premise that an intelligent search requires the use of adaptive memory, which is organized to take advantage of the search history in relation to the structure of the problem being solved. Rather than "memoryless" designs as in simulated annealing or "rigid memory" designs as in branch and bound, tabu search relies on flexible memory structures that incorporate short term and long term information about solutions and trajectories. This memory is organized by monitoring particular attributes of solutions that change during the search, and uses characteristics such as recency, frequency, influence and logic to control the occurrence of such attributes in future solutions generated. More advanced forms of the tabu search can be used as a complete solution approach by incorporating a long term exploration search space. To present the tabu search strategy, we denote by $N(history, x)$ the history information of the solution x [6]. The generic tabu search is described by the following pseudocode.

---

begin
Stop:= false
history:=$\emptyset$
iteration:=0
y:=start
while stopping criterion is not satisfied **do**:
Choose z^* such that
$f(z^*) = \min\{f(z) : z \in N(history, y)\}$

```
        y = z*
        iteration:=iteration + 1
        adjust history
    end do
    Stop:= true
end
```

4. A Tabu Based Pattern Search Heuristic

In this section, we present a version of the tabu search that is specifically adapted to execute the implementation of the pattern search in the context of continuous function optimization. The tabu search strategy can be adapted for solving continuous optimization problems by discretizing domains [3] and using scatter search [12]. The crucial elements for the implementation of tabu search include the definition of neighborhood structure, the management of memory structure that incorporates short term and long term search information, and the adaptation of the strategies of aspiration condition.

We introduce a neighborhood structure in the continuous search space. Generally, the exploration move of the pattern search is an alternative coordinate search along the axes. For a fixed solution $x \in \Re^n$, the coordinate search is computationally expensive if the dimension n is large. We consider a hypercube centered about the current iterate. Because the pattern search only requires a simple decrease of the objective function, we do not evaluate all vertices of the hypercube at each iterate step. We randomly choose n_x vertices. These vertices are neighbors of the current solution x. The size of the neighborhood $N(x)$ is n_x, and remains constant during the search procedure. In contrast with the conventional pattern search, we modify the search step length as a constant. Fig. 2 shows the three dimensions hypercube around the current iterate.

The principle of the tabu restriction is easily used for the continuous space by selecting problem variables as a basis for defining move attributes [12]. Let x and $x^{'}$ denote a pair of successive solutions, where $x^{'} \in \Re^n$ was generated from x by a recent move. Like in discrete settings, we try to avoid a move reversal for the continuous search by preventing a current move that would create a solution $x^{''}$ containing selected attributes of x. For defining the move attributes, we try to select a subset of variables that change from one solution to another. From the implementation point of view it is better to focus on a subset of components rather than all components of the solution x. In our

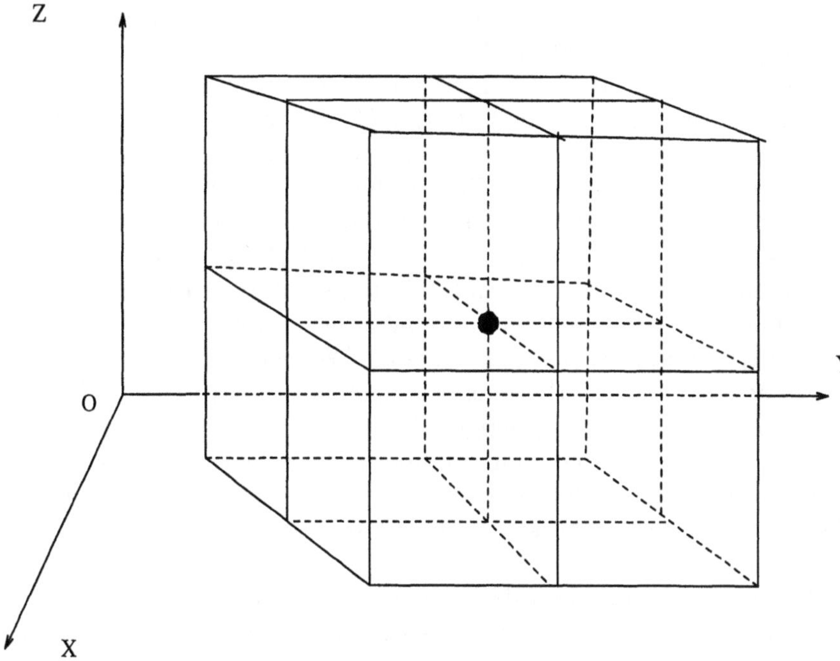

Figure 2. Three dimensional hypercube neighborhood of the solution

algorithm, we choose two components of x whose values change the most from x to $x^{'}$ in the neighbor $N(x)$ of the solution x. We impose a tabu restriction that prohibits at least one of these two components from changing in the reverse direction by an amount greater than half of its current change.

By avoiding repeated visits to recently explored areas, the tabu condition encourages the search towards the regions far from the current solutions. To select some components, whose values are changing most but their frequency subject to tabu is smaller, is to favor the diversification search strategy. We can also use the pattern move as the diversification search strategy, that is after a fixed number of iterations, if the objective function values does not decrease, we apply several pattern move steps.

The aspiration condition provides a supplementary basis for controlling the search, by allowing a move to be selected if the resulting solution is sufficiently good or novel, in spite of being classified tabu. A simple aspiration is to allow a tabu move to be selected if it leads to a solution better than the best one computed so far, or the best one computed within a particular region or duration of the search. We implement the aspiration condition by simply designing the function

AF. We say the aspiration condition is satisfied and the tabu status of the one move is canceled, if $f(x') \leq AF$.

We applied the proposed algorithm to the distance geometry problem. The algorithm has been written in C and implemented on a SUN/4 workstation. The preliminary test data contain 57 atoms and 218 pairwise distances which results 171 variables [17, 18]. First, we compute the sum of all distances (d_{ij}) in order to find an initial solution. In the worst case, all atoms can be positioned along a line. Hence, we generated a random number as the initial solution in the range of the sum of all distances. The random number generator (ranlib.c) available from `http://netlib.att.com` was used. We predefined the search step length as constant and the maximal number of iteration as 5 000. We can also let the search step length change after a certain number of iterations and use the maximum number of iterations, during which the objective function no longer decreases, as the stopping criterion. For each test problem n_x was set to be $n/2$ where n denotes the number of variables.

Table 1 presents computational results with the above test data. The numbers of the first column express the ranked numbers of the atoms. The x, y, and z denote the three dimensional coordinates (x, y, z) of the atom. Because the coordinates (x, y, z) are translation invariant, that is we can always make the coordinate transformation for all atoms, $(x, y, z) \longrightarrow (x + \Delta, y + \Delta, z + \Delta)$, where Δ is the parameter of transformation step, negative coordinates among the solution do not influence the correctness of the optimal solution for the distance geometry problem. When the initial solution for the distance geometry problem is generated, we can restrict it in the interval $(0, L)$ or $(-L/2, L/2)$, where L is a parameter for the range of the optimal solution. We consider the distance geometry problem as an unconstrained optimization problem, that is we do not consider the solution distribution because we can use coordinate transformations. The efficiency of the proposed approach depends on the distribution of the initial solutions. More testing is underway for larger molecules.

5. Concluding Remarks

Our study has shown that conventional minimization methods for nonlinear programming such as direct search methods, and heuristics such as the tabu search, can be efficiently used to design new algorithms for global continuous optimization problems. The proposed algorithm has several other attractive features. It is problem-dependent and the size of neighborhood and the step length are determined by the problem being

Table I. Experimental Results

atoms	x	y	z
1	-3.53	-1.60	-4.07
2	-1.69	0.64	0.29
3	-0.75	0.59	-2.73
4	1.69	0.54	2.27
5	-2.67	-0.98	-0.99
6	-1.36	-0.05	2.69
7	3.01	-2.29	-1.19
8	3.42	-0.04	-1.22
9	0.80	-1.01	-2.50
10	-2.27	1.83	-0.51
11	2.12	1.25	-0.46
12	0.68	-1.90	-0.41
13	-1.70	0.84	1.89
14	-0.66	2.04	0.59
15	-0.09	1.81	1.42
16	1.20	1.87	-2.78
17	-1.45	-0.57	-1.49
18	-2.20	-2.51	-1.49
19	1.25	-0.32	-1.05
20	-0.80	0.34	-0.73
21	-1.27	0.63	0.40
22	-1.70	-2.89	1.92
23	0.03	0.51	-0.12
24	-2.36	3.12	-2.07
25	-0.29	0.42	-0.23
26	-1.02	2.05	-1.51
27	1.66	-1.37	0.39
28	-1.26	0.08	-0.23
29	-1.95	-2.02	-1.83

atoms	x	y	z
30	-1.85	-0.51	1.31
31	0.53	-0.76	-0.73
32	-1.60	0.36	2.19
33	-2.32	1.01	-1.17
34	-0.66	0.22	0.66
35	-3.71	0.35	-1.09
36	-0.76	1.11	1.36
37	-1.35	-0.96	0.21
38	-1.30	3.67	-2.63
39	1.54	-1.40	0.07
40	1.88	1.08	1.12
41	1.53	-3.75	-0.81
42	-1.10	-0.49	-0.68
43	0.20	2.29	2.18
44	-0.17	0.64	2.31
45	2.37	0.13	1.07
46	1.53	0.01	0.36
47	-0.73	-1.62	-2.32
48	1.46	0.43	0.32
49	-2.83	1.86	1.06
50	2.32	-0.07	-2.10
51	3.18	2.81	0.82
52	-1.72	0.49	-1.73
53	-2.42	0.41	2.88
54	0.88	-0.33	1.14
55	0.32	2.22	-0.49
56	2.30	0.14	-0.13
57	0.00	0.00	0.00

solved. The proposed global optimization method is not specific to the distance geometry problem but will work for any function. In addition, the proposed algorithm is easy to implement. Further computational results with a variety of test problems are underway.

Acknowledgements. We would like to thank Professor F. Glover for directing us to many references about tabu search, and Dr. A. B. Hendricson for providing his test data.

References

1. Box, G. and K. Wilson, *On the experimental attainment of optimum conditions,* J. Royal Statistical Society, Series B, 13 (1951), pp. 1-45.
2. Crippen, M. G. and T. F. Havel, *Distance Geometry and Molecular Conformation,* John Wiley & Sons, 1988.
3. Cvijović, D. and J. Klinowski, *Taboo search: An approach to the multiple minima problem,* Science 267 (1995), pp. 664-666.
4. Dennis, J. and V. Torczon, *Direct search methods on parallel machines,* SIAM J. Optimization, 1 (1991), pp. 448-474.
5. Dress, A. W. M. and T. F. Havel, *Bound smoothing under chirality constraints,* SIAM J. Discr. Math. 4 (1991), pp. 535–549.
6. Fleurent, C., F. Glover, P. Michelon and Z. Valli, *A scatter search approach for unconstrained continuous optimization,* Manuscript.
7. Floudas, C. and P. M. Pardalos, *Recent Advances in Global Optimization,* Princeton Univ. Press, Princeton, NJ, 1991.
8. Feo, T. A., and M. G. C. Resende, *Greedy randomized adaptive search procedures,* J. Global. Opt. 6 (1995), pp. 106-133.
9. Glover, F., *Heuristic for integer programming using surrogate constrains,* Decision Sciences 8 (1977), pp. 156–166.
10. Glover, F., *Tabu Search, Part 1,* ORSA. J. Comput. 1 (1989), pp. 190-206.
11. Glover, F., *Tabu Search, Part 2,* ORSA. J. Comput. 2 (1990), pp. 4-32.
12. Glover, F., *Tabu search for nonlinear and parameter optimization,* manuscript, Graduate School of Business, University of Colorado, Boulder, Colorado (1991)
13. Glover, F., *Tabu search fundamentals and uses,* manuscript , Graduate School of Business, University of Colorado, Boulder, Colorado (1995).
14. Goldberg, D. E., *Genetic Algorithms in Search, Optimization and Machine Learning,* Addison-Wesley, (1989).
15. Havel, F. T., *An evaluation of computational strategies for use in the determination of protein structure from distance geometry constrains Obtained by Nuclear Magnetic Resonance,* Prog. Mol. Biol., 56 (1991), pp. 43-78.
16. Havel, F. T., *Distance geometry,* To appear in "The Encyclopedia of NMR", eds. D. M. Grant and R. K. Harris, John Wiley & Sons.
17. Hendricson, A. B., *The molecular problem: Determining conformation from pairwise distances,* Ph.D. thesis, Cornell University, Ithaca, New York (1991).
18. Hendricson, A. B., *The molecular problem: Exploiting structure in global optimization,* SIAM J. Optimization, 5(1995), pp. 835-857.
19. Hooke, R. and T. Jeeves, *Direct search solution of numerical and statistical problems,* J. Assoc. Comp. Mach., 8 (1961), pp. 212-229.
20. Horst, R., P. M. Pardalos and N. V. Thoai, *Introduction to Global Optimization,* Kluwer Academic Publishers, 1995.
21. Kirkpatrick, S. , C. D. Gelatt and Jr. M. P. Vecchi, *Optimization by simulated annealing,* Science, 220 (1983), pp. 671-680.
22. Morè, J. J. and Wu, Z., *ϵ-Optimal solution to distance geometry problems via global continuation*, Preprint Technical Report (1995), Argonne National Laboratory.
23. Nelder, J. A. and Mead, R., *A simplex method for finding stationary values of a function of several variables,* Computer J., 5 (1965), pp. 147-151.
24. Pardalos, P. M., D. Shalloway and G. Xue, *Global Minimization of Nonconvex Energy Functions: Molecular Conformation and Protein Folding,* DIMACS Series Vol. 23 (1996), American Mathematical Society.

Programming Under Probabilistic Constraint with Discrete Random Variable

András Prékopa
RUTCOR, Rutgers Center for Operations Research, Rutgers University, New Brunswick, NJ, 08903.

Béla Vizvári
Dept. of Operations Research, Eötvös Loránd University, H-1088 Budapest, Múzeum krt. 6-8, RUTCOR

Tamás Badics
RUTCOR, Rutgers Center for Operations Research, Rutgers University, New Brunswick, NJ, 08903.

Abstract. The most important static stochastic programming models, that can be formulated in connection with a linear programming problem, where some of the right-hand side values are random variables, are: the simple recourse model, the probabilistic constrained model and the combination of the two. In this paper we present algorithmic solution to the second and third models under the assumption that the random variables have a discrete joint distribution. The solution method uses the concept of a p-level efficient point (pLEP) intoduced by the first author (1990) and works in such a way that first all pLEP's are enumerated, then a cutting plane method does the rest of the job.

1. Introduction

Stochastic programming problems are formulated in such a way that we start from an underlying problem or base problem, in which we observe that some of the parameters are random variables; then, we reformulate the problem, by the use of some decision principle, and obtain a stochastic programming formulation. Assume that the base problem is:

$$\begin{aligned} &\text{Min}\, \mathbf{c}^T\mathbf{x} \\ &\text{subject to} \\ &\mathbf{Ax} = \mathbf{b} \\ &\mathbf{Tx} \geq \xi \\ &\mathbf{x} \geq \mathbf{0}, \end{aligned} \tag{1}$$

where $\mathbf{A}$ is an $m \times n$, $\mathbf{T}$ is an $r \times n$ matrix, and $\xi = (\xi_1, ..., \xi_r)^T$ is a random vector. Starting from problem (1), usually we formulate a recourse, or a probabilistic constrained stochastic programming problem, or a problem which combines the two formulations. In the present

F. Giannessi et al. (eds.), New Trends in Mathematical Programming, 235–255.

paper we look at the latter two problems under the assumption that ξ has a discrete distribution with a finite number of possible values, and propose algorithmic solutions for them.

Let $q_1, ..., q_r$ be nonnegative constants and designate by $T_1, ..., T_r$ the rows of the matrix $\mathbf{T}$. Then, our stochastic programming problem is the following:

$$\begin{aligned} &\text{Min}\left(\mathbf{c}^T\mathbf{x} + \textstyle\sum_{i=1}^r q_i E[\xi_i - T^i\mathbf{x}]_+\right) \\ &\text{subject to} \\ &\qquad \mathbf{Ax} = \mathbf{b} \\ &P(\mathbf{Tx} \geq \xi) \geq p \\ &\qquad\quad \mathbf{x} \geq \mathbf{0}, \end{aligned} \tag{2}$$

where p is a given probability $(0 < p < 1)$.

If $F_i(z)$ designates the probability distribution function of ξ_i, i.e., $F_i(z) = P(\xi_i \leq z)$, $z \in \mathbf{R}$, $i = 1, ..., r$, then problem (2) can be written in the form (see, e.g., Prékopa (1990)):

$$\begin{aligned} &\text{Min}\left(\mathbf{c}^T\mathbf{x} + \textstyle\sum_{i=1}^r q_i\left(\mu_i - T^i\mathbf{x} + \int_{-\infty}^{T_i\mathbf{x}} F_i(z)dz\right)\right) \\ &\text{subject to} \\ &\qquad \mathbf{Ax} = \mathbf{b} \\ &P(\mathbf{Tx} \geq \xi) \geq p \\ &\qquad\quad \mathbf{x} \geq \mathbf{0}. \end{aligned} \tag{3}$$

Let $z_{j1} < ... < z_{jk_j}$ be the possible values of ξ_j, $j = 1, ..., r$. We supplement each set of possible values by two further values, which are z_{j0} and z_{j,k_j+1} in case of ξ_j, and assume that they satisfy the following relations: $z_{j0} < z_{j1}$, $z_{jk_j} < z_{j,k_j+1}$. These values should be chosen in such a way that if $\mathbf{x}$ satisfies $\mathbf{Ax} = \mathbf{b}$, $\mathbf{x} \geq \mathbf{0}$, then it should also satisfy the inequalities

$$z_{j0} \leq T_j\mathbf{x} \leq z_{jk_j+1},\ j = 1, ..., r.$$

The set $\mathcal{Z} = \{(z_{1j_1}, ..., z_{rj_r}) \mid 0 \leq j_i \leq k_i + 1,\ i = 1, ..., r\}$ clearly contains all possible values of the random vector ξ. Due to stochastic dependency of the components of ξ, some elements of $\mathcal{Z}$ may have 0 probability, but for notational convinienice we take $\mathcal{Z}$ as the set of possible values of ξ. Let $F(\mathbf{z})$ designate the probability distribution function of ξ, i.e., $F(\mathbf{z}) = P(\xi \leq \mathbf{z})$, $\mathbf{z} \in \mathbf{R}^r$.

A point $\mathbf{z} \in \mathcal{Z}$ is called a p-level efficient point (pLEP) of the probability distribution of ξ, if $F(\mathbf{z}) \geq p$ and there is no $\mathbf{y} \in \mathcal{Z}$ satisfying $\mathbf{y} \leq \mathbf{z}$, $\mathbf{y} \neq \mathbf{z}$, $F(\mathbf{y}) \geq p$. Let $\mathcal{E} = \{\mathbf{z}^{(1)}, ..., \mathbf{z}^{(N)}\}$ be the set of all pLEP's.

If we choose $q_1 = ... = q_r = 0$ in problem (3), then an equivalent form of the problem is the following:

$$\text{Min}\mathbf{c}^T\mathbf{x} \tag{4a}$$

subject to

$$\mathbf{Ax} = \mathbf{b} \tag{4b}$$

$$\mathbf{x} \geq \mathbf{0}, \tag{4c}$$

$$\mathbf{Tx} \geq \mathbf{z}^{(i)}, \text{ for at least one } i \in \{1, ..., N\}. \tag{4d}$$

In the general case we intruduce the notations:

$$c_{ij} = -q_i z_{ij} + q_i \int_{z_{i0}}^{z_{ij}} F_i(z)dz, \; j = 0, ..., k_i + 1; \; i = 1, ..., r, \tag{5}$$

and apply the λ-representation for the piecewise linear convex functions

$$f_i(y_i) = -q_i y_i + q_i \int_{z_{i0}}^{z_{ij}} F_i(z)dz, \; z_{i0} \leq y_i \leq z_{ik_i+1}, \; i = 1, ..., r. \tag{6}$$

This means that for each $i = 1, ..., r$ we have:

$$\begin{aligned} f_i(y_i) = \; & \text{Min} \sum_{j=0}^{k_i+1} f_i(z_{ij})\lambda_{ij} \\ & \text{subject to} \\ & \sum_{j=0}^{k_i+1} z_{ij}\lambda_{ij} = y_i \\ & \sum_{j=0}^{k_i+1} \lambda_{ij} = 1 \\ & \lambda_{ij} \geq 0, \quad j = 0, ..., k_i + 1. \end{aligned} \tag{7}$$

Using these and the pLEP's, problem (3) can be written in the form (the constant term $\sum_{i=1}^r q_i\mu_i$ is left out from the objective function):

$$\text{Min} \left(\mathbf{c}^T\mathbf{x} + \sum_{i=1}^{r} \sum_{j=0}^{k_i+1} c_{ij}\lambda_{ij} \right) \tag{8a}$$

subject to

$$\mathbf{A}\mathbf{x} = \mathbf{b} \tag{8b}$$

$$\mathbf{T}_i\mathbf{x} - \sum_{j=0}^{k_i+1} z_{ij}\lambda_{ij} = 0 \tag{8c}$$

$$\sum_{j=0}^{k_i+1} \lambda_{ij} = 1 \tag{8d}$$

$$\mathbf{x} \geq \mathbf{0} \tag{8e}$$

$$\lambda_{ij} \geq 0, j = 0, ..., k_i + 1;\ \ i = 1, ..., r \tag{8f}$$

$$\mathbf{T}\mathbf{x} \geq \mathbf{z}^{(i)} \text{ for at least one } i \in \{1, ..., N\}. \tag{8g}$$

We present algorithmic solutions to problems (4), and (8). To this end, we have to study the set $\mathcal{E}$ from two points of view: how to enumerate them algorithmically and what are their geometric properties.

2. Algorithmic Enumeration of the p-Level Efficient Points

The algorithm is presented in a recursive form. When we enumerate the pLEP's in $\mathbf{R}^r$, we assume that an enumeration technique in $\mathbf{R}^{r-1}$ is available for functions which are not necessarily probability distribution functions in the sense that the sum of all probabilities we are dealing with may be smaller than 1. Thus, it is convenient to define the function $F(z_1, ..., z_r)$, the pLEP's of which are to be determined, right at the outset in a somewhat more general manner.

We assume that to each element u in $\mathcal{Z} = \mathbb{Z}_1 \times ... \times \mathbb{Z}_r$ a probability p_u is assigned such that $\sum_{\mathbf{u}\in\mathcal{Z}} p_{\mathbf{u}} \leq 1$, and then define

$$F(\mathbf{z}) = \sum_{\substack{\mathbf{u}\in\mathcal{Z} \\ \mathbf{u}\leq\mathbf{z}}} p_{\mathbf{u}}, \ \ \mathbf{z} \in \mathbf{R}^n. \tag{9}$$

For the function (9) the set $\mathcal{E}$ of all pLEP's can similarly be defined as we have done it for the probability distribution functions in Section 1. In the present case, however, the set $\mathcal{E}$ may be empty for some p values. In case of $r = 1$, the pLEP is that element h of $\mathbb{Z}_1$, which satisfies

$$h = \text{argmin}\{l \mid F(z_{1l}) \geq p\}, \tag{10}$$

provided that such an h exists. In the general case the steps of the algorithm are as presented below. We assume that $F(z_{1,k_1+1}, ..., z_{r,k_r+1}) \geq p$.

Step 0. Initialize $k \leftarrow 0$. Go to **Step 1.**

Step 1. Let

$$\begin{aligned} z_{1,j_1} &= \operatorname{argmin}\{y \mid F(y, z_{2,k_2+1}, ..., z_{r,k_r+1}) \geq p\} \\ z_{2,j_2} &= \operatorname{argmin}\{y \mid F(z_{1,j_1}, y, z_{3,k_3+1}, ..., z_{r,k_r+1}) \geq p\} \\ &\cdot \\ &\cdot \\ &\cdot \\ z_{r,j_r} &= \operatorname{argmin}\{y \mid F(z_{1,j_1}, ..., z_{r-1,j_{r-1}}, y) \geq p\}. \end{aligned}$$

Go to **Step 2.**

Step 2. Let $\mathcal{E} \leftarrow \{z_{1,j_1}, ..., z_{r,j_r}\}$. Go to **Step 3.**

Step 3. Let $k \leftarrow k + 1$. If $j_1 + k > k_1 + 1$, then go to **Step 5.** If $j_1 + k \leq k_1 + 1$, then go to **Step 4.**

Step 4. Enumerate all pLEP's of the function $F(z_{1,j_1+k}, \mathbf{y})$ of the variable $\mathbf{y} \in \mathbf{R}^{r-1}$, and eliminate those which dominate at least one element in $\mathcal{E}$ ($\mathbf{y}$ dominates $\mathbf{z}$, if $\mathbf{y} \geq \mathbf{z}$, and $\mathbf{y} \neq \mathbf{z}$). If $\mathcal{H}$ is the set of the remaining pLEP's, which may be empty, then let $\mathcal{E} \leftarrow \mathcal{E} \cup \mathcal{H}$. Go to **Step 3.**

Step 5. Stop, $\mathcal{E}$ is the set of all pLEP's of the function $F(\mathbf{z})$, $\mathbf{z} \in \mathcal{Z}$.

Example 1. Let $\mathcal{Z}_1 = \mathcal{Z}_2 = \{0, 1, 2, 3, 4, 5, 6, 7, 8, 9\}$, $p_{ih} = 0.019$, if $0 \leq i \leq 4$, $0 \leq h \leq 5$, or $h = 8, 9; p_{ih} = 0.038$, if $0 \leq i \leq 4$, $h = 6$; $p_{ih} = 0$, if $0 \leq i \leq 4$, $h = 7$; $p_{ih} = 0.001$ if $5 \leq i \leq 9$, $0 \leq h \leq 9$, and $p = 0.6$. In **Step 1** we determine

$$\begin{aligned} 3 &= \operatorname{argmin}\{y \mid F(y, 9) \geq 0.6\} \\ 6 &= \operatorname{argmin}\{y \mid F(3, y) \geq 0.6\}. \end{aligned}$$

Thus, $(z_{1,j_1}, z_{2,j_2}) = (3, 6)$, and at the end of **Step 2** we have $k = 0$, $\mathcal{E} = \{(3, 6)\}$.

In **Step 3** we take $k = 1$, and go to **Step 4**, where we obtain $\mathcal{H} = \{(4, 6)\}$. Then we eliminate $(4, 6)$, and define $\mathcal{E} = \{(3, 6)\}$.

Then we go to **Step 3** but $\mathcal{H}$ will be empty for $k = 2, 3, 4, 5$. In case of $k = 6$ we obtain $\mathcal{H} = \{(9, 5)\}$, and the procedure terminates. The set of all pLEP's is $\mathcal{E} = \{(3, 6), (9, 5)\}$ (see Figure 1).

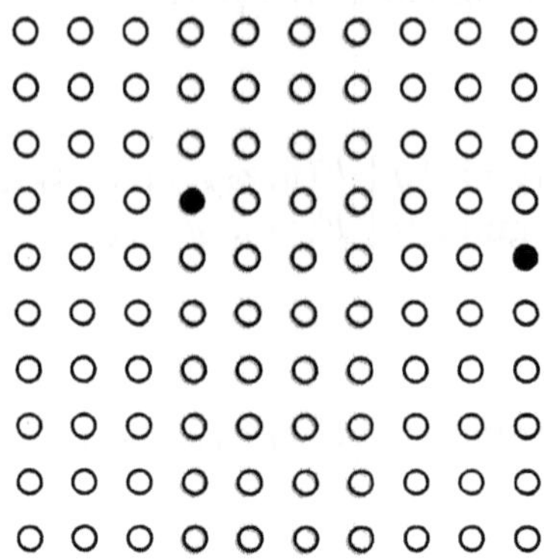

Figure 1. The lattice points are assigned to the set of integers $\{(i,h) \mid 0 \le i,\ h \le 9\}$. The corresponding probabilities are those mentioned in Example 1 of Section 2. The marked points are the 0.6-level efficient points.

3. Properties of the p-Level Efficient Points

First we remark that the notion of a pLEP can be defined for arbitrary probability distributions. In addition to discrete distributions with finite supports we will discuss discrete distributions, where the support is the set $\mathbb{Z}_+^r$ of all lattice points of the nonnegative orthant of $\mathbf{R}^r$ or a subset of it.

A theorem of Vizvári (1977, Thm 4.5, p.28) asserts that any subset $\mathcal{H}$ of $\mathbb{Z}_+^r$ which has the property that if $\mathbf{x}, \mathbf{y} \in \mathcal{H},\ \mathbf{x} \neq \mathbf{y}$, and neither $\mathbf{x} \le \mathbf{y}$ nor $\mathbf{y} \le \mathbf{x}$ holds, is necessarily finite. This implies that for any $0 < p < 1$, and for any probability distribution with support in $\mathbb{Z}_+^r$ the set of all pLEP's is finite. The question that what are the sets that may be pLEP sets is answered by the following theorem.

Theorem 1. Let $\mathcal{H}$ be any nonempty subset of $\mathbb{Z}$ (or $\mathbb{Z}_+^r$) such that if $\mathbf{x}, \mathbf{y} \in \mathcal{H},\ \mathbf{x} \neq \mathbf{y}$, then neither $\mathbf{x} \le \mathbf{y}$ nor $\mathbf{y} \le \mathbf{x}$ holds. Then, for any $0 < p < 1$ there exists a probability distribution with support $\mathbb{Z}$ (or $\mathbb{Z}_+^r$), the set of all pLEP's of which is $\mathcal{H}$.

Proof. Let N be the number of elements of $\mathcal{H}$. If $N = 1$ then the probability p is assigned to the single point $\mathbf{x}$ of $\mathcal{H}$ and the probability $1-p$ is assigned to another point $\mathbf{z}$ of $\mathbb{Z}$ (or $\mathbb{Z}_+^r$) satisfying $\mathbf{x} \neq \mathbf{z},\ \mathbf{x} \le \mathbf{z}$. Assume that $N \ge 2$. At first let us suppose that $\frac{1}{N} \le p < 1$. Then

$$\varepsilon = \frac{1-p}{N-1} \le p.$$

Let $\mathbf{z} \in \mathbb{Z}$ (or $\mathbf{z} \in \mathbb{Z}^r_+$) be a point in $\mathbf{R}^r$ that satisfies $\mathbf{z} \leq \mathbf{x}$, for every $\mathbf{x} \in \mathcal{H}$. We define the probability distribution by assigning probability $p - \varepsilon$ to $\mathbf{z}$, and probability ε to each element of $\mathcal{H}$. Assume that $0 < p < \frac{1}{N}$. Let $\mathbf{z} \in \mathbb{Z}$ (or $\mathbf{z} \in \mathbb{Z}^r_+$) be a point in $\mathbf{R}^r$ that satisfies $\mathbf{z} \geq \mathbf{x}$, for every $\mathbf{x} \in \mathcal{H}$. Then we define the probability distribution by assigning probability $1 - Np$ to $\mathbf{z}$, and probability p to each element of $\mathcal{H}$.

□

Definition 1. A set $\{z^{(i)}, i = 1, ..., N\}$ is said to be a discrete convex set if $z^j \notin \text{riconv}\{z^{(i)}, i = 1, ..., N\}$, $j = 1, ..., N$.

In most cases we have encountered so far the pLEP sets turned out to be convex.

Below we present three examples to show that the set of pLEP's can be nonconvex, even in case of some well-known discrete probability distributions.

Example 1. Let $r = 2$ and define the birariate probability distribution on $\mathbb{Z}^2_+$ as follows

$$p_{ik} := \frac{1}{(i+1)(i+2)} \frac{1}{(k+1)(k+2)}, \quad i \geq 0, \ k \geq 0. \tag{11}$$

The probability distribution function of this distribution is $F(z_1, z_2) = F(z_1)F(z_2)$ if z_1, z_2 are nonegative integers, where $F(z) = \frac{z}{z+1}$, $z \in \mathbb{Z}^1_+$.

pLEP's (i, k)	Probability levels $F(i, k)$
(99,10)	0.9
(32,13)	0.9004329
(27,14)	0.9
(24,15)	0.9
(22,16)	0.9002557545
(21,17)	0.9015151515
(19,18)	0.9

Table 1: The set of 0.9-level efficients points and the corresponding probabilities in case of the discrete distribution given by (11). The points obtained by tranposing the two co-ordinates are pLEP's, too.

Let $p = 0.9$. The set of pLEP's together with the corresponding probability distribution function values are presented in Table 1. The set $\mathcal{E}$ is nonconvex, because the points $(21,17)$, and $(17,21)$ are interior points of the convex hull of the pLEP's. By symmetry it is enough to show it for one of these points. It can easily be checked that $(21,17) = \lambda_1(22,16) + \lambda_2(19,18) + \lambda_3(10,99)$ where

$$\lambda_1 = \frac{153}{225}, \quad \lambda_2 = \frac{71}{225}, \quad \lambda_3 = \frac{1}{225}.$$

Example 2. Bivariate geometric distribution. Let again $r = 2$, and define the bivariate probability distribution as follows:

$$p_{ik} := (1-q)^2 q^{i+k}, \; i,k \geq 0. \tag{12}$$

Thus, each marginal distribution is geometric, with probabilities $(1-q)p^i$, $i = 0,1,\ldots$.

pLEP's (i,k)	Probability levels $F(i,k)$
(44,143)	0.9000016434
(45,99)	0.9001705286
(46,87)	0.9002818328
(47,80)	0.9003904620
(48,75)	0.9003710351
(49,71)	0.9000762339
(50,69)	0.9013348881
(51,66)	0.9006190340
(52,64)	0.9007330886
(53,62)	0.9003042365
(54,61)	0.9013589450
(55,60)	0.9021475573
(56,58)	0.9003777116
(57,57)	0.9005120454

Table 2: The set of 0.9-level efficient points and the corresponding probabilities, in case of the discrete distribution given by (12). The points obtained by tranposing the two co-ordinates are pLEP's, too.

If we choose $q = 0.95$, and $p = 0.9$, then there are 27 pLEP's. We present them, together with the corresponding probability levels (values of the probability distribution function), in Table 2. The set $\mathcal{E}$ is nonconvex, because the points $(55,60)$ and $(60,55)$ are interior points of the convex hull of $\mathcal{E}$. By symmetry it is enough to show it for one of these points. We can easily check that $(55, 60) = \lambda_1(54, 61) + \lambda_2(56, 58) + \lambda_3(143, 44)$, where

$$\lambda_1 = \frac{2720}{3961}, \quad \lambda_2 = \frac{1224}{3961}, \quad \lambda_3 = \frac{17}{3961}.$$

Example 3. Bivariate Poisson distribution. Let $r = 2$, and define

$$p_{ik} := \frac{\lambda^i}{i!}e^{-\lambda}\frac{\lambda^k}{k!}e^{-\lambda}, \; i, k \geq 0. \tag{13}$$

Each marginal distribution is a Poisson distribution, with parameter λ.

For the case of $\lambda = 100$, and $p = 0.01$ the set of pLEP's, and the corresponding probability distribution function values are tabulated in Table 3. The points $(84, 91)$, $(91, 84)$ are interior points of the convex hull of $\mathcal{E}$. In fact, the point $(84, 91)$ is in the interior of the triangle determined by the points $(83, 92)$, $(85, 89)$, $(133, 70)$. The assertion for $(91, 84)$ follows by symmetry.

pLEP's (i, k)	Probability levels $F(i, k)$
(70,133)	0.01000114075
(75,107)	0.01030997036
(79,102)	0.01055318148
(80,98)	0.01012056273
(81,96)	0.01071687362
(82,94)	0.01088975998
(83,92)	0.01059854598
(84,91)	0.01144575315
(85,89)	0.01035405859
(86,88)	0.01066081614
(87,87)	0.01076508548

Table 3: The set of 0.01-level efficient points and the corresponding probabilities, in case of the discrete distribution given by (13). The points obtained by tranposing the two co-ordinates are pLEP's, too.

In all three examples there exist continuous, and logconcave probability distribution functions such that at the nonnegative lattice points the discrete and the continunous distribution functions coincide. Note that if $F(z)$ is a logconcave probability distribution function, then it is also quasi-concave, and thus, the set $\{z \mid F(z) \geq p\}$ is convex for any $0 < p < 1$. For facts concerning logconcavity see Prékopa (1995).

In example 1 the continuous distribution function is:

$$G(z_1, z_2) = \frac{z_1}{z_1+1}\frac{z_2}{z_2+1}. \tag{14}$$

In example 2 the continuous distribution function is:

$$G(z_1, z_2) = (1 - e^{-0.0513z_1})(1 - e^{-0.0513z_2}). \tag{15}$$

Finally, in example 3 the continuous distribution function is (see Prékopa (1995)):

$$\begin{aligned} G(z_1, z_2) &= \int_\lambda^\infty \frac{x^{z_1}}{\Gamma(z_1+1)} e^{-x} dx \int_\lambda^\infty \frac{y^{z_2}}{\Gamma(z_2+1)} e^{-y} dy \\ &= \frac{\int_\lambda^\infty x^{z_1} e^{-x} dx}{\int_0^\infty x^{z_1} e^{-x} dx} \frac{\int_\lambda^\infty y^{z_2} e^{-y} dy}{\int_0^\infty y^{z_2} e^{-y} dy}, \quad z_1 \geq -1, \; z_2 \geq -1. \end{aligned} \tag{16}$$

The logconcavity of the function (14) is trivial. Note that the probability density function of $G(z) = \frac{z}{z+1}$ is not logconcave, but logconvex.

The function (15) is also trivially logconcave. The probability density function of the exponential distribution function $(1 - e^{-\lambda z})' = \lambda e^{-\lambda z}$ is both logconcave, and logconvex: it is loglinear.

For the logconcavity of the function (16) see Prékopa (1995). There is no information about the logconcavity of the probability density function

$$\begin{aligned} \frac{d}{dz} G(z) &= \frac{d}{dz} \frac{\int_\lambda^\infty x^z e^{-x} dx}{\int_0^\infty x^z e^{-x} dx} \\ &= G(z) \left(\frac{\int_\lambda^\infty x^z \ln x e^{-x} dx}{\int_\lambda^\infty x^z e^{-x} dx} - \frac{\int_0^\infty x^z \ln x e^{-x} dx}{\int_\lambda^\infty x^z e^{-x} dx} \right). \end{aligned}$$

We know, however, that the elements of the Poisson distribution: $p_i = \frac{\lambda^i}{i!} e^{-\lambda}, \; i = 0, 1, 2, ...$ form a logconcave sequence.

Thus, in all of the three examples the bivariate p-quantile, defined by,

$$\{(z_1, z_2) \mid G(z_1, z_2) = p\}$$

is a convex curve and

$$\{(z_1, z_2) \mid G(z_1, z_2) \geq p\}$$

is a convex set. This suggests that if there exists a continous and logconcave probability distribution function that coincides with the discrete probability distribution function at the lattice points, then the set of pLEP's may not be far away from convexity.

4. Cutting Plane Method for the Solution of Relaxations of Problems (4) and (8)

First we consider problem (4). A relaxation of it is obtained by replacing the constraints

$$\begin{aligned} \mathbf{Tx} \geq \textstyle\sum_{i=1}^{N} \mathbf{z}^{(i)}\mu_i \\ \textstyle\sum_{i=1}^{N} \mu_i = 1 \\ \mu_i \geq 0, i \in \{1, ..., N\} \end{aligned} \tag{17}$$

for the constraint (4d). If in (17) we introduce surplus variables to make the first constraint equality and use (17) in that form, then the relaxation of problem (4) is the following:

$$\text{Min } \mathbf{c}^T\mathbf{x} \tag{18a}$$

subject to

$$\mathbf{Ax} = \mathbf{b} \tag{18b}$$

$$\mathbf{Tx} - \mathbf{u} - \sum_{i=1}^{N} \mathbf{z}^{(i)}\mu_i = \mathbf{0} \tag{18c}$$

$$\sum_{i=1}^{N} \mu_i = 1 \tag{18d}$$

$$\mathbf{x} \geq \mathbf{0},\ \mathbf{u} \geq \mathbf{0},\ \mu \geq 0, \tag{18e}$$

where $\mu = (\mu_1, ..., \mu_N)^T$. The matrix of the problem is presented in Figure 2.

Clearly, we have the relation:

$$\begin{aligned} &\{\mathbf{x} \mid \mathbf{Ax} = b;\ \mathbf{x} \geq \mathbf{0};\ \exists i \in \{1, ..., N\}, \mathbf{Tx} \geq \mathbf{z}^i\} \\ &\subset \{(\mathbf{x} \mid (\mathbf{x}, \mathbf{u}, \mu) \text{ is feasible in (18a-e)}\} \end{aligned} \tag{19}$$

which shows that problem (18a-e) is in fact a relaxation of problem (4).

We enumerate the pLEP's and then solve problem (18a-d) by a cutting plane method. We omit the constraints (18c-d), and subsequently generate cuts towards the constraints (17).

$\mathbf{A}$	$\mathbf{0}$	$\mathbf{0}$
$\mathbf{T}$	$-\mathbf{I}$	$-\mathbf{z}^{(1)} \ldots - \mathbf{z}^{(N)}$
$\mathbf{0}$	$\mathbf{0}$	$1 \ \ldots \ 1$

Figure 2. The matrix of the equality constraints in problem (18a-d).

Let $\overline{\mathbf{x}}$ be a relative interior point of conv$\{\mathbf{z}^{(i)}, i = 1, ..., N\}$. We can take, e.g.,

$$\overline{\mathbf{z}} = \frac{1}{N} \sum_{i=1}^{N} \mathbf{z}^{(i)}.$$

Since the set of pLEP's can be concentrated on an affine manifold with dimension smaller than r, first we look at the system of linear equations (for the components of $\mathbf{w}$):

$$\mathbf{w}^T(\mathbf{z}^{(i)} - \overline{\mathbf{z}}) = 0, \quad i = 1, ..., N. \tag{20}$$

If $\mathbf{w}_1, ..., \mathbf{w}_k$ is a maximum number of linearly independent solutions of (20), then we append the contraints

$$\mathbf{w}_l^T(\mathbf{T}\mathbf{x} - \mathbf{u}) = 0, \quad l = 1, ..., k \tag{21}$$

to the constraint $\mathbf{A}\mathbf{x} = \mathbf{b}$, and keep them together throughout the procedure. For the sake of simplicity of the discussion, right at the outset we assume that the constraint $\mathbf{A}\mathbf{x} = \mathbf{b}$ already contains the constraints (21).

To create a cutting plane in the kth iteration, we formulate the auxiliary problem

$$\begin{array}{c} \text{Min}\, \mathbf{e}^T\mu = \alpha \\ \text{subject to} \\ \sum_{i=1}^{N}(\mathbf{z}^{(i)} - \overline{\mathbf{z}})\mu_i = \mathbf{T}\mathbf{x}^k - \mathbf{u}^k - \overline{\mathbf{z}} \\ \mu \geq \mathbf{0}, \end{array} \tag{22}$$

where $\mathbf{e} = (1, 1, ..., 1)^T$, $(\mathbf{x}^k, u^k)$ is the current optimal solution, and $\mu = (\mu_1, ..., \mu_N)$ is the decision vector. If problem (18a-d) has a feasible solution, then so does problem (22).

If $\alpha \leq 1$, then we terminate the procedure. The current optimal solution $(\mathbf{x}^k, u^k)$ is an optimal solution to problem (18a-d). If $\alpha > 1$, then we proceed in the following way.

Let $\mathbf{z}^{(i_1)} - \overline{\mathbf{z}}, ..., \mathbf{z}^{(i_{r-k})} - \overline{\mathbf{z}}$ be an optimal basis to problem (22). Then, find a $\mathbf{w} \neq \mathbf{0}$ such that

$$\begin{aligned} \mathbf{w}^T \mathbf{w}_i &= 0, \quad i = 1, ..., h \\ \mathbf{w}^T (\mathbf{z}^{(i_j)} - \mathbf{z}^{(i_1)}) &= 0, \quad j = 2, ..., r-h. \end{aligned} \tag{23}$$

The $r - 1$ equations (23) determine $\mathbf{w}$ up to a constant factor. Assume that $\mathbf{w}$ is determined in such a way that

$$\mathbf{w}^T (\mathbf{T}\mathbf{x}^k - \mathbf{u}^k - \overline{\mathbf{z}}) < 0. \tag{24}$$

Then, define $\mathbf{w}^{k+1} = \mathbf{w}$, and the cut:

$$(\mathbf{w}^{k+1})^T (\mathbf{T}\mathbf{x} - \mathbf{u} - \overline{\mathbf{z}}) \geq 0. \tag{25}$$

The algorithm that we designate by A1 can be summerized as follows.

Algorithm A1

STEP 1. Enumerate all pLEP's. Initialize $k \leftarrow 0$, and go to STEP 2.
STEP 2. Solve the following LP:

$$\begin{gathered} \text{Min } \mathbf{c}^T \mathbf{x} \\ \text{subject to} \\ \mathbf{A}\mathbf{x} = \mathbf{b} \\ (\mathbf{w}^i)^T (\mathbf{T}\mathbf{x} - \mathbf{u} - \overline{\mathbf{z}}) \geq 0, \quad i = 1, ..., k \\ \mathbf{x} \geq \mathbf{0}, \ \mathbf{u} \geq \mathbf{0}. \end{gathered} \tag{26}$$

If $k = 0$, then ignore the constraints involving the cuts. Let $(\mathbf{x}^k, \mathbf{u}^k)$ be an optimal solution to problem (26).
STEP 3. Solve the LP (22). If $\alpha \leq 1$, then STOP, the current $(\mathbf{x}^k, \mathbf{u}^k)$ is an optimal solution to problem (18a-d). If $\alpha > 1$, then go to STEP 4.
STEP 4. Create the cut (25), set $k \leftarrow k + 1$, and go to STEP 2.

If the set of pLEP's is a convex discrete set, then all cuts generate facets of the convex polytope in the affine manifold given by

$$\left\{ \sum_{i=1}^{h} (\mathbf{w}_i - \overline{\mathbf{z}}) \beta_i \mid \sum_{i=1}^{h} \beta_i = 1 \right\}.$$

Since the number of facets is finite, it follows that Algorithm A1 terminates in a finite number of steps, and provides us with an optimal solution.

The execution of the above cutting plane method can be done in a similar fashion as we do it in case of Gomory's method for the solution of the integer variable linear programming problem. This means that each time a new cut is created, a surplus variable is assigned to it (x_{n+k} in case of the kth cut) with nonnegativity and integrality restrictions and then, the problem is reoptimized by the use of the (lexicographic) dual method. In addition, once a variable x_{n+k} becomes basic at a later iteration, then we delete x_{n+k} together with its constraint from the problem.

The above remark also shows that Algorithm A1 can be combined with Gomory's algorithm to solve problem (18a-d) with the additional restriction that all components of $\mathbf{x}$ and $\mathbf{u}$ be integers. In this case we are dealing with cuts of the type (25) as well as Gomory cuts. Each time a Gomory cut is created, we reoptimize a problem of the type (18a-d).

Now we turn our attention to problem (8). The relaxation, after the introduction of the surplus vector $\mathbf{u}$, takes the form:

$$\text{Min}\left(\mathbf{c}^T\mathbf{x} \quad + \quad \sum_{i=1}^{r}\sum_{j=0}^{k_i+1} c_{ij}\lambda_{ij}\right) \tag{27a}$$

subject to

$$\mathbf{A}\mathbf{x} = \mathbf{b} \tag{27b}$$

$$\mathbf{T}_i\mathbf{x} - \sum_{j=0}^{k_i+1} z_{ij}\lambda_{ij} = \mathbf{0} \tag{27c}$$

$$\sum_{j=0}^{k_i+1} \lambda_{ij} = 1 \tag{27d}$$

$$\mathbf{T}\mathbf{x} - \mathbf{u} - \sum_{j=1}^{N} \mathbf{z}^{(j)}\mu_j = \mathbf{0} \tag{27e}$$

$$\sum_{j=1}^{N} \mu_j = 1 \tag{27f}$$

$$\mathbf{x} \geq \mathbf{0};\ \lambda_{ij} \geq 0,\ j = 0, ..., k_i + 1,\ i = 1, ..., r;\ \mathbf{u} \geq \mathbf{0};\ \mu \geq \mathbf{0}. \tag{27g}$$

The matrix of the equality constraints is illustrated in Figure 3.

A	**0**	**0**	**0**
T	$-z_{10} \ \cdots \ -z_{1,k_1+1}$ $\ddots$ $-z_{r0} \ \cdots \ -z_{r,k_r+1}$	**0**	**0**
0	$1 \ \cdots \ 1$ $\ddots$ $1 \ \cdots \ 1$	**0**	**0**
T	**0**	$-\mathbf{I}$	$\mathbf{z}^{(1)} \ldots \mathbf{z}^{(N)}$
0	**0**	**0**	$1 \ \ldots \ 1$

Figure 3. The matrix of the equality constraints in problem (27a-f).

Algorithm A1 solves problem (27a-f), too. If we compare Figures 2 and 3, we see that replacing the matrix

A	**0**
T	$-z_{10} \ \cdots \ -z_{1,k_1+1}$ $\cdot \ \cdot \ \cdot$ $-z_{r0} \ \cdots \ -z_{r,k_r+1}$
0	$1 \ \cdots \ 1$ $\cdot \ \cdot \ \cdot$ $1 \ \cdots \ 1$

taken from Figure 3, for the matrix **A** in Figure 2, we obtain the matrix of the equality constraints of problem (27a-g). Thus, a code which is suitable to solve problem (18a-e), also solves problem (27a-f) (Of course, the objective function coefficients and the right-hand side values need replacement as well). However, problem (27a-f) can be solved more efficiently if its special structure is exploited in full details. The dual type method of Prékopa (1990) and the improved dual type method of Findler, Prékopa and Fábián (1995) offer promising possibilities in this respect.

5. An Exact Solution of Problem (8)

The method is outlined in Prékopa (1990), and is based on the dual method presented in that paper. That dual method solves any problem of the type (8a-e), i.e., problem (8) without the restriction that $\mathbf{Tx}$ should dominate at least one pLEP.

Problem (8) is nonlinear. If, however, constraint (8g) is replaced by the single constraint that $\mathbf{Tx} \geq \mathbf{z}^{(i)}$ for a given i, then the problem becomes an LP. We may append the new constraint to the constraint $\mathbf{Ax} = \mathbf{b}$, but there is a more practical way to take it into account which decreases (rather than increases) the size of the problem: we simply ignore all elements in $\mathbf{z} \in \mathcal{Z}$ for which $\mathbf{z} \not\geq \mathbf{z}^{(i)}$. This means that we delete all terms from (8a), (8c-d) and all inequalities from (8e) which correspond to deleted elements of $\mathcal{Z}$. The obtained problem is of the type (8a-e) and can be solved by the dual method. This is because $\{\mathbf{z} \in \mathcal{Z} \mid \mathbf{z} \geq \mathbf{z}^{(i)}\}$ is a rectangular set, i.e., the Cartesian product of sets in $\mathbf{R}$, as is $\mathcal{Z}$ itself.

If we subdivide the set

$$\bigcup_{i=1}^{N} \left\{\mathbf{z} \mid \mathbf{z} \geq \mathbf{z}^{(i)}\right\} \tag{28}$$

into a finite number of disjoint rectangular sets, then, in view of the above remark, we can solve problem (8a-f) by the application of the dual method as many times as the number of the subdividing sets. In fact, each application of the dual method produces an optimal solution and an optimum value. That optimal solution which corresponds to the smallest optimum value, is the optimal solution of problem (8a-f). Note that a nondisjoint subdivision of the set (28) is inefficient because in that case $\mathbf{Tx}$ is allowed to visit parts of the set (28) more than once.

In view of the above remarks, we mainly have to be concerned with the subdivision of the set (28) and below we present a method to do it.

Method to subdivide the set (28) into pairwise disjoint rectangular sets

In connection with a set $C \subset \mathbf{R}^r$ we will use the notation:

$$\mathbf{z} + C = C + \mathbf{z} = \{\mathbf{z} + \mathbf{y} \mid y \in C\}.$$

Let $H_1, ..., H_M$ be arbitrary subsets of $\mathbf{R}^r$. Then, the following formula holds true:

$$H_1 \cup ... \cup H_M = H_1 \cup \left(\bigcup_{i=2}^{M} \left(H_i \cap \overline{H}_1 \cap ... \cap \overline{H}_{i-1} \right) \right), \tag{29}$$

where H_1, $H_i \cap \overline{H}_1 \cap ... \cap \overline{H}_{i-1}$, $i = 2, ..., M$ are pairwise disjoint sets.

Let us define the sets

$$\begin{aligned} C_0(\mathbf{z}) &= \{\mathbf{y} \in \mathbf{R}^r \mid \mathbf{y} \geq \mathbf{z}\} \\ C_i(\mathbf{z}) &= \{\mathbf{y} \in \mathbf{R}^r \mid y_i < z_i\}, \quad i = 1, ..., r. \end{aligned} \tag{30}$$

If we apply formula (29) for the case $M = r$, $H_i = C_i(\mathbf{z})$, $i = 1, ..., r$, then we obtain

$$\bigcup_{i=1}^{r} C_i(\mathbf{z}) = C_1(\mathbf{z}) \cup \left(\bigcup_{i=2}^{r} \left(C_i(\mathbf{z}) \cap \overline{C_1(\mathbf{z})} \cap ... \cap \overline{C_{i-1}(\mathbf{z})} \right) \right). \tag{31}$$

The sets

$$\begin{aligned} D_1(\mathbf{z}) &= C_1(\mathbf{z}) \\ D_i(\mathbf{z}) &= C_i(\mathbf{z}) \cap \overline{C_1(\mathbf{z})} \cap ... \cap \overline{C_{i-1}(\mathbf{z})} \end{aligned} \tag{32}$$

are paiwise disjoint, and rectangular, i.e., each of them is the Cartesian product of onedimensional intervals. We remark that

$$\bigcup_{i=1}^{r} C_i(\mathbf{z}) = \bigcup_{i=1}^{r} D_i(\mathbf{z}) = \overline{C_0(\mathbf{z})}. \tag{33}$$

Now, we use formula (29) for $M = N$, and $H_i = C_0(\mathbf{x}^{(i)})$, $i = 1, ..., N$. We obtain

$$\bigcup_{i=1}^{N} C_0(\mathbf{z}^{(i)}) = C_0(\mathbf{z}^{(1)}) \cup \left(\bigcup_{i=2}^{N} \left(C_0(\mathbf{z}^{(i)}) \cap \overline{C_0(\mathbf{z}^{(1)})} \cap ... \cap \overline{C_0(\mathbf{z}^{(i-1)})} \right) \right). \tag{34}$$

Taking into account (31)-(34), we derive our final formula:

$$\bigcup_{i=1}^{N} C_0(\mathbf{z}^{(i)}) = C_0(\mathbf{z}^{(1)}) \cup \left(\bigcup_{i=2}^{N} \left(C_0(\mathbf{z}^{(i)}) \cap \bigcap_{j=1}^{i-1} \left(\bigcup_{l=1}^{r} D_l(\mathbf{z}^{(j)}) \right) \right) \right). \tag{35}$$

Formula (35) suggests the following algorithm to subdivide the set (28).

Algorithm A2

STEP 0. Initialize $i \leftarrow 1$, $k \leftarrow 1$, $F_1^{(1)} = C_0(\mathbf{z}^{(1)})$.
STEP 1. If the current sets are $F_1^i, ..., F_k^i$, then form the sets $F_j^{(i)} \cap C_h(\mathbf{z}^{(i)})$, $j = 1, ..., k$; $h = 0, ..., r$. Let l be the number of nonempty sets among them.
STEP 2. Set $i \leftarrow i+1$, $k \leftarrow l$. If $i = N$, then STOP, $F_1 = F_1^{(N)}, ..., F_k = F_k^{(N)}$ are the subdividing rectangular sets. Otherwise go to STEP 1.

Having created the above subdivision, problems (4) and (8) can be solved in such a way that we replace the constraint

$$\mathbf{Tx} \geq \mathbf{z}^{(i)} \text{ for at least one } i \in \{1, ..., N\}$$

by the constraint

$$\mathbf{Tx} \in F_h \text{ for at least one } h \in \{1, ..., k\}. \tag{36}$$

For each h we solve the LP (4a-c), $\mathbf{Tx} \in F_h$, or (8a-e), $\mathbf{Tx} \in F_h$ and take that optimal solution which produces the smallest objective function value. This will be the optimal solution to problem (4) or (8).

If we solve problem (8a-e), $\mathbf{Tx} \in F_h$, then we ignore those z_{ij}, λ_{ij} in the problem, for which $z_{ij} \notin F_h$. In other words, the set $\mathcal{Z}$ is now defined as the intersection of the $\mathcal{Z}$, defined in the Introduction, with the set F_h. For each $h = 1, ..., k$ the problem can be solved by the dual type method of Prékopa (1990), or the improved dual type method of Fiedler, Prékopa and Fábián (1995).

6. Numerical Example

The following numerical example is based on the paper by Prékopa and Boros (1991) about transportation network reliability calculation. The problem presented in this section is a transportation network capacity design problem, where the network reliability (i.e., the probability of the existence of a feasible flow=1 - loss of load Probability) is prescribed to be at least $p = 0.9975$.

Let (N, A) designate the network, where N is the set of nodes and A is the set of arcs. Our capacity design problem is of the following type:

$$\begin{gathered}\text{Min}\left\{\textstyle\sum_{i\in N} c_i x_i + \sum_{(i,j)\in A} c_{ij} y_{ij}\right\} \\ \text{subject to} \\ X_i^{(l)} \leq x_i \leq X_i^{(u)} \\ Y_{ij}^{(l)} \leq y_{ij} \leq Y_{ij}^{(u)} \\ P(\text{a feasible flow exists}) \geq p,\end{gathered} \tag{37}$$

where x_i, y_{ij} are node and arc capacities, respectively, for which lower and upper bounds are prescribed. The c_i, c_{ij} constants represent capacity unit prices.

We take Example 3 of the paper by Prékopa and Boros as a basis to formulate the design problem.

The random event that a feasible flow exists can be formulated in terms of "cuts" which are presented in Table VII of that paper. Each cut is an inequality involving random demand values. The first, second and seventh cuts (cut 15, cut 11,12 and cut 1,2) represent inequalities that hold if and only if the random variables involved assume their largest possible values. Recall that the system reliability is prescribed to be at least $p = 0.9975$. This implies that these three cut inequalities can be separated from the others and can be handled as determinitic constraints in the problem. The other ten cut inequalities should hold by probability p.

We enumerated the set of all pLEP's and then formulated and solved the problem of the type (18a-e).

The original problem

Minimize

$$x_1 + x_2 + x_3 + x_4 + x_5 + x_6 + x_7 + x_8 + x_9 + x_{10} + x_{11} + x_{12} + \\ + x_{13} + x_{14} + x_{15} + y_{1,2} + y_{2,3} + y_{3,4} + y_{3,5} + y_{3,13} + y_{4,5} + y_{5,6} + \\ + y_{5,8} + y_{5,13} + y_{5,14} + y_{6,7} + y_{7,8} + y_{7,10} + y_{8,9} + y_{8,11} + y_{9,10} + \\ + y_{9,11} + y_{9,15} + y_{10,11} + y_{11,12} + y_{13,14} + y_{14,15}$$

Subject to

cut 15: $x_{15} + y_{9,15} + y_{14,15} \geq 4400$

cut 11,12: $x_{11} + x_{12} + y_{8,11} + y_{9,11} + y_{10,11} \geq 5395$

cut 1,2: $x_1 + x_2 + y_{2,3} \geq 3945$

Bounds

$0 \leq x_1 \leq 3125$	$0 \leq x_2 \leq 3030$	$0 \leq x_3 \leq 3020$
$0 \leq x_4 \leq 3200$	$0 \leq x_5 \leq 3770$	$0 \leq x_6 \leq 2500$
$0 \leq x_7 \leq 3800$	$0 \leq x_8 \leq 3900$	$0 \leq x_9 \leq 3415$
$0 \leq x_{10} \leq 1500$	$0 \leq x_{11} \leq 2880$	$0 \leq x_{12} \leq 1525$
$0 \leq x_{13} \leq 2800$	$0 \leq x_{14} \leq 3500$	$0 \leq x_{15} \leq 2500$
$2000 \leq y_{1,2} \leq 3000$	$1500 \leq y_{5,6} \leq 2250$	$1500 \leq y_{2,3} \leq 2250$
$1800 \leq y_{3,4} \leq 2700$	$1800 \leq y_{3,5} \leq 2700$	$1800 \leq y_{3,13} \leq 2700$
$1800 \leq y_{4,5} \leq 2700$	$2000 \leq y_{5,8} \leq 3000$	$1800 \leq y_{5,13} \leq 2700$
$1450 \leq y_{5,14} \leq 2175$	$2000 \leq y_{6,7} \leq 3000$	$2000 \leq y_{7,8} \leq 3000$
$1200 \leq y_{7,10} \leq 1800$	$3000 \leq y_{8,9} \leq 4500$	$1500 \leq y_{8,11} \leq 2250$
$1225 \leq y_{9,10} \leq 1835$	$1500 \leq y_{9,11} \leq 2250$	$1500 \leq y_{9,15} \leq 2250$
$2450 \leq y_{10,11} \leq 3675$	$1675 \leq y_{11,12} \leq 2500$	$1200 \leq y_{13,14} \leq 1800$
$2000 \leq y_{14,15} \leq 3000$		

The 10×37 matrix T

```
0 0 0 0 0 0 0 0 0 1 1 1 0 0 0 0 0 0 0 0 0 0 0 0 0 0 0 1 0 1 1 1 0 0 0 0 0
0 0 0 0 0 0 1 1 1 1 1 1 0 0 1 0 0 0 0 0 0 0 1 0 0 1 0 0 0 0 0 0 0 0 0 0 1
0 0 1 1 1 0 0 0 0 0 0 0 1 1 1 0 1 0 0 0 0 1 1 0 0 0 0 0 0 0 0 0 1 0 0 0 0
0 0 1 1 1 1 1 1 1 1 1 1 1 1 1 0 1 0 0 0 0 0 0 0 0 0 0 0 0 0 0 0 0 0 0 0 0
1 1 1 0 0 0 0 0 0 0 0 0 1 1 1 0 0 1 1 0 0 0 0 1 1 0 0 0 0 0 0 0 1 0 0 0 0
1 1 1 1 1 0 0 0 0 0 0 0 1 0 0 0 0 0 0 0 0 1 1 0 1 0 0 0 0 0 0 0 0 0 0 1 0
1 1 1 1 1 0 0 0 0 0 0 0 1 1 0 0 0 0 0 0 0 1 1 0 0 0 0 0 0 0 0 0 0 0 0 0 1
1 1 1 1 1 0 0 0 0 0 0 0 1 1 1 0 0 0 0 0 0 1 1 0 0 0 0 0 0 0 0 0 1 0 0 0 0
1 1 1 1 1 1 1 1 1 1 1 1 1 0 0 0 0 0 0 0 0 0 0 1 0 0 0 0 0 0 0 0 1 0 0 1 0
1 1 1 1 1 1 1 1 1 1 1 1 1 1 1 0 0 0 0 0 0 0 0 0 0 0 0 0 0 0 0 0 0 0 0 0 0
```

The detailed description of the problem of type (18a-e) can be found in Prékopa et al. (1994).

The optimal solution of the problem of the type (18a-e)
Solution value = 78273.333333
The dimension of **x** is 126. **x** =

1683.33	3030.00	3020.00	3200.00	3770.00	2500.00	3800.00
450.00	3415.00	1500.00	2880.00	1525.00	2800.00	3500.00
2500.00	2000.00	1500.00	1800.00	1800.00	1800.00	1800.00
1500.00	2000.00	1800.00	1450.00	2000.00	2000.00	1200.00
3000.00	1500.00	1225.00	1500.00	1500.00	2450.00	1675.00
1200.00	2000.00	0.00	0.00	0.00	0.00	0.00
0.00	0.00	0.00	0.00	0.00	0.00	0.00
0.00	0.00	0.00	0.00	0.00	0.00	0.00
0.00	0.00	0.00	0.00	0.00	0.00	0.00
0.14	0.00	0.00	0.00	0.00	0.00	0.00
0.35	0.00	0.00	0.00	0.00	0.00	0.00
0.00	0.00	0.00	0.00	0.00	0.00	0.30
0.00	0.00	0.00	0.00	0.00	0.00	0.00
0.00	0.00	0.00	0.00	0.00	0.15	0.00
0.00	0.00	0.00	0.00	0.00	0.00	0.00
0.00	0.00	0.00	0.00	0.00	0.00	0.00
0.00	0.00	0.00	0.00	0.00	0.00	0.00
0.00	0.06	0.00	0.00	0.00	0.00	0.00

Tx is NOT DOMINATED by any pLEP. The dimension of **Tx** is 10.
Tx = (11330.00, 22070.00, 25290.00, 36360.00, 24883.33, 23653.33, 26503.33, 28503.33, 37723.33, 39573.33)

Acknowledgement

Research supported by the Air Force, grant numbers: AFORS-89-0512B, F49620-93-1-0041.

References

1. Fiedler, O., A. Prékopa, and Cs. Fábián (1995). "On a Dual Method for a Specially Structured Linear Programming Problem", *RUTCOR Research Report*, 25–95.
2. Prékopa, A. (1995). "Stochastic Programming", *Kluwer Scientific Publishers*, Boston.
3. Prékopa, A. (1990). "Dual Method for the Solution of a One-Stage Stochastic Programming Problem with Random RHS Obeying a Discrete Probability Distribution", *ZOR-Methods and Models of Operations Research*, 34, 441–461.
4. Prékopa, A., and E. Boros (1991). "On the Existence of a Feasible Flow in a Stochastic Transportation Network", *Operations Research* 39, 119–129.
5. Prékopa, A., and W. Li (1995). "Solution of and Bounding in a Linearly Constrained Optimization Problem with Convex, Polyhedral Objective Function", *Mathematical Programming* 70, 1–16.
6. Prékopa A., B. Vizvári, T. Badics (1994) "Programming Under Probabilistic Constraint with Discrete Random Variables", *RUTCOR, Rutgers University, Research Report*, RRR 10–96.
7. Rockafellar, R. T. (1972). "Convex Analysis", Princeton University Press, Princeton, N.J.
8. Sen, S. (1992). "Relaxations for the Probabilistically Constrained Programs with Discrete Random Variables", *Operations Research Letters* 11, 81–86.
9. Vizvári, B. (1987). "Beitræge zum Frobenius Problem", *Dr. Sc. Nat. Dissertation*, Technische Hochschule "Carl Schorlemmer", Leuna-Merseburg, Germany.

Variable metric methods along geodetics

T. Rapcsák
Laboratory of Operations Research and Decision Systems,
Computer and Automation Institute, Hungarian Academy of Sciences,
H-1518 Budapest, P.O. Box 63 Hungary

Abstract. In the paper, the behaviour of variable metric methods along geodesics is analyzed. First, a general framework is given in the case of Riemannian submanifolds in R^n, then two general convergence theorems for a wide class of nonlinear optimization methods are proved to find a stationary point or a local optimum point of a smooth function defined on a compact set of a Riemannian manifold and the rate of a convergence is studied. These methods and theorems should be extended in such a way that penalty methods be generalized in the case of inequality constraints defined on Riemannian manifolds.

Key words: Variable metric algorithms, Riemannian manifolds, convergence, rate of convergence, penalty methods

1. Introduction

This paper is devoted to the analysis of variable metric methods along geodesics. These methods are iterative, meaning that the algorithms generate a series of points, each point calculated on the basis of the points preceding it, and they are descent which means that each new point is generated by the algorithms and that the corresponding value of some function, evaluated at the most recent point, decreases in value. The theory of iterative algorithms can be divided into three parts. The first is concerned with the creation of the algorithms based on the structure of problems and the efficiency of computers. The second is the verification whether a given algorithm generates a sequence converging to a solution. This aspect is referred to as global convergence analysis, since the question is whether an algorithm starting from an arbitrary initial point, be it far from the solutions, converges to a solution. Thus, an algorithm is said to be globally convergent if for arbitrary starting points, the algorithm is guaranteed to generate a sequence of points converging to a solution. Many of the most important algorithms for solving nonlinear optimization problems are not globally convergent in the purest form, and thus occasionally generate sequences that either do not converge at all, or converge to points that are not solutions. The third is referred to as local convergence analysis and is concerned with the rate at which the generated sequence of points converges to a solution.

Consider a smooth optimization problem in the form of

F. Giannessi et al. (eds.), New Trends in Mathematical Programming, 257–275.

$$\min f(\mathbf{x})$$
$$\mathbf{x} \in A \subseteq M \subseteq \mathcal{M}, \tag{1.1}$$

where f is a C^2 function, M is a C^2 k-dimensional Riemannian manifold and $\mathcal{M}$ is the n-dimensional differentiable manifold R^n endowed with a Riemannian metric $G_1(\mathbf{x})$, $\mathbf{x} \in R^n$, which induces the metric $G(\mathbf{x})$ of M.

In the case of classical nonlinear optimization problems with full rank Jacobian matrices of equality constraints at the feasible points, the sets of problem (1.1) can be given by

$$\begin{aligned} M &= \{\mathbf{x} \in R^n | \quad h_j(\mathbf{x}) = 0, \quad j = 1, \dots, n-k\}, \\ A &= \{\mathbf{x} \in M | \quad g_i(\mathbf{x}) \geq 0, \quad i = 1, \dots, m\}, \\ \mathcal{M} &= R^n , \end{aligned} \tag{1.2}$$

where h_j, g_i, $j = 1, \dots, n-k$, $i = 1, \dots, m$, are twice continuously differentiable functions and $G(\mathbf{x})$, $\mathbf{x} \in M$, is induced by the $n \times n$ identity matrix I at every point.

This paper contains two general convergence theorems for a wide class of nonlinear optimization methods to find a stationary point or a local optimum point of problem (1.1), then the rate of convergence is studied. The first convergence theorem was published in Rapcsák and Thang (1995), but without a detailed proof. These methods and theorems should be extended in such a way that penalty methods be generalized in the case of inequality constraints defined on Riemannian manifolds.

2. General framework for variable metric methods on Riemannian submanifolds in R^n

In this part, a general framework for variable metric methods along geodesics will be studied, developing further the results of Luenberger (1972) and Gabay (1982). This general framework related to the nonlinear coordinate representations of nonlinear optimization algorithms was elaborated in explicit form based on the immersion of the manifold M into $\mathcal{M}$ (Rapcsák and Thang, 1995). The main contribution was that, instead of one Riemannian manifold, two were considered, taking the structure of smooth optimization problems into account. So, explicit expressions can be formulated for basic optimization algorithms. The question seems to be important because the performance of nonlinear optimization methods may strongly depend on coordinate representations (Gabay, 1982). The following definition is standard in differential geometry (e.g., Kobayashi and Nomizu II., 1969):

DEFINITION 2.1 Let G be the Riemannian metric on a differentiable manifold M, and $f : M \to R$ a differentiable function. Then, the gradient of the function f on M at a point m is defined to be the tangent vector $\nabla^G f(m)^T \in TM_m$ such that

$$< \nabla^G f(m), \mathbf{v} >= Df(m)\mathbf{v}, \qquad \text{for all} \qquad \mathbf{v} \in TM_m \,, \tag{2.1}$$

where $< \nabla^G f(m), \mathbf{v} >$ is the scalar product with respect to G at m and $Df(m)\mathbf{v}$ is the derivative of f at m in the direction of $\mathbf{v}$.

Remark 2.1 If $M = R^n$, then $\nabla^G f^T = G^{-1}\nabla f^T$.

Now, it will be proved that, in the case of optimization problem (1.1) originated from classical nonlinear optimization problems, the transpose of the gradient $\nabla^G f^T$ is just the orthogonal projection of the vector $G_1^{-1}\nabla f^T$ to the subspace TM with respect to the Riemannian metric G_1 at every point. This statement can be found in Bayer and Lagarias (1989) with a different proof, and in Rapcsák and Thang (1995).

THEOREM 2.1 *Let G_1 and G be the Riemannian metrics on the differentiable manifolds $\mathcal{M}$ and M of problem (1.1), respectively, and $f : M \to R$ a differentiable function. If the differentiable manifold M can be given by equality constraints (1.2) with full rank Jacobian matrices at every feasible point, then*

$$\nabla^G f^T = (I - G_1^{-1} J\mathbf{h}^T (J\mathbf{h} G_1^{-1} J\mathbf{h}^T)^{-1} J\mathbf{h}) G_1^{-1} \nabla f^T \,, \tag{2.2}$$

where $\nabla f = (\frac{\partial f}{\partial x_1}, \dots, \frac{\partial f}{\partial x_n})$, $J\mathbf{h}$ is the Jacobian matrix of the mapping $\mathbf{h} : R^n \to R^{n-k}$ and I, G^{-1} and $J\mathbf{h}$ are $n \times n$ and $(n-k) \times n$ matrices at every point, respectively.

Proof. In our case, $TM = \mathcal{N}(J\mathbf{h})$, where $\mathcal{N}(J\mathbf{h})$ denotes the null space of the Jacobian matrix $J\mathbf{h}$. By Definition 2.1,

$$(\nabla^G f G_1 - \nabla f)\mathbf{v} = 0, \qquad \text{for all} \qquad \mathbf{v} \in \mathcal{N}(J\mathbf{h}) \,, \tag{2.3}$$

which means that $(\nabla^G f G_1 - \nabla f)^T \in \mathcal{R}(J\mathbf{h}^T)$, where $\mathcal{R}(J\mathbf{h}^T)$ denotes the range of the transpose of the Jacobian matrix $J\mathbf{h}$. So, there exists $\mathbf{u} \in R^k$ such that

$$(\nabla^G f G_1 - \nabla f)^T = J\mathbf{h}^T \mathbf{u} \,,$$

i.e.,

$$\nabla^G f^T - G_1^{-1} J\mathbf{h}^T \mathbf{u} = G_1^{-1} \nabla f^T . \tag{2.4}$$

If we choose

$$J\mathbf{h}^- = G_1^{-1} J\mathbf{h}^T (J\mathbf{h} G_1^{-1} J\mathbf{h}^T)^{-1} , \tag{2.5}$$

then by Theorem 3.2 (Rapcsák and Thang, 1995), we have that

$$\mathcal{R}(J\mathbf{h}^-) = \mathcal{R}(G_1^{-1} J\mathbf{h}^T) \quad \text{and} \quad R^n = \mathcal{R}(J\mathbf{h}^-) \oplus \mathcal{N}(J\mathbf{h}) . \tag{2.6}$$

The subspaces of the above decomposition are orthogonal with respect to the Riemannian metric G_1. However,

$$-G_1^{-1} J\mathbf{h}^T \mathbf{u} \in \mathcal{R}(J\mathbf{h}^-) \quad \text{and} \quad \nabla^G f^T \in \mathcal{N}(J\mathbf{h}) ,$$

thus, by Lemma 3.1 (Rapcsák and Thang, 1995),

$$\nabla^G f^T = (I - G_1^{-1} J\mathbf{h}^T (J\mathbf{h} G_1^{-1} J\mathbf{h}^T)^{-1} J\mathbf{h}) G_1^{-1} \nabla f^T, \tag{2.7}$$

which is the statement. □

Now, a general framework for variable metric methods along geodesics can be suggested to find a stationary point or a local optimum of problem (1.1) in the case of $A = M$. These algorithms represent a generalization of unconstrained optimization methods and constrained optimization methods under equalities and proceed as follows.

Starting from an initial feasible solution $\mathbf{x}_0 \in M \subseteq \mathcal{M}$, let $\mathbf{x}_k \in M \subseteq \mathcal{M}$ be the feasible solution of the kth iteration step, D_k an $n \times n$ symmetric matrix defining a linear map $D_k : TM_{\mathbf{x}_k} \to TM_{\mathbf{x}_k}$, and a positive definite quadratic form on $TM_{\mathbf{x}_k}$ with a uniform lower bound at every iteration point (i.e., $\mathbf{v}_k D_k \mathbf{v}_k \geq \varepsilon |\mathbf{v}_k|^2$, $\mathbf{v}_k \in TM_{\mathbf{x}_k}$, $\varepsilon > 0$ for all k), G_1 the Riemannian metric of manifold $\mathcal{M}$, G the induced Riemannian metric of M; furthermore, let D_k and the Riemannian metric G_1 commute on $TM_{\mathbf{x}_k}$ at every iteration point.

1. Compute the direction $\mathbf{p}_k \in R^n$ as follows:

$$\mathbf{p}_k = -D_k^2 \nabla^G f(\mathbf{x}_k)^T, \qquad \nabla^G f(\mathbf{x}_k)^T \in R^n . \tag{2.8}$$

2. Let

$$\mathbf{x}_{k+1} = \gamma_{\mathbf{x}_k}(t_k, \mathbf{p}_k) , \tag{2.9}$$

where $\gamma_{\mathbf{x}_k}(t_k, \mathbf{p}_k)$ means the arc of the geodesic starting from $\mathbf{x}_k$ with tangent $\mathbf{p}_k$ and the stepsize t_k is determined by an exact geodesic search to find the first local minimum along the above geodesic, i.e.,

$$t_k = \arg\min\{ f(\gamma_{\mathbf{x}_k}(t, \mathbf{p}_k)) | \; t \in R_+ \} \tag{2.10}$$

or chosen according to the Armijo principle (e.g., Ortega and Rheinboldt, 1970), i.e., given $\alpha \in (0, \frac{1}{2})$, let $t_k = 2^{-l_k}$, with l_k the smallest integer such that

$$f(\gamma_{\mathbf{x}_k}(2^{-l_k}, \mathbf{p}_k)) \leq f(\mathbf{x}_k) - \alpha 2^{-l_k} |\nabla^G f(\mathbf{x}_k) G_1(\mathbf{x}_k) D_k^2 \nabla^G f(\mathbf{x}_k)^T| \, . \tag{2.11}$$

This general framework contains several well-known nonlinear optimization algorithms by choosing $D_k = I$ for all k.

In the case of

$$J\mathbf{h}(\mathbf{x}) = 0, \quad \text{and} \quad G_1(\mathbf{x}) = I, \qquad \forall \mathbf{x} \in R^n \, ,$$

the steepest descent methods are obtained.

In the case of

$$J\mathbf{h}(\mathbf{x}) = 0 \quad \text{and} \quad G_1(\mathbf{x}) = Hf(\mathbf{x}), \qquad \forall \mathbf{x} \in R^n \, ,$$

where Hf is the Hessian matrix of the objective function, Newton-type methods are obtained.

In the case of

$$J\mathbf{h}(\mathbf{x}) = A \, ,$$

where A is a $k \times n$ full rank matrix, and

$$G_1(\mathbf{x}) = \begin{pmatrix} \frac{1}{x_1^2} & & \\ & \ddots & \\ & & \frac{1}{x_n^2} \end{pmatrix}, \qquad \forall \mathbf{x} \in R_+^n \, ,$$

where R_+^n denotes the positive orthant, the affine scaling vector field for the linear optimization of a special form is provided (Karmarkar, 1990). This is one of the main research directions related to interior point methods, on which the affine scaling algorithms are based.

In the case of

$$G_1(\mathbf{x}) = I, \qquad \forall \mathbf{x} \in R^n \, ,$$

gradient projection methods along geodesics are obtained (Luenberger, 1972; Neto and Oliveira, 1995; Ferreira and Oliveira, 1995).

In the case of

$$G_1(\mathbf{x}) = Hf(\mathbf{x}), \qquad \forall \mathbf{x} \in R^n \, ,$$

where Hf is the Hessian matrix of the objective function, Newton-type methods along geodesics are obtained (Gabay, 1982).

In the case of

$$J\mathbf{h}(\mathbf{x}) = A \, ,$$

where A is a $k \times n$ full rank matrix, and

$$G_1(\mathbf{x}) = Hf(\mathbf{x}), \qquad \forall \mathbf{x} \in R^n_+ ,$$

where

$$f(\mathbf{x}) = \mathbf{c}^T\mathbf{x} - \mu \sum_{i=1}^{n} \log(x_i) , \qquad \forall \mathbf{x} \in R^n_+ ,$$

is the logarithmic barrier function of the standard linear optimization problem, the directions of the projective vector field and method can be obtained by a projected Newton barrier method with a special choice of the barrier parameter (Gill, Murray, Saunders, Tomlin and Wright, 1986).

In this terminology, the reduced gradient means a nonlinear coordinate representation of the projected gradient on the tangent space of M with respect to the Euclidean metric.

It is obvious that quasi-Newton, SQP and conjugate gradient methods can be involved as well by choosing D_k^2 as the matrix of updating formulas at the kth iteration point or it is possible to extend the matrices of updating formulas to a Riemannian metric.

3. Convergence of variable metric methods along geodesics

Variable metric algorithms along geodesics in forms (2.8), (2.9) and (2.10), where $D_k^2 = I$, were first introduced and analyzed by Luenberger (1972), who explicitly used Riemannian metrics induced by the Euclidean one. In his paper, Luenberger established the global convergence of the algorithm to a critical point of f on M and estimated the speed of convergence in a neighbourhood of a critical point which is a strict local minimum. As has been mentioned, the projected gradient methods along geodesics was deduced. Lichnewsky (1979) obtained similar results and studied one of its conjugate gradient versions as well. A general framework given by Gabay (1982) makes the generalization of the global convergence theorem possible for methods in R^n (e.g., Polak, 1971; Ortega and Rheinboldt, 1970). Udriste (1994) dealt with the convergence of descent and Newton methods along geodesics on finite-dimensional complete Riemannian manifolds. The convergence of the variants of Luenberger's method was investigated by Neto and Oliveira (1995) and Ferreira and Oliveira (1995).

Riemannian metrics were used in an implicit way, e.g., in quadratic optimization by Fletcher (1971) and in differential equation approaches by Tanabe (1980) and Yamashita (1980). Recently, Riemannian metrics have been used in algorithms elaborated for linear and convex optimization. Gonzaga (1990) used the affine metric and a potential function leading to polynomial algorithms with an $O(n^2L)$ or $O(nL)$ bound for linear optimization. The Riemannian metric originating from the projective transformation, first proposed by Karmarkar (1990), was used for linear optimization as well. Iri and Imai (1986) introduced a special metric based on the scaled Hessian matrix of the multiplicative potential function to obtain an $O(nL)$ method for linear optimization. Rapcsák and Thang (1996) studied a class of Riemannian metrics containing, among other things, the affine, projective and Iri-Imai metrics, each element of which generates a polynomial interior point method for linear optimization. Riemannian metrics based on the Hessian matrix of the logarithmic barrier function were investigated by den Hertog, Roos and Terlaky (1992), Jarre (1992), Mehrotra and Sun (1990), Nesterov and Nemirovsky (1989), providing $O(\sqrt{n}L)$ algorithms for linear, quadratic and a class of convex optimization problems.

Now, two general convergence theorems will be proved for variable metric methods along geodesics, developing further Gabay's results (1982). The first theorem was published in Rapcsák and Thang (1995), but the proof did not contain all the details. First, we recall the Kirszbraun theorem (e.g., Schwartz, 1969).

THEOREM 3.1 *(Kirszbraun) Let H be a Hilbert space, A any subset of H, and $\Phi : A \to H$. Suppose that*

$$|\Phi(x_1) - \Phi(x_2)| < L|x_1 - x_2|, \qquad \forall x_1, x_2 \in A\,,$$

where L is a constant. Then, Φ can be extended to all of H in such a way that the extension satisfies the same Lipschitz condition.

We need to introduce the following notation. Let W_k denote the connected component containing the $\mathbf{x}_k$ of the level set

$$\{\mathbf{x} \in M \subseteq \mathcal{M} |\ f(\mathbf{x}) \le f(\mathbf{x}_k)\}\,.$$

THEOREM 3.2 *If f is continuously differentiable, W_0 is a compact set in a complete Riemannian submanifold $M \subseteq \mathcal{M}$, the sequence $\{\mathbf{x}_k\}$ is generated by (2.8), (2.9) and (2.10), and if a mapping is defined by $D_k \nabla^G f(\mathbf{x}_k)^T \in TM_{\mathbf{x}_k}$, for all k at the iteration points and satisfies there the Lipschitz condition, D_k are positive definite on $TM_{\mathbf{x}_k}$ with a uniform lower bound for all k, and D_k and G_1 commute on $TM_{\mathbf{x}_k}$ at every iteration point, then the sequence $\{\mathbf{x}_k\}$ is either finite terminating at a critical point, or infinite, and every accumulation point is critical. If the critical values of f are distinct, the whole sequence $\{\mathbf{x}_k\}$ converges to a critical point.*

Proof. If $\mathbf{x}_k$ is a critical point, then $\nabla^G f(\mathbf{x}_k) = \mathbf{0}$, and the algorithm does not generate new iteration points, so introducing a stopping test, the procedure can be terminated at iteration k.

Assume now that $\mathbf{x}_k$ is not a critical point. Hence, $\mathbf{p}_k$ is a descent direction because

$$\begin{aligned} \nabla^G f(\mathbf{x}_k)G_1(\mathbf{x}_k)\mathbf{p}_k = -\nabla^G f(\mathbf{x}_k)G_1(\mathbf{x}_k)D_k^2\nabla^G f(\mathbf{x}_k)^T = \\ -\nabla^G f(\mathbf{x}_k)D_kG_1(\mathbf{x}_k)D_k\nabla^G f(\mathbf{x}_k)^T < 0\,. \end{aligned} \tag{3.1}$$

Denote the arc of geodesic starting from $\mathbf{x}_k$ with tangent $\mathbf{p}_k$ by $\gamma_{\mathbf{x}_k}(t, \mathbf{p}_k)$. The compactness of W_0 implies the compactness of all the W_k because the sequence $\{f(\mathbf{x}_k)\}$ is nonincreasing. By the Hopf-Rinow theorem, the manifold M is geodesically complete, thus if W_k is a compact set, $\mathbf{x}_k$ and $\mathbf{p}_k$ are given, then there exists a geodesic $\gamma_{\mathbf{x}_k}(t, \mathbf{p}_k)$ such that it can either be extended for all $t \in R_+$ taking its value in W_k, or can be defined on a closed interval, and one of its values belongs to the boundary of W_k. Let us consider the function $f(\gamma_{\mathbf{x}_k}(t, \mathbf{p}_k))$ and let

$$\bar{t}_k = \limsup \mathcal{J}_k,$$

where the set $\mathcal{J}_k$ is equal to

$$\begin{aligned} \mathcal{J}_k = \{t \in R_+| \\ \gamma_{\mathbf{x}_k}(t, \mathbf{p}_k) \text{ is defined and } f(\gamma_{\mathbf{x}_k}(t, \mathbf{p}_k)) < f(\gamma_{\mathbf{x}_k}(0, \mathbf{p}_k)) = f(\mathbf{x}_k)\}\,. \end{aligned}$$

The set $\mathcal{J}_k$ is nonempty by (3.1) and either $\bar{t}_k = +\infty$, $\gamma_{\mathbf{x}_k}(t, \mathbf{p}_k) \in W_k$ for all $t \in [0, \infty)$, or $\bar{t}_k$ is finite, and $\gamma_{\mathbf{x}_k}(t, \mathbf{p}_k) \in W_k$ for all $t \in [0, \bar{t}_k]$. In the first case, assume that the single variable function has no local minimum. Then, the function is monotone decreasing and bounded from below, so $t_k \to +\infty$ determining a point in W_k. It turns out that, in both cases, the stepsize rule (2.10) is well defined and $t_k \in (0, \bar{t}_k]$, i.e., the function $f(\gamma_{\mathbf{x}_k}(t, \mathbf{p}_k))$ has a minimum (or infimum) at the point t_k, and $f(\gamma_{\mathbf{x}_k}(t_k, \mathbf{p}_k)) = f(\mathbf{x}_{k+1})$.

If $\alpha \in (0, \frac{1}{2})$, the equation

$$\nabla^G f(\gamma_{\mathbf{x}_k}(t, \mathbf{p}_k))G_1(\gamma_{\mathbf{x}_k}(t, \mathbf{p}_k))\frac{d\gamma_{\mathbf{x}_k}(t, \mathbf{p}_k)}{dt} = \alpha\nabla^G f(\mathbf{x}_k)G_1(\mathbf{x}_k)\mathbf{p}_k \tag{3.2}$$

has the smallest solution $\hat{t}_k \in (0, t_k)$, because the continuous function

$$\nabla^G f(\gamma_{\mathbf{x}_k}(t, \mathbf{p}_k))G_1(\gamma_{\mathbf{x}_k}(t, \mathbf{p}_k))\frac{d\gamma_{\mathbf{x}_k}(t, \mathbf{p}_k)}{dt}$$

takes every value between $\nabla^G f(\mathbf{x}_k)G_1(\mathbf{x}_k)\mathbf{p}_k$ and 0, and

$$\nabla^G f(\gamma_{\mathbf{x}_k}(t, \mathbf{p}_k))G_1(\gamma_{\mathbf{x}_k}(t, \mathbf{p}_k))\frac{d\gamma_{\mathbf{x}_k}(t, \mathbf{p}_k)}{dt} < \alpha\nabla^G f(\mathbf{x}_k)G_1(\mathbf{x}_k)\mathbf{p}_k \tag{3.3}$$

for all $t \in [0, \hat{t}_k)$.

The sequence $\{f(\mathbf{x}_k)\}$ is monotone decreasing and bounded from below, since the continuous function f attains its minimum on the compact W_0. Hence, it converges to a limit. Let us extend the mapping $D_k \nabla^G f(\mathbf{x}_k)$ defined at the iteration points to a continuous vector-valued function $D(\mathbf{x})$ defined in R^n. This is always possible by the Kirszbraun theorem. Two cases are distinguished:

a) There exists a subsequence $\{\hat{t}_{k_i}\}$ of $\hat{t}_k$ converging to zero. By passing further to a subsequence, if necessary, we can assume that the subsequences $\{\hat{\mathbf{x}}_{k_i}\}$ related to $\{\hat{t}_{k_i}\}$ and $\{\mathbf{x}_{k_i}\}$ related to $\{t_{k_i}\}$ converge to $\hat{\mathbf{x}} \in W_0$. From equation (3.2) and the continuity of the Riemannian metric G_1, the gradient field $\nabla^G f$ and the vector-valued function D, we obtain that

$$D(\hat{\mathbf{x}})^T G_1(\hat{\mathbf{x}}) D(\hat{\mathbf{x}}) = \alpha D(\hat{\mathbf{x}})^T G_1(\hat{\mathbf{x}}) D(\hat{\mathbf{x}}) \,, \tag{3.4}$$

which implies that $D(\hat{\mathbf{x}}) = 0$. If $\nabla^G f(\hat{\mathbf{x}}) \neq 0$, then $\nabla^G f(\mathbf{x}_{k_i}) \neq 0$ for all i large enough, i.e., at least one component of the gradients is greater than a constant, thus because of the uniform lower bound of the positive definite quadratic forms defined by D_k on $TM_{\mathbf{x}_k}$ at the iteration points, $\nabla^G f(\hat{\mathbf{x}}) D(\hat{\mathbf{x}}) > 0$ is a contradiction. So, $\nabla^G f(\hat{\mathbf{x}}) = 0$ and $\hat{\mathbf{x}}$ is a critical point.

b) There exists a $\beta > 0$ such that $\hat{t}_{k_i} \geq \beta$ for all k_i. Let $\{\hat{\mathbf{x}}_{k_i}\}$ related to $\{\hat{t}_{k_i}\}$ and $\{\mathbf{x}_{k_i}\}$ related to $\{t_{k_i}\}$ be the subsequences defined above converging to $\hat{\mathbf{x}} \in W_0$. Suppose that $\hat{\mathbf{x}}$ is not a critical point. Hence,

$$|D(\hat{\mathbf{x}})^T G_1(\hat{\mathbf{x}}) D(\hat{\mathbf{x}})| = \delta > 0 \,. \tag{3.5}$$

The continuity of the Riemannian metric G_1, the gradient field $\nabla^G f$ and the vector-valued function D imply that

$$\left| \frac{df(\gamma_{\mathbf{x}_{k_i}}(0, \mathbf{p}_{k_i}))}{dt} \right| = |\nabla^G f(\mathbf{x}_{k_i}) G_1(\mathbf{x}_{k_i}) \mathbf{p}_{k_i}| > \delta/2 \tag{3.6}$$

for all k_i greater than a constant. Hence, by the mean value theorem and (3.3),

$$\begin{gathered} f(\mathbf{x}_{k_{i+1}}) \leq f(\mathbf{x}_{k_i+1}) \leq f(\hat{\mathbf{x}}_{k_i}) = \\ f(\mathbf{x}_{k_i}) + \hat{t}_{k_i} \nabla^G f(\gamma_{\mathbf{x}_{k_i}}(\tilde{t}, \mathbf{p}_{k_i})) G_1(\gamma_{\mathbf{x}_{k_i}}(\tilde{t}, \mathbf{p}_{k_i})) \frac{d\gamma_{\mathbf{x}_{k_i}}(\tilde{t}, \mathbf{p}_{k_i})}{dt} < \\ f(\mathbf{x}_{k_i}) + \hat{t}_{k_i} \alpha \nabla^G f(\mathbf{x}_{k_i}) G_1(\mathbf{x}_{k_i}) \mathbf{p}_{k_i} < f(\mathbf{x}_{k_i}) - \frac{1}{2} \alpha \beta \delta \,, \qquad \tilde{t} \in (0, \hat{t}_{k_i}) \,, \end{gathered} \tag{3.7}$$

which contradicts the fact that $\{f(\mathbf{x}_{k_i})\}$ converges to $f(\hat{\mathbf{x}})$. Thus, $\hat{\mathbf{x}}$ is a critical point.

Finally, suppose that $\mathbf{x}^*$ and $\mathbf{x}^{**}$ are distinct accumulation points of the sequence $\{\mathbf{x}_k\}$ in W_0. By the first part of Theorem 3.2, $\mathbf{x}^*$ and $\mathbf{x}^{**}$ are critical points of f. Since $\{f(\mathbf{x}_k)\}$ converges, we must have $f(\mathbf{x}^*) = f(\mathbf{x}^{**})$, which is impossible if the critical values of f are distinct. □

Remark 3.1 In order to determine a descent direction (3.1), we use the square of a symmetric matrix which is strongly positive definite on the tangent space at every iteration point. Another possibility for studying the convergence of variable methods along geodesics is to consider the product of the matrix of the Riemannian metric and the matrix of the updating formulas at the iteration points. It is well-known that, in general, the product of two real, positive definite matrices is not positive definite as the following example shows.

EXAMPLE 3.1 Let us consider two matrices in the form of

$$\begin{pmatrix} 1 & \beta \\ \beta & 1 \end{pmatrix} \quad \text{and} \quad \begin{pmatrix} 1 & \gamma \\ 0 & 1 \end{pmatrix},$$

then the product of these matrices is not positive definite for some parameter values, e.g., in the case of $\beta = \frac{3}{4}$, $\gamma = -\frac{16}{3}$,

$$\begin{pmatrix} 1 & \frac{3}{4} \\ \frac{3}{4} & 1 \end{pmatrix} \begin{pmatrix} 1 & -\frac{16}{3} \\ 0 & 1 \end{pmatrix} = \begin{pmatrix} 1 & -\frac{55}{12} \\ \frac{3}{4} & -3 \end{pmatrix}.$$

COROLLARY 3.1 *The proof of Theorem 3.2 shows that the statement holds if we replace the stepsize rule by (3.2), i.e., if $\alpha \in (0, \frac{1}{2})$, then the equation*

$$\nabla^G f(\gamma_{\mathbf{x}_k}(t, \mathbf{p}_k)) G_1(\gamma_{\mathbf{x}_k}(t, \mathbf{p}_k)) \frac{d\gamma_{\mathbf{x}_k}(t, \mathbf{p}_k)}{dt} = \alpha \nabla^G f(\mathbf{x}_k) G_1(\mathbf{x}_k) \mathbf{p}_k \tag{3.8}$$

has the smallest solution $\hat{t}_k \in (0, t_k)$ at every iteration point.

THEOREM 3.3 *If f is continuously differentiable, W_0 is a compact set in a complete Riemannian submanifold $M \subseteq \mathcal{M}$, the sequence $\{\mathbf{x}_k\}$ is generated by (2.8), (2.9) and (2.11), and if a mapping is defined by $D_k \nabla^G f(\mathbf{x}_k)^T \in TM_{\mathbf{x}_k}$, for all k at the iteration points and satisfies there the Lipschitz condition, D_k are positive definite on $TM_{\mathbf{x}_k}$ with a uniform lower bound for all k, and D_k and G_1 commute on $TM_{\mathbf{x}_k}$ at every iteration point, then the sequence $\{\mathbf{x}_k\}$ is either finite terminating at a critical point, or infinite, and every accumulation point is critical. If the critical values of f are distinct, the whole sequence $\{\mathbf{x}_k\}$ converges to a critical point.*

Proof. If $\mathbf{x}_k$ is a critical point, then $\nabla^G f(\mathbf{x}_k) = 0$, and the algorithm does not generate new iteration points, so introducing a stopping test, the procedure can be terminated at iteration k.

Assume now that $\mathbf{x}_k$ is not a critical point. Hence, $\mathbf{p}_k$ is a descent direction because of (3.1). Denote the arc of geodesic starting from $\mathbf{x}_k$ with tangent $\mathbf{p}_k$ by $\gamma_{\mathbf{x}_k}(t, \mathbf{p}_k)$. The compactness of W_0 implies the compactness of all the W_k because the sequence $\{f(\mathbf{x}_k)\}$ is nonincreasing. By the Hopf-Rinow theorem, the manifold M is geodesically complete, thus if W_k is a compact set, $\mathbf{x}_k$ and $\mathbf{p}_k$ are given, then there exists a geodesic $\gamma_{\mathbf{x}_k}(t, \mathbf{p}_k)$ such that it can either be extended for all $t \in R_+$ taking its value in W_k, or can be defined on a closed interval and one of its values belongs to the boundary of W_k. The direction $\mathbf{p}_k$ is a descent one, thus the stepsize rule (2.11) is well defined.

The sequence $\{f(\mathbf{x}_k)\}$ is monotone decreasing and bounded from below, since the continuous function f attains its minimum on the compact W_0. Hence, it converges to a limit. Let us extend the mapping $D_k \nabla^G f(\mathbf{x}_k)$ defined at the iteration points to a continuous vector-valued function $D(\mathbf{x})$ defined in R^n. This is always possible by the Kirszbraun theorem. Two cases are distinguished:

a) There exists a subsequence $\{t_{k_i} = 2^{-l_{k_i}}\}$ of t_k converging to zero. By passing further to a subsequence, if necessary, we can assume that the subsequence $\{\mathbf{x}_{k_i}\}$ related to $\{t_{k_i}\}$ converges to $\hat{\mathbf{x}} \in W_0$.

By using the Taylor expansion in stepsize rule (2.11), it follows that

$$\alpha 2^{-l_{k_i}} |\nabla^G f(\mathbf{x}_{k_i}) D_k G_1(\mathbf{x}_{k_i}) D_k \nabla^G f(\mathbf{x}_{k_i})^T| \leq$$
$$f(\mathbf{x}_{k_i}) - f(\gamma_{\mathbf{x}_{k_i}}(2^{-l_{k_i}}, \mathbf{p}_{k_i})) =$$
$$f(\mathbf{x}_{k_i}) - f(\mathbf{x}_{k_i}) - 2^{-l_{k_i}} \nabla^G f(\gamma_{\mathbf{x}_{k_i}}(\tilde{t}, \mathbf{p}_{k_i})) G_1(\gamma_{\mathbf{x}_{k_i}}(\tilde{t}, \mathbf{p}_{k_i})) \frac{d\gamma_{\mathbf{x}_{k_i}}(\tilde{t}, \mathbf{p}_{k_i})}{dt} =$$
$$-2^{-l_{k_i}} \nabla^G f(\gamma_{\mathbf{x}_{k_i}}(\tilde{t}, \mathbf{p}_{k_i})) G_1(\gamma_{\mathbf{x}_{k_i}}(\tilde{t}, \mathbf{p}_{k_i})) \frac{d\gamma_{\mathbf{x}_{k_i}}(\tilde{t}, \mathbf{p}_{k_i})}{dt}, \qquad \tilde{t} \in (0, 2^{-l_{k_i}}) \tag{3.9}$$

i.e.,

$$\alpha |\nabla^G f(\mathbf{x}_{k_i}) D_k G_1(\mathbf{x}_{k_i}) D_k \nabla^G f(\mathbf{x}_{k_i})^T| \leq$$
$$-\nabla^G f(\gamma_{\mathbf{x}_{k_i}}(\tilde{t}, \mathbf{p}_{k_i})) G_1(\gamma_{\mathbf{x}_{k_i}}(\tilde{t}, \mathbf{p}_{k_i})) \frac{d\gamma_{\mathbf{x}_{k_i}}(\tilde{t}, \mathbf{p}_{k_i})}{dt}, \qquad \tilde{t} \in (0, 2^{-l_{k_i}}). \tag{3.10}$$

From inequality (3.10) and the continuity of the Riemannian metric G_1, the gradient field $\nabla^G f$ and the vector-valued function D, we obtain that

$$(1 + \alpha) D(\hat{\mathbf{x}})^T G_1(\hat{\mathbf{x}}) D(\hat{\mathbf{x}}) \leq 0\,, \tag{3.11}$$

which implies that $D(\hat{\mathbf{x}}) = \mathbf{0}$. If $\nabla^G f(\hat{\mathbf{x}}) \neq 0$, then $\nabla^G f(\mathbf{x}_{k_i}) \neq 0$ for all i large enough, i.e., at least one component of the gradients is greater than a constant, thus because of the uniform lower bound of the positive definite quadratic forms defined by D_k on $TM_{\mathbf{x}_k}$ at the iteration points, $\nabla^G f(\hat{\mathbf{x}})D(\hat{\mathbf{x}}) > 0$ is a contradiction. So, $\nabla^G f(\hat{\mathbf{x}}) = 0$ and $\hat{\mathbf{x}}$ is a critical point.

b) There exists a $\beta > 0$ such that $t_{k_i} \geq \beta$ for all k_i. Let $\{\mathbf{x}_{k_i}\}$ related to $\{t_{k_i}\}$ be the subsequence defined above which converge to $\hat{\mathbf{x}} \in W_0$. Because of the stepsize rule, we have that

$$\alpha 2^{-l_{k_i}} |\nabla^G f(\mathbf{x}_{k_i}) D_k G_1(\mathbf{x}_{k_i}) D_k \nabla^G f(\mathbf{x}_{k_i})^T| \leq f(\mathbf{x}_{k_i}) - f(\gamma_{\mathbf{x}_{k_i}}(2^{-l_{k_i}}, \mathbf{p}_{k_i})) \ . \tag{3.12}$$

The right size of the sequence in inequality (3.12) tends to zero, $2^{-l_{k_i}} \geq \beta$, and the continuity of the Riemannian metric G_1, the gradient field $\nabla^G f$ and the vector-valued function D imply that

$$D(\hat{\mathbf{x}})^T G_1(\hat{\mathbf{x}}) D(\hat{\mathbf{x}}) = 0 \ , \tag{3.13}$$

from which it follows that $D(\hat{\mathbf{x}}) = 0$, and so, $\nabla^G f(\hat{\mathbf{x}}) = 0$.

Finally, suppose that $\mathbf{x}^*$ and $\mathbf{x}^{**}$ are distinct accumulation points of the sequence $\{\mathbf{x}_k\}$ in W_0. By the first part of Theorem 3.3, $\mathbf{x}^*$ and $\mathbf{x}^{**}$ are critical points of f. Since $\{f(\mathbf{x}_k)\}$ converges, we must have $f(\mathbf{x}^*) = f(\mathbf{x}^{**})$, which is impossible if the critical values of f are distinct. □

4. Rate of convergence for variable metric methods along geodesics

The speed of convergence is a rich and elementary theory dealing with convergence rates. By this, one can predict with confidence the relative effectiveness of a wide class of algorithms. First, the notion of the order of convergence is introduced in order to measure the speed of convergence.

Consider a sequence of real numbers $\{t_k\}_{k=0}^{+\infty}$ converging to the limit t^*.

DEFINITION 4.1 Let the sequence $\{t_k\}$ converge to t^*. The order of convergence of $\{t_k\}$ is defined as the supremum of the nonnegative numbers α satisfying

$$0 \leq \limsup_{k \to \infty} \frac{|t_{k+1} - t_{k^*}|}{|t_k - t_{k^*}|^{\alpha}} < +\infty \ . \tag{4.1}$$

In order to ensure that the definition is applicable to any sequence, it is stated in terms of limit superior rather than just limit, and $\frac{0}{0}$ – which occurs if $t_k = t^*$ for all k – is regarded as finite.

In order to measure convergence in optimization theory, error functions are commonly used as the same function that defines the objective function of the original optimization problem. This means that we measure convergence by the speed the objective function converges to its minimum. Alternatively, a distance function between the iteration points and the optimum point is used sometimes, and it measures convergence by the speed the distance from the solution point decreases to zero.

THEOREM 4.1 *Let* $\mathbf{x}^*$ *be a local optimum of problem (1.1). If* $\{\mathbf{x}_k\}$ *is a sequence generated by a geodesic descent method that converges to* $\mathbf{x}^*$*, then the sequence of the objective values* $\{f(\mathbf{x}_k)\}$ *converges to* $f(\mathbf{x}^*)$ *with an order one and a ratio not greater than*

$$\limsup_{k\to\infty} \frac{\mathbf{u}_1'(0)^T D^2 f(\mathbf{x}^*)\mathbf{u}_1'(0)}{\mathbf{u}'(0)^T D^2 f(\mathbf{x}^*)\mathbf{u}'(0)} \leq 1 , \tag{4.2}$$

where at $\mathbf{x}^*$ $\mathbf{u}_1'(0)$ *and* $\mathbf{u}'(0)$ *mean the tangents of the unique geodesics defined in a geodesic convex neighbourhood, between* $\mathbf{x}_{k+1}$ *and* $\mathbf{x}^*$ *as well as between* $\mathbf{x}_k$ *and* $\mathbf{x}^*$*, respectively, and* $D^2 f$ *the second covariant derivative of the function* f *on* M.

Proof. Let us consider a geodesic convex neighbourhood of $\mathbf{x}^*$ and, without loss of generality, assume that the sequence $\{\mathbf{x}_k\}$ is contained in this neighbourhood. Consider the unique geodesic

$$\mathbf{x}_1(\mathbf{u}_1(s)), \quad \mathbf{x}_1(\mathbf{u}_1(0)) = \mathbf{x}^*, \qquad s \in [0, b_1] , \tag{4.3}$$

between $\mathbf{x}_{k+1}$ and $\mathbf{x}^*$ as well as the unique geodesic

$$\mathbf{x}(\mathbf{u}(s)), \quad \mathbf{x}(\mathbf{u}(0)) = \mathbf{x}^*, \qquad s \in [0, b] , \tag{4.4}$$

between $\mathbf{x}_k$ and $\mathbf{x}^*$.

By the tensor expansion of the function f in a geodesic convex neighbourhood of the manifold M (Theorem 4.1 in Rapcsák, 1995), we have that

$$\begin{aligned} f(\mathbf{x}_{k+1}) &= f(\mathbf{x}^*) + \frac{1}{2}\mathbf{u}_1'(0)^T D^2 f(\mathbf{x}^*)\mathbf{u}_1'(0) + \varepsilon_1(0; s) , \\ f(\mathbf{x}_k) &= f(\mathbf{x}^*) + \frac{1}{2}\mathbf{u}'(0)^T D^2 f(\mathbf{x}^*)\mathbf{u}'(0) + \varepsilon(0; s) , \end{aligned} \tag{4.5}$$

where $\mathbf{u}_1'(0)$ and $\mathbf{u}'(0)$ mean the tangents of the geodesics at $\mathbf{x}^*$, respectively. Because of $f(\mathbf{x}_{k+1}) < f(\mathbf{x}_k)$,

$$\frac{f(\mathbf{x}_{k+1}) - f(\mathbf{x}^*)}{f(\mathbf{x}_k) - f(\mathbf{x}^*)} = \frac{\mathbf{u}_1'(0)^T D^2 f(\mathbf{x}^*)\mathbf{u}_1'(0) + \varepsilon_1(0; s)}{\mathbf{u}'(0)^T D^2 f(\mathbf{x}^*)\mathbf{u}'(0) + \varepsilon(0; s)} < 1 ,$$

which gives the statement. □

EXAMPLE 4.1 It will be shown that, in the case of a quadratic function and the Euclidean metric, the result of Theorem 4.1 coincides with the classical one based on the Kantorovich inequality (e.g., Luenberger, 1973).

Let us consider a quadratic function in the form of

$$f(\mathbf{x}) = \frac{1}{2}\mathbf{x}^T Q\mathbf{x} - \mathbf{b}^T\mathbf{x}, \qquad \mathbf{x}, \mathbf{b} \in R^n ,$$

where Q is a positive definite symmetric $n \times n$ matrix. Since Q is positive definite, all of its eigenvalues are positive. Assume that the smallest eigenvalue is $\lambda_1 > 0$, the largest one λ_n and the unique minimum point of f is $\mathbf{x}^*$. Then $M = R^n$, $G(\mathbf{x}) = I$, $\mathbf{x} \in R^n$,

$$D^2 f(\mathbf{x}^*) = Q, \quad \mathbf{u}_1'(0) = \mathbf{x}_{k+1} - \mathbf{x}^*, \quad \mathbf{u}'(0) = \mathbf{x}_k - \mathbf{x}^*.$$

Because of

$$\nabla f(\mathbf{x}_k) = (\mathbf{x}_k - \mathbf{x}^*)^T Q ,$$

$$\mathbf{x}_{k+1} = \mathbf{x}_k - \alpha \nabla f(\mathbf{x}_k) ,$$

$$\alpha = \frac{\nabla f(\mathbf{x}_k)^T \nabla f(\mathbf{x}_k)}{\nabla f(\mathbf{x}_k)^T Q \nabla f(\mathbf{x}_k)} ,$$

we obtain that

$$(\mathbf{x}_k - \mathbf{x}^*)^T Q (\mathbf{x}_k - \mathbf{x}^*) = \nabla f(\mathbf{x}_k) Q^{-1} \nabla f(\mathbf{x}_k)^T ,$$

$$(\mathbf{x}_{k+1} - \mathbf{x}^*)^T Q (\mathbf{x}_{k+1} - \mathbf{x}^*) = \nabla f(\mathbf{x}_k) Q^{-1} \nabla f(\mathbf{x}_k)^T - \frac{(\nabla f(\mathbf{x}_k) \nabla f(\mathbf{x}_k)^T)^2}{\nabla f(\mathbf{x}_k) Q \nabla f(\mathbf{x}_k)^T} .$$

Thus,

$$\frac{\mathbf{u}_1'(0)^T D^2 f(\mathbf{x}^*) \mathbf{u}_1'(0)}{\mathbf{u}'(0)^T D^2 f(\mathbf{x}^*) \mathbf{u}'(0)} = \frac{(\mathbf{x}_{k+1} - \mathbf{x}^*)^T Q (\mathbf{x}_{k+1} - \mathbf{x}^*)}{(\mathbf{x}_k - \mathbf{x}^*)^T Q (\mathbf{x}_k - \mathbf{x}^*)} =$$

$$1 - \frac{(\nabla f(\mathbf{x}_k) \nabla f(\mathbf{x}_k)^T)^2}{\nabla f(\mathbf{x}_k) Q \nabla f(\mathbf{x}_k)^T \nabla f(\mathbf{x}_k) Q^{-1} \nabla f(\mathbf{x}_k)^T} ,$$

and by Kantorovich inequality (e.g., Luenberger, 1973), stating that if Q is a positive definite symmetric $n \times n$ matrix, then for any vector $\mathbf{x} \in R^n$ the following inequality holds:

$$\frac{(\mathbf{x}^T \mathbf{x})^2}{(\mathbf{x}^T Q \mathbf{x})(\mathbf{x}^T Q^{-1} \mathbf{x})} \geq \frac{4\lambda_1 \lambda_n}{(\lambda_1 + \lambda_n)^2} , \tag{4.6}$$

from which we have the statement

$$\frac{\mathbf{u}_1'(0)^T D^2 f(\mathbf{x}^*) \mathbf{u}_1'(0)}{\mathbf{u}'(0)^T D^2 f(\mathbf{x}^*) \mathbf{u}'(0)} \leq \left(\frac{\lambda_n - \lambda_1}{\lambda_n + \lambda_1} \right)^2 .$$

5. Variable metric methods along geodesics under inequality constraints

Based on the convergence Theorems 3.2 and 3.3, the penalty methods seem to be applicable for extension to solve problem (1.1) in the case of $A \neq M$, i.e., if there exist not only equality, but inequality constraints as well. Thus, it is possible to preserve a more general and flexible treatment of the nonconvex character of smooth optimization problems.

The idea of penalty function methods is to replace problem (1.1) by a serie of unconstrained minimization problems in the form of

$$\begin{aligned} &\min f(\mathbf{x}) + \mu_k P(\mathbf{x}) \\ &\mathbf{x} \in M \subseteq \mathcal{M} \,, \end{aligned} \tag{5.1}$$

where $\{\mu_k\}$, $k = 1, 2, \ldots$ is a sequence tending to infinity such that for each k

$$0 < \mu_k < \mu_{k+1} \,, \tag{5.2}$$

and P is a continuous function on M satisfying the following two conditions:

$$P(\mathbf{x}) \geq 0, \qquad \forall \mathbf{x} \in M \,, \tag{5.3}$$

$$P(\mathbf{x}) = 0 \qquad \text{iff} \qquad \mathbf{x} \in A \,. \tag{5.4}$$

For large μ it is clear that the minimum point of problem (1.1) is in a region where the function P is small. Thus, for increasing μ, it is expected that the corresponding solution point will approach to the feasible region A, and if A is closed, it will minimize f on A, i.e., if $\mu_k \to +\infty$, $k = 1, \ldots, \infty$, then the solution points of the penalty problems will converge to a solution of the constrained problem. If we assume that for all k, the unconstrained problems (5.1) have a solution each, then the following classic statements (e.g., Luenberger, 1973) can be proved:

LEMMA 5.1

$$f(\mathbf{x}_k) + \mu_k P(\mathbf{x}_k) \leq f(\mathbf{x}_{k+1}) + \mu_{k+1} P(\mathbf{x}_{k+1}) \,, \qquad \forall k \,, \tag{5.5}$$

$$P(\mathbf{x}_k) \geq P(\mathbf{x}_{k+1}) \,, \qquad \forall k \,, \tag{5.6}$$

$$f(\mathbf{x}_k) \leq f(\mathbf{x}_{k+1}) \,, \qquad \forall k \,. \tag{5.7}$$

LEMMA 5.2 *Let* $\mathbf{x}^*$ *be a solution of problem (1.1). Then, for each* k,

$$f(\mathbf{x}^*) \geq f(\mathbf{x}_k) + \mu_k P(\mathbf{x}_k) \geq f(\mathbf{x}_k) \,. \tag{5.8}$$

The global convergence of the above penalty methods follows from the two previous lemmas.

THEOREM 5.1 *Let $\{\mathbf{x}_k\}$ be a sequence generated by a penalty method. Then, any limit point of the sequence is a solution of problem (1.1).*

For ensuring the local-global property of unconstrained optimization problems, the assumption of geodesic convexity of the penalty functions can be used. To solve continuously differentiable unconstrained optimization problems a general framework for globally convergent variable metric methods along geodesics was introduced in Section 3, containing a generalization of the steepest descent, Newton, quasi-Newton, gradient projection, reduced gradient, projected gradient, SQP methods, etc. In numerical realizations, tensor approximations can be applied.

By the assumption of Theorems 3.2 and 3.3, f is a continuously differentiable function. In penalty function approach, there is a continuously differentiable subclass for inequality constraints (e.g., Bertsekas, 1982) where the convergence analysis is developed for the convex case. A general image problem approach to penalty methods was published by Pappalardo (1990) based on the relationships between the weak and strong separation functions and the penalty methods discussed by Giannessi (1984). The following penalty function with two parameters, suggested by Forgó (1969) for classical nonlinear optimization problems with inequality constraints, is a modification of the exponential one:

$$f(\mathbf{x}) + 1/b_k \sum_{i=1}^{m} e^{-a_k g_i(\mathbf{x})}, \qquad \mathbf{x} \in M \subseteq \mathcal{M}\,,$$

$$0 < a_k < a_{k+1}\,, \quad 0 < b_k < b_{k+1}\,, \qquad \lim_{k\to\infty} a_k = \lim_{k\to\infty} b_k = +\infty\,,$$

$$\lim_{k\to\infty} a_k/\log(b_k) = +\infty\,.$$

It is also a weak separation function for every pair of parameters. In the case of convex optimization problems with unbounded feasible region, nonempty and bounded solution set, the algorithm converges to an optimal solution, and moreover, if the interior of the feasible domain is nonempty, then a pair of parameters can be selected such that an ϵ-approximate solution is reached after one global minimization (Rapcsák, 1975). It follows from the proof that if the global minimums of this penalty function exist for every feasible pair of parameters, then they converge to a solution of the optimization problem. By extending the ϵ-exactness property to our case, the combinations of this result and

Theorems 3.2 and 3.3 should result in globally convergent algorithms for determining ϵ-approximate solutions.

Interior point or barrier methods are applicable to problems where the constraint set has a nonempty interior that is arbitrarily close to any point of the set. Intuitively, this means that the set has an interior and it is possible to get any boundary point by approaching it from the interior. Such a set is referred to in the literature as robust. This kind of set often arises in conjunction with inequality constraints in R^n. Interior point methods work by establishing a barrier on the boundary of the feasible region that prevents a search procedure from leaving the region. These methods have received enormous attention recently because of their close relationship with the polynomial approach to linear programming. Based on the above convergence theorems, the interior point methods seem to be applicable for extension to solve problem (1.1) with inequality constraints as well.

The characterization of problems where variable metric methods along geodesic provide polynomial algorithms is an open question. In the paper of Rapcsák and Thang (1996), the behaviour of interior point algorithms in the case of linear optimization is analyzed by using a variable metric method approach. A class of polynomial variable metric algorithms is given achieving $O(\frac{n}{\beta}L)$ iterations for solving a canonical form linear optimization problem with respect to a wide class of Riemannian metrics, where n is the number of dimensions and β a fixed value. It is shown that the vector fields of several interior point algorithms for linear optimization is the negative Riemannian gradient vector field of a linear, a potential or a logarithmic barrier function for suitable Riemannian metrics.

6. Acknowledgement

Research partially supported by the Hungarian National Research Foundation, Grant No. OTKA-T016413.

References

1. Bertsekas, D. P., Constrained optimization and Lagrange multiplier methods, Academic Press, New York, London, 1982.
2. den Hertog, D., Roos, C. and Terlaky, T., On the classical logarithmic barrier method for a class of smooth convex programming problems, Journal of Optimization Theory and Applications 73 (1992) 1–25.
3. Ferreira, O. P. and Oliveira, P. R., Subgradient algorithm on Riemannian manifolds, Publicacoes Técnicas, Rio de Janeiro (1995).
4. Fletcher, R., A general quadratic programming algorithm, Journal of the Institute of Mathematics and Its Applications 7 (1971) 76–91.

5. Forgó, F., A method for solving nonlinear programming problems approximately, Szigma 1 (1969) 67–75. (in Hungarian)
6. Gabay, D., Minimizing a differentiable function over a differentiable manifold, Journal of Optimization Theory and Applications 37 (1982) 177–219.
7. Giannessi, F., Theorems of the alternative and optimality conditions, Journal of Optimization Theory and Applications 42 (1984) 331–365.
8. Gill, P. E., Murray, W., Saunders, M. A., Tomlin, J. A. and Wright, M. H., On projected Newton barrier methods for linear programming and an equivalence to Karmarkar's projective method, Mathematical Programming 36 (1986) 183–209.
9. Gonzaga, C. C., Polynomial affine algorithms for linear programming, Mathematical Programming 49 (1990) 7–21.
10. Iri, M. and Imai, H., A multiplicative barrier function method for linear programming, Algorithmica 1 (1986) 455–482.
11. Jarre, F., Interior-points methods for convex programming, Applied Mathematics and Optimization 26 (1992) 287–311.
12. Karmarkar, N., Riemannian geometry underlying interior points methods for linear programming, Contemporary Mathematics 114 (1990) 51–76.
13. Kobayashi, S. and Nomizu, K., Foundations of differential geometry, Interscience Publishers, New-York, London, Sydney, 1969.
14. Lichnewsky, A., Minimization de fonctionnelle définies sur une variété par la method du gradient conjugué, Thèse de Doctorat d'État, Université de Paris-Sud, Paris, France, 1979.
15. Luenberger, D. G., The gradient projection methods along geodesics, Management Science 18 (1972) 620–631.
16. Luenberger, D. G., Introduction to linear and nonlinear programming, Addison-Wesley Publishing Company, Reading, 1973.
17. Mehrotra, S. and Sun, J., An interior point algorithm for solving smooth convex programs based on Newton's method, Mathematical Developments Arising from Linear Programming, J. C. Lagarias and M. J. Todd, eds., Contemporary Mathematics 114 (1990) 265–284.
18. Nesterov, Y. E. and Nemirovsky, A. S., Self-concordant functions and polynomial time methods in convex programming, Report, Central Economic and Mathematical Institute, USSR Academy of Science, Moscow, USSR, 1989.
19. Neto, J. X. C. and Oliveira, P. R., Geodesic methods in Riemannian manifolds, Publicacoes Técnicas, Rio de Janeiro (1995).
20. Ortega, J. M. and Rheinboldt, N. C., Iterative solution of nonlinear equations in several variables, Academic Press, New York, New York, 1970.
21. Pappalardo, M., Image space approach to penalty methods, Journal of Optimization Theory and Applications 64 (1990) 141–152.
22. Polak, E., Computational methods in optimization, A unified approach, Academic Press, New York, New York, 1971.
23. Rapcsák, T., An exterior point algorithms for solving convex nonlinear programming problems, Alkalmazott Matematikai Lapok 1 (1975) 357–354. (in Hungarian)
24. Rapcsák, T., Geodesic convexity in nonlinear optimization, Journal of Optimization Theory and Applications 69 (1991) 169–183.
25. Rapcsák, T. and Thang, T. T., Nonlinear coordinate representations of smooth optimization problems, Journal of Optimization Theory and Applications 86 (1995) 459–489.
26. Rapcsák, T., Tensor applications of smooth nonlinear complementarity systems, in: Variational inequalities and network equilibrium problems, eds.: F. Giannessi and A. Maugeri, Plenum Press (1995) 235–249.
27. Rapcsák, T. and Thang, T. T., A class of polynomial variable metric algorithms for linear optimization, Mathematical Programming 74 (1996) 319–331.

28. Schwartz, J. T., Nonlinear functional analysis, Gordon and Breach Science Publishers, New York, 1969.
29. Tanabe, K., A geometric method in nonlinear programming, Journal of Optimization Theory and Applications 30 (1980) 181–210.
30. Udriste, C., Convex functions and optimization methods on Riemannian manifolds, Kluwer Academic Publishers, Boston, 1994.
31. Yamashita, H., A differential equation approach to nonlinear programming, Mathematical Programming 18 (1980) 155–168.

Criteria for Generalized Monotonicity

Siegfried Schaible
Graduate School of Management
University of California
Riverside, CA 92521
U.S.A.

Abstract. Characterizations of different kinds of generalized monotonicity are surveyed for the following subclasses of maps: affine maps, differentiable maps, locally Lipschitz maps.

Key words: Generalized convexity, generalized monotonicity; affine maps; differentiable maps; locally Lipschitz maps

1. Introduction

In the analysis and solution of complementarity problems and variational inequalities it is commonly assumed that the defining map is *monotone*. This is not suprising since in the special case of an underlying optimization problem usually *convexity* is assumed, and convexity of the objective function corresponds to monotonicity of its gradient.

For several decades much effort has been devoted to generalizing convexity in various ways, often with the view of nonconvex optimization in mind [2]. On the other hand, only recently a systematic study of generalizations of monotonicity has emerged. Since the article [15] in 1990 about onehundred publications have appeared. They deal with either concepts and characterizations of generalized monotonicity or with applications in variational inequalities and related models; e.g. [34] for a survey of major results until 1994.

The present paper surveys criteria for generalized monotonicity. An updated survey on applications of different kinds of generalized monotone maps in variational inequalities and more general models is in preparation [10].

In this survey we bring together characterizations of generalized monotonicity for different subclasses of maps. The need for such criteria is obvious, given that the defining inequalities are often hard to verify. The presentation is restricted to results derived in projects the author was involved in himself. For other contributions we refer to the references in the publications reviewed and in the previous survey [34].

The paper is organized as follows. In Section 2 we provide a brief review of some basic generalized monotonicity concepts and their relationships, using corresponding generalized convexity concepts and their

F. Giannessi et al. (eds.), New Trends in Mathematical Programming, 277–288.

geometric properties as the background. This is followed by a presentation of criteria for generalized monotonicity in case of differentiable, affine and nondifferentiable (locally Lipschitz) maps in Section 3, 4 and 5, respectively. The connections between these various results will become clear in retrospect.

2. Nine kinds of (generalized) monotonicity

Nine basic kinds of convex/generalized convex functions are [2]:

- convex (CX), strictly convex (S.CX), strongly convex (Str.CX);
- pseudoconvex (PCX), strictly pseudoconvex (S.PCX), strongly pseudoconvex (Str.PCX);
- quasiconvex (QCX), semistrictly quasiconvex (Ss.QCX) and strictly quasiconvex (S.QCX).

These functions are related to each other as follows in the continuous case:

$$\begin{array}{ccccc} & & & & (QCX) \\ & & & & \Uparrow \\ (CX) & \Rightarrow & (PCX) & \Rightarrow & (Ss.QCX) \\ \Uparrow & & \Uparrow & & \Uparrow \\ (S.CX) & \Rightarrow & (S.PCX) & \Rightarrow & (S.QCX) \\ \Uparrow & & \Uparrow & & \\ (Str.CX) & \Rightarrow & (Str.PCX) & & \end{array}$$

Different kinds of generalized convexity preserve different properties of convex functions. E.g., the characteristic of a pseudoconvex function is that a stationary point is a global minimum. Similarly, for a semistrictly quasiconvex function a local is a global minimum and for a quasiconvex function the lower level sets are convex. In contrast to convex functions, inflection points are admissible for all types of generalized convex functions.

The qualifier 'strictly' indicates that a global minimum is unique while 'strongly' indicates at least quadratic increase around a global minimum.

We note that in [2] the terminology of quasiconvex and pseudoconvex functions was harmonized, resulting in renaming former 'strongly quasiconvex' functions as strictly quasiconvex and 'strictly quasiconvex' functions as semistrictly quasiconvex.

It is well known that a differentiable convex function is characterized by a monotone gradient. Correspondingly, a strictly convex and strongly convex function is characterized by a strictly and strongly monotone gradient, respectively. Accordingly, *generalized monotonicity concepts* have been introduced in such a way that in case of a gradientmap $F = \nabla f$ generalized monotonicity of F corresponds to some kind of generalized convexity of the underlying function f.

Below we list the definitions of (generalized) monotone maps.

Consider $F : C \to R^n$,where $C \subseteq R^n$.

F is *monotone (M)* on C if for all $x, y \in C$

$$(y-x)^T(F(y)-F(x)) \geq 0. \tag{2.1}$$

F is *strictly monotone (S.M)* on C if for all $x, y \in C$, $x \neq y$

$$(y-x)^T(F(y)-F(x)) > 0. \tag{2.2}$$

F is *strongly monotone (Str.M)* on C if there exists $\beta > 0$ such that for all $x, y \in C$, $x \neq y$

$$(y-x)^T(F(y)-F(x)) \geq \beta \|y-x\|^2 . \tag{2.3}$$

F is *pseudomonotone (PM)* on C if for all $x, y \in C$

$$(y-x)^T F(x) \geq 0 \text{ implies } (y-x)^T F(y) \geq 0 \tag{2.4}$$

which is equivalent to

$$(y-x)^T F(x) > 0 \text{ implies } (y-x)^T F(y) > 0.$$

F is *strictly pseudomonotone (S.PM)* on C if for all $x, y \in C$, $x \neq y$

$$(y-x)^T F(x) \geq 0 \text{ implies } (y-x)^T F(y) > 0. \tag{2.5}$$

F is *quasimonotone (QM)* on C if for all $x, y \in C$

$$(y-x)^T F(x) > 0 \text{ implies } (y-x)^T F(y) \geq 0. \tag{2.6}$$

F is *strongly pseudomonotone (Str.PM)* on C (open) if for all $x \in C$ and $v \in R^n$ such that $\|v\| = 1$, $v^T F(x) = 0$, there exist $\beta > 0$, $\varepsilon > 0$ such that

$$x + \varepsilon v \in C \text{ and } v^T F(x+tv) \geq \beta t,\ 0 \leq t < \varepsilon. \tag{2.7}$$

This definition in [8] replaces the more restrictive one in [15].

F is *strictly quasimonotone (S.QM)* on C if F is quasimonotone on C and for all $x, y \in C$, $x \neq y$ there exists $z = tx + (1-t)y, t \in (0,1)$ such that

$$(y-x)^T F(z) \neq 0. \tag{2.8}$$

F is *semistrictly quasimonotone (Ss.QM)* on C if F is quasimonotone on C and for all $x, y \in C$, $x \neq y$

$$(y-x)^T F(x) > 0 \text{ implies } (y-x)^T F(z) > 0 \tag{2.9}$$

for some $z = tx + (1-t)y, t \in (0, \frac{1}{2})$.

If F is continuous, quasimonotonicity of F does not have to be required explicitely for strictly/semistrictly quasimonotone maps since it is implied by (2.8), (2.9), respectively. In terms of references for the concepts above, please consult [14] for pseudomonotone maps, [15] for quasimonotone and strictly pseudomonotone maps and [8] for strongly pseudomonotone, strictly quasimonotone and semistrictly quasimonotone maps. We also refer to related, independent studies referenced in these publications and in [34].

The authors of [8,14,15] derived the following diagram for general maps which are not necessarily gradientmaps.

$$\begin{array}{ccccc} & & & & \text{(QM)} \\ & & & & \Uparrow \\ \text{(M)} & \Rightarrow & \text{(PM)} & \Rightarrow & \text{(Ss.QM)} \\ \Uparrow & & \Uparrow & & \Uparrow \\ \text{(S.M)} & \Rightarrow & \text{(S.PM)} & \Rightarrow & \text{(S.QM)} \\ \Uparrow & & \Uparrow & & \\ \text{(Str.M)} & \Rightarrow & \text{(Str.PM)} & & \end{array}$$

Now consider the special case of a gradientmap $F = \nabla f$, where $f : C \rightarrow R$ is differentiable on the open convex set $C \subseteq R^n$. It can be shown [8,14,15]:

THEOREM 2.1 *The map $F = \nabla f$ is monotone, strictly monotone, strongly monotone, pseudomonotone, strictly pseudomonotone, strongly pseudomonotone, quasimonotone, semistrictly quasimonotone, strictly quasimonotone if and only if the function f is convex, strictly convex, strongly convex, pseudoconvex, strictly pseudoconvex, strongly pseudoconvex, quasiconvex, semistrictly quasiconvex, strictly quasiconvex, respectively.*

We point out that in case of a strongly pseudomonotone map the above equivalence is no longer true if f is only once, but not twice differentiable [8,9], i.e., there are once differentiable strongly pseudoconvex functions for which the gradient is not strongly pseudomonotone.

We note that in case of semistrictly quasiconvex functions Theorem 2.1 provides the first successful characterization in terms of the gradient. Before, the existence of such a characterization was doubted [25].

We would like to mention at this point that there are several studies where similar results are obtained for nondifferentiable functions by replacing the gradient with some generalized derivative; e.g. [1, 6, 11, 12, 18–22, 28, 29, 30].

We further remark that, given the geometric properties of generalized convex functions [2], it is not difficult to derive the geometric properties describing generalized monotonicity of gradientmaps; e.g. [3,17].

Finally we mention the following result which shows how new generalized monotone maps can be constructed from existing ones [31]. Consider $z = Ax + b$, where A is an $m \times n$ matrix and $b \in R^m$. Let $D \subseteq R^m$ and $C = \{x \in R^n \mid Ax + b \in D\}$. We have:

PROPOSITION 2.2 *The map $F(x) = A^T G(Ax + b)$ is quasimonotone (pseudomonotone) on C if G is quasimonotone (pseudomonotone) on D. F is strictly pseudomonotone on C if G is strictly pseudomonotone on D and A has full rank.*

3. The differentiable case

In this section we assume that $F : C \to R^n$ is differentiable and $C \subseteq R^n$ is an open convex set. Let $J_F(x)$ denote the Jacobian of F. First-order characterizations of generalized monotone maps have been established in [16,17]. In case of gradientmaps they extend classical second-order characterizations of generalized convex functions.

Consider the following conditions, given $x \in C, v \in R^n, v \neq 0$:

(A) $\qquad v^T F(x) = 0 \quad \text{implies} \quad v^T J_F(x) v \geq 0\,,$

(A$^+$) $\qquad v^T F(x) = 0 \quad \text{implies} \quad v^T J_F(x) v > 0\,,$

(B) $v^T F(x) = v^T J_F(x) v = 0$ and $v^T F(x + \hat{t}v) > 0$ for some $\hat{t} < 0$ implies that there exists $\tilde{t} > 0$ such that $x + \tilde{t}v \in C, v^T F(x+tv) \geq 0$ for all $0 \leq t \leq \tilde{t}$,

(C) $v^T F(x) = v^T J_F(x) v = 0$ implies that there exists $\tilde{t} > 0$ such that $x + \tilde{t}v \in C, v^T F(x + tv) \geq 0$ for all $0 \leq t \leq \tilde{t}$.

It can be shown [16,17]:

THEOREM 3.1 *Let $F : C \to R^n$ be differentiable on the open convex set $C \subseteq R^n$.*

(i) F is quasimonotone if and only if (A) and (B) hold for all $x \in C$ and $v \in R^n$,

(ii) F is pseudomonotone if and only if (A) and (C) hold for all $x \in C$ and $v \in R^n$,

(iii) F is strictly pseudomonotone if (A^+) holds for all $x \in C$ and $v \in R^n$.

More recently, it was shown in [4] that for continuously differentiable maps the condition $v^T F(x) = 0$ in (B) and (C) can be replaced by the less restrictive assumption $F(x) = 0$, and (i) and (ii) are still true. An immediate corollary of this stronger characterization of quasimonotone and pseudomonotone maps is

PROPOSITION 3.2 *Let $F : C \to R^n$ be continuously differentiable on an open convex set $C \subseteq R^n$ such that $F(x) \neq 0$ on C. Then F is pseudomonotone if and only if F is quasimonotone if and only if (A) holds for all $x \in C$ and $v \in R^n$.*

4. The affine case

In this section we focus on the special case of affine maps. Let $F(x) = Mx + q$, where M is an nxn matrix and $q \in R^n$. Consider F on an open convex set $C \subseteq R^n$. For general differentiable maps we have $F = \nabla f$ if and only if $J_F(x)$ is symmetric for all x. Hence for an affine map $F(x) = Mx + q$ we have $F = \nabla f$ if and only if M is symmetric. In this case $f(x) = \frac{1}{2} x^T M x + q^T x$. Therefore first-order characterizations of affine maps correspond to second order characterizations of quadratic functions.

For affine maps conditions (B) and (C) are always satisfied. Hence, specializing Theorem 3.1, we have

THEOREM 4.1 *The map $F(x) = Mx+q$ is quasimonotone on an open convex set $C \subseteq R^n$ if and only if F is pseudomonotone on C if and only if for all $x \in C$ and $v \in R^n$*

$$v^T(Mx+q) = 0 \quad \text{implies} \quad v^TMv \geq 0.$$

From this we see:

PROPOSITION 4.2 *If there exists $\bar{x} \in R^n$ such that $M\bar{x}+q=0$, then $F(x) = Mx+q$ is quasimonotone on a neighborhood $N(\bar{x})$ of $\bar{x}$ if and only if F is monotone on R^n, i.e. M is positive semidefinite.*

This implies:

COROLLARY 4.3 *If F is quasimonotone but not monotone on an open convex set $C \subseteq R^n$, then $F(x) \neq 0$ for all $x \in C$.*

In addition to Proposition 4.2 we have [17]:

PROPOSITION 4.4 *If $F(x) = Mx+q$ is quasimonotone on R^n, then F is monotone on R^n, i.e., M is positive semidefinite.*

Proposition 2.2 can be used to construct new generalized monotone affine maps with the help of a given one. Specializing further to the case of the nonnegative orthant $D = R^m_+$, we obtain [31]:

PROPOSITION 4.5 *For any nonnegative $m \times n$ matrix A, the map $F(x) = (A^TMA)x$ is quasimonotone (pseudomonotone) on R^n_+, if $G(z) = Mz$ is quasimonotone (pseudomonotone) on R^m_+.*

Furthermore, the authors of [31] study admissible translations of generalized monotone linear maps. For a given linear map $F(x) = Mx$, which is quasimonotone on R^n_+, the properties of the set

$$Q_M = \{q \in R^n \mid Mx+q \text{ is quasimonotone on } R^n_+\}$$

are analyzed. The special case $M^T = M$ was treated previously in [33]. For general matrices it can be shown [31]:

PROPOSITION 4.6 *For a general $n \times n$ matrix M we have*

(i) Q_M *is a closed convex cone;*

(ii) $MR^n_+ \subseteq Q_M$, *hence* $\dim Q_M \geq \text{rank} M$.

In particular, Q_M has a nonempty interior if M is nonsingular.

Very recently a matrix-theoretic characterization of generalized monotone affine maps was obtained [5]. The departure point for its derivation is Theorem 4.1. We need the following notation to describe the results.

For the affine map $F(x) = Mx + q$ we consider

$$B = \frac{1}{2}(M + M^T), P = \frac{1}{2}M^T B^\dagger M,$$

where $B^\dagger$ is the Moore-Penrose pseudo-inverse of B, n_+, n_- and n_o is the number of positive, negative and zero eigenvalues of B, respectively,

$$r = \dim(\text{kern}(M)),$$

$$f(x) = (Mx + q)^T B^\dagger (Mx + q),$$

$$S = \{x \in R^n \mid f(x) \leq 0\},$$

$$T = \left\{x \in R^n \mid x^T P x \leq 0\right\},$$

$C \subseteq R^n$ is convex where int$C \neq \emptyset$.

We have [5]:

THEOREM 4.7 *F is quasimonotone on C (and pseudomonotone on int$C$) if and only if one of the following conditions holds:*

(i) $n_- = 0$, i.e., B is positive semidefinite and F is monotone on R^n;

(ii1) $n_- = 1, r = n_o + 1, -q \notin M(\text{int}C), q \in B(R^n) \supseteq M(R^n)$, P is positive semidefinite, S is a closed convex set and $C \subseteq S$;

(ii2) $n_- = 1, r = n_o, -q \notin M(\text{int}C), q \in B(R^n) = M(R^n), T = T_+ \cup (-T_+)$ where T_+ is a closed convex cone, int$T_+ \neq \emptyset$, and for $\bar{x}$ such that $M\bar{x} = q$ either $C \subseteq -\bar{x} + T_+$ or $C \subseteq -\bar{x} - T_+$.

Hence the maximal domain of quasimonotonicity is:

R^n in case (i),

S in case (ii1) and

$-\bar{x}+T_+$ or $-\bar{x}-T_+$ in case (ii2).

We note that in the last case quasimonotonicity of F on C implies quasiconvexity of f on C.

From Theorem 4.7 a characterization of quasimonotone (pseudomonotone) affine maps on convex cones can be derived. Through further specialization a characterization on the nonnegative orthant is obtained [5]:

COROLLARY 4.8 *The map $F(x) = Mx + q$ is quasimonotone on R^n_+ (and pseudomonotone on intR^n_+) if and only if one of the following conditions holds:*

(i) $n_- = 0$, *i.e., B is positive semidefinite and F is monotone on R^n;*

(ii1) $n_- = 1, n_+ = 1, M$ *has rank* $1, q \in B(R^n) \supseteq M(R^n), q^T B^\dagger q \leq 0$ *and* $-M^T B^\dagger q \in R^n_+$;

(ii2) $n_- = 1, r = n_o, -P$ *is copositive,* $q \in B(R^n) = M(R^n), T = T_+ \cup (-T_+)$, *where T_+ is a closed convex cone,* int$T_+ \neq \emptyset$, *and for $\bar{x}$ such that $M\bar{x} = q$ either $R^n_+ \subseteq T_+$ and $\bar{x} \in T_+$ or $R^n_+ \subseteq -T_+$ and $\bar{x} \in -T_+$.*

We point out that in the special case $M^T = M$, case (ii1) does not occur and Theorem 4.7 reduces to classical characterizations of generalized convex quadratic functions [7,26,27,32,33]; see also [2]. Case (ii1) does not occur either if M is nonsingular. Hence it arises only if M is not symmetric and singular.

5. The nondifferentiable case

In this section we present characterizations of certain nondifferentiable generalized monotone maps [24].

Let $F : C \to R^n$ be locally Lipschitz where $C \subseteq R^n$ is open convex. The criteria below make use of the generalized Jacobian in the sense of Clarke. Let

$$\partial F(x) = \text{conv}\left\{\lim DF(x_i) \mid x_i \to x,\ F \text{ is differentiable at } x_i \in C\right\},$$

where $DF(x_i)$ is the Jacobian at x_i and 'conv' denotes the convex hull. We further introduce for $x \in C$ and $v \in R^n$

$$D_+F(x;v) = \sup\left\{v^T A v \mid A \in \partial F(x)\right\},$$

$$D_-F(x;v) = \inf\left\{v^T A v \mid A \in \partial F(x)\right\}.$$

In generalization of Theorem 3.1 (i) we have

THEOREM 5.1 *The locally Lipschitz map F is quasimonotone on C if and only if for all $x \in C, v \in R^n$*

(A') $v^T F(x) = 0$ implies $D_+ F(x; v) \geq 0$, and

(B') $v^T F(x) = 0$, $0 \in \left\{ v^T A v \mid A \in \partial F(x) \right\}$ and $v^T F(x + \hat{t}v) > 0$ for some $\hat{t} < 0$ imply that there exists $\tilde{t} > 0$ such that $v^T F(x + tv) \geq 0$ for all $t \in [0, \tilde{t}]$.

In light of [4], a stronger sufficient condition can be obtained which however is no longer necessary, in contrast to the differentiable case [24]:

PROPOSITION 5.2 *The map F is quasimonotone on C if for all $x \in C, v \in R^n, v \neq 0$*

(A") $v^T F(x) = 0$ implies $D_- F(x; v) \geq 0$, and

(B") $F(x) = 0, D_-(x; v) = 0$ and $v^T F(x + \hat{t}v) > 0$ for some $\hat{t} < 0$ imply that there exists $\tilde{t} > 0$ such that $v^T F(x + tv) \geq 0$ for all $t \in [0, \tilde{t}]$.

In analogy to the differentiable case (see Theorem 3.1), corresponding characterizations can be obtained for pseudomonotone maps, replacing (B') ((B")) by a stronger condition (C') ((C")) where the existence of $\hat{t} < 0$ such that $v^T F(x + \hat{t}v) > 0$ is not included; see condition (C) in Theorem 3.1.

Furthermore, criteria for strict pseudomonotonicity and strong pseudomonotonicity are derived in [24]. In particular, we have:

PROPOSITION 5.3 *The map F is strictly pseudomonotone on C if for all $x \in C, v \in R^n$*

(A'$^+$) $v^T F(x) = 0$ implies $D_- F(x; v) > 0$.

Very recently, the generalized monotonicity criteria for maps above have been extended to the class of general continuous maps [13]. In this study Clarke's generalized Jacobian is replaced by an "approximate Jacobian".

6. Conclusion

In this survey we have presented various characterizations of generalized monotone maps. Details are shown mainly for quasimonotone and pseudomonotone maps. In retrospect, it becomes clear how the main characterization in the differentiable case (Theorem 3.1) specializes in the affine case (Theorem 4.1) and how it extends in the nondifferentiable case (Theorem 5.1).

References

1. Aussel, D., Corvallec, J. N. and M. Lassonde, "Subdifferential characterization of quasiconvexity and convexity", Journal of Convex Analysis, **1** (1994) 195-201.
2. Avriel, M., Diewert, W. E., Schaible, S. and I. Zang, "Generalized concavity", Plenum Publishing Corporation, New York, 1988.
3. Castagnoli, E. and P. Mazzoleni, "Order-preserving functions and generalized convexity", Rivista di Matematica per le Scienze Economiche e Sociali, **14** (1991) 33-46.
4. Crouzeix, J.-P. and J. A. Ferland, "Criteria for differentiable generalized monotone maps", Mathematical Programming, **75** (1996) 399–406.
5. Crouzeix, J.-P. and S. Schaible, "Generalized monotone affine maps ", SIAM Journal on Matrix Analysis and Applications, **17** (1996) 992–997.
6. Ellaia, R. and A. Hassouni, "Characterization of nonsmooth functions through their generalized gradients", Optimization, **22** (1991) 401-416.
7. Ferland, J. A., "Quasi-convex and pseudo-convex functions on solid convex sets", Technical Report 71-4, Department of Operations Research, Stanford University, 1971.
8. Hadjisavvas, N. and S. Schaible, "On strong pseudomonotonicity and (semi)strict quasimonotonicity", Journal of Optimization Theory and Applications, **79** (1993) 139-155.
9. Hadjisavvas, N. and S. Schaible, "Errata corrige. On strong pseudomonotonicity and (semi)strict quasimonotonicity", Journal of Optimization Theory and Applications, **85** (1995) 741-742.
10. Hadjisavvas, N. and S. Schaible, "Quasimonotonicity and pseudomonotonicity in variational inequalities and equilibrium problems", in: Crouzeix, J.-P., Martinez-Legaz, J.-E. and M. Volle (eds.) "Generalized convexity and monotonicity: recent developments", Kluwer Academic Publishers, Dordrecht, London, Boston, 1997, forthcoming.
11. Hassouni, A., "Sous-differentiels des fonctions quasi-convexes", Thèse de 3ème cycle, Universite Paul Sabatier, Toulouse III, 1983.
12. Hassouni, A., "Operateurs quasimonotones, applications a certains problemes variationnels", Thèse de Ph.D., Universite Paul Sabatier, Toulouse III, 1993.
13. Jeyakumar, V., Luc, D. T. and S. Schaible "Characterizations of generalized monotone nonsmooth continuous maps using approximate Jacobians", Working Paper 96-22, A. G. Anderson Graduate School of Management, University of California, Riverside, October 1996.
14. Karamardian, S., "Complementarity over cones with monotone and pseudomonotone maps", Journal of Optimization Theory and Applications, **18** (1976) 445-454.
15. Karamardian, S. and S. Schaible, "Seven kinds of monotone maps", Journal of Optimization Theory and Applications, **66** (1990) 37-46.
16. Karamardian, S. and S. Schaible, "First-order characterizations of generalized monotone maps", Working Paper 90-5, Graduate School of Management, University of California, Riverside, 1989.

17. Karamardian, S., Schaible, S. and J.-P. Crouzeix, "Characterizations of generalized monotone maps", Journal of Optimization Theory and Applications, **76** (1993) 399-413.
18. Komlósi, S., "On generalized upper quasidifferentiability", in: Giannessi, F., (ed.), "Nonsmooth optimization methods and applications", Gordon and Breach, Amsterdam, 1992. 189-200.
19. Komlósi, S., "Generalized monotonicity in nonsmooth analysis", in: Komlósi, S., Rapcsák, T. and S. Schaible, "Generalized Convexity", Springer Verlag, Berlin-Heidelberg-New York, 1994, 263-275.
20. Komlósi, S., "Generalized monotonicity and generalized convexity", Journal of Optimization Theory and Applications, **84** (1995) 361-376.
21. Komlósi, S., "Monotonicity and quasimonotonicity in nonlinear analysis", in: Du, D. Z., Qi, L. and R. S. Womersley (eds.), "Recent advances in nonsmooth optimization", World Scientific Publishing Co, Singapore, 1995, 193-214.
22. Luc, D. T., "Characterizations of quasiconvex functions", Bulletin of the Australian Mathematical Society, **48** (1993) 393-405.
23. Luc, D. T., "On generalized convex nonsmooth functions", Bulletin of the Australian Mathematical Society, **49** (1994) 139-149.
24. Luc, D. T. and S. Schaible, "Generalized monotone nonsmooth maps", Journal of Convex Analysis, **3**, No. 2, (1996) forthcoming.
25. Mangasarian, O. L., "Nonlinear Programming ", McGraw Hill, New York, 1969.
26. Martos, B., "Subdefinite matrices and quadratic forms", SIAM Journal of Applied Mathematics, **17** (1969) 1215-1223.
27. Martos, B., "Quadratic programming with a quasiconvex objective function", Operations Research, **19** (1971) 82-97.
28. Penot, J.-P., "Generalized convexity in the light of nonsmooth analysis", in: Duriez, R. and C. Michelot (eds.), Lecture Notes in Mathematical Systems and Economics **429**, Springer Verlag, Berlin-Heidelberg-New York, 1995, 269-290.
29. Penot, J.-P. and P. H. Quang, "Generalized convexity of functions and generalized monotonicity of set-valued maps", Journal of Optimization Theory and Applications, **92** (1997) 343-356.
30. Penot, J.-P. and P. H. Sach, "Generalized monotonicity of subdifferentials and generalized convexity", Journal of Optimization Theory and Applications, forthcoming.
31. Pini, R. and S. Schaible, "Invariance properties of generalized monotonicity", Optimization, **28** (1994) 211-222.
32. Schaible, S., "Beiträge zur quasikonvexen Programmierung", Doctoral Dissertation, Universität Köln, 1971.
33. Schaible, S., "Quasiconvex, pseudoconvex and strictly pseudoconvex quadratic functions", Journal of Optimization Theory and Applications, **35** (1981) 303-338.
34. Schaible, S., "Generalized monotonicity – concepts and uses", in: Giannessi, F. and A. Maugeri (eds.), "Variational inequalities and network equilibrium problems ", Plenum Publishing Corporation, New York, 1995, 289-299.

On the $k - out - of - r - from - n : F$ System with Unequal Element Probabilities*

T. Szántai and A. Habib †
Institute of Mathematics, Technical University of Budapest
Műegyetem rkp. 3, Budapest, 1111 HUNGARY
E-mail: szantai@vma.bme.hu

Abstract. In this paper the consecutive *k-out-of-r-from-n:F* reliability system with unequal element probabilities will be investigated. When the element probabilities are equal one can determine Boole-Bonferroni bounds of order up to four. In this case this is possible only for order two. However we also can produce Hunter-Worsley lower bound on the overall system reliability and this bound many times can be sharp enough. When the calculated bounds are not sharp enough we propose the application of a Monte-Carlo simulation variance reduction technique which takes more computation time but it can be applied in those cases too when all other estimations are slow. Numerical results and test examples are given.

Key words: Reliability systems, Boole-Bonferroni bounds, Monte-Carlo simulation

1. Introduction

Given the events (or cut-sets) that cause system-failure, the probability of system failure is the probability of the union of these events. One way to calculate this probability is to use the well-known inclusion-exclusion formula. The computation burden can be excessive if there are many events, thus necessitating the calculation of upper and lower bounds instead of the exact system-failure probability.

The simplest way to obtain upper viz. lower bound of the exact system-failure probability is to stop the calculation of the inclusion-exclusion formula after the sum of all single event probabilities viz. after the sum of all event pair intersection probabilities. These bounds are the original Boole-Bonferroni bounds. They have been improved by many researchers. Prékopa [9], [10] worked out a general framework for determining exact formulae of the best possible bounds involving the first few sums of the inclusion-exclusion formula and proposed a dual type simplex algorithm for calculating the best possible bounds in the most general case. Hunter [5] and later Fong & Buzacott [1] worked out a general scheme to get improved bounds by using infor-

* This work was partly supported by grants from the National Scientific Research Fund, Hungary, T014102.

† On leave from Dept. of Math.; Faculty of Science; Menoufia University, Shibeen El-Kom; EGYPT.

F. Giannessi et al. (eds.), New Trends in Mathematical Programming, 289–303.

mation contained in the individual probability values in the sums of the inclusion-exclusion formula.

Sfakianakis, Kounias & Hillaris [12] proposed first the use of improved Bonferroni type bounds for calculating reliability of a consecutive $k - out - of - r - from - n : F$ system when the reliability is the same for each component. They use the best possible upper and lower bounds when applying the first three sums of inclusion-exclusion formula.

T. Szántai and Aziz Habib [2] proposed the application of the Hunter-Worsley bound which was not used before in the framework of reliability of the system. They generalized the algorithm by Sfakianakis, Kounias & Hillaris [12] for calculating the fourth sum of the inclusion-exclusion formula and gave best Bonferroni type bounds using these new terms, too.

T. Szántai [13] developed earlier a simulation technique for approximation of multivariate probability distribution function values. We show that this technique can be applied with success for the estimation of the reliability of the consecutive $k - out - of - r - from - n : F$ systems.

In this paper we are dealing with consecutive $k - out - of - r - from - n : F$ systems with unequal element (component) reliability. These systems were investigated by Papastavridis & Koutras [8] and Jun Cai [6]. Some good lower and upper bounds were given for the reliability of the system. These bounds were not Bonferroni type. The main problem in this type of bounds is the calculation of all event pair intersections. We developed algorithms not only for calculation of the single event probability but for calculation of all event pair intersection probabilities, too. So we were able to obtain the best Bonferroni type bounds applying the first two sums of inclusion-exclusion formula. The Hunter-Worsley bound and the simulation technique by T. Szántai became applicable, too.

2. The consecutive $k - out - of - r - from - n : F$ system

Let be

$e_1, e_2, \ldots, e_n$ the elements of the system,
$p_1, p_2, \ldots, p_n$ the probabilities of working the elements,
$q_1, q_2, \ldots, q_n$ the probabilities of failing the elements.

The system fails if and only if at least k out of r consecutive components fail, where $1 \leq k \leq r \leq n$. It is a generalization of both systems consecutive $k - out - of - n : F$ system when $k = r$ and the simple $k - out - of - n : F$ system when $r = n$.

Let be $N = n - r + 1$ and for $i = 1, 2, \dots, N$ let us introduce the following notations

$$\begin{aligned} E_i^r &= \{e_i, e_{i+1}, \dots, e_{i+r-n}\}, \\ A_i^{k,r} &= \text{at least } k \text{ out of elements in the set } E_i^r \text{ are failing}, \\ B_i^{k,r} &= \text{exactly } k \text{ out of elements in the set } E_i^r \text{ are failing}. \end{aligned}$$

The figure below demonstrates the sets $E_1^r, \dots, E_N^r$:

$$\underbrace{e_1, \dots, e_r}_{E_1^r},\ \underbrace{e_2, \dots, e_{r+1}}_{E_2^r},\ \underset{\dots}{\dots}\ ,\underbrace{e_N, \dots, e_n}_{E_N^r}$$

Then the probability of the system failing is

$$\Pr(A_1^{k,r} + A_2^{k,r} + \ldots + A_N^{k,r}).$$

The calculation of this probability is a hard problem instead we calculate

$$S_1 = \sum_{1 \le i \le N} \Pr(A_i^{k,r}), \quad S_2 = \sum_{1 \le i < j \le N} \Pr(A_i^{k,r} A_j^{k,r}),$$

and give lower & upper bounds and approximation by simulation on the system failing probability.

3. Calculation of single event probabilities involved in S_1

If we determine first the auxiliary variables

$$t_1 = \frac{p_1}{q_1}, \dots, t_n = \frac{p_n}{q_n}$$

then for all $i = 1, 2, \dots, N$ we have

$$\Pr(A_i^{k,r}) = a_i^{0,0} + \sum_{i_1=l_1}^{u_1} \left(a_i^{1,i_1} + \sum_{i_2=l_2}^{u_2} \left(a_i^{2,i_2} + \ldots + \sum_{i_j=l_j}^{u_j} a_i^{j,i_j} \right) \right) \tag{3a}$$

where $j = r - k$, and

$$\begin{aligned} a_i^{0,0} &= q_i q_{i+1} \cdots q_{i+r-1},\ i = 1, 2, \dots, N \\ a_i^{1,i_1} &= t_{i_1} a_i^{0,0}, & l_1 &= i, & u_1 &= i + r - 1 \\ a_i^{2,i_2} &= t_{i_2} a_i^{1,i_1}, & l_2 &= i_1 + 1, & u_2 &= i + r - 1 \\ &\vdots & &\vdots & &\vdots \\ a_i^{j,i_j} &= t_{i_j} a_i^{j-1,i_{j-1}}, & l_j &= i_{j-1} + 1, & u_j &= i + r - 1. \end{aligned}$$

We remark that when we apply the formula (3a) for calculation of the probability $\Pr(A_i^{k,r})$ then we can get easily the probabilities of events $B_i^{r,r}, \ldots, B_i^{k,r}$. If we use different registers for the partial sums in the embedded summation we get these probabilities in the following way:

$$\Pr(A_i^{k,r}) = a_i^{0,0} \quad + \sum_{i_1=l_1}^{u_1} (a_i^{1,i_1} \quad + \sum_{i_2=l_2}^{u_2} (a_i^{2,i_2} \quad + \ldots + \sum_{i_j=l_j}^{u_j} a_i^{j,i_j}))$$

$$\begin{array}{l} \uparrow \\ \Pr(B_i^{r,r}) \\ \qquad \uparrow \\ \qquad \Pr(B_i^{r-1,r}) \\ \qquad\qquad \uparrow \\ \qquad\qquad \Pr(B_i^{r-2,r}) \\ \qquad\qquad\qquad \ddots \qquad \uparrow \\ \qquad\qquad\qquad\qquad \Pr(B_i^{k,r}) \end{array} \tag{3b}$$

These probabilities will play an important role in the calculation of the pair intersection probabilities in the next section.

4. Calculation of pair intersection probabilities involved in $\mathbf{S}_2$

We have to calculate the probabilities

$$\Pr(A_i^{k,r} A_j^{k,r}),\ i = 1, 2, \ \ldots\ , N-1;\ j = i+1, i+2,\ \ldots, N$$

There are two cases of these probability calculations as follows:

i) *Case* $j - i > r - 1$

In this case there are no common elements in the sets E_i^r and E_j^r. We simply have

$$\Pr(A_i^{k,r} A_j^{k,r}) = \Pr(A_i^{k,r}) \Pr(A_j^{k,r})$$

ii) *Case* $j - i \leq r - 1$

In this case the sets E_i^r and E_j^r have common elements so the events $A_i^{k,r}$ and $A_j^{k,r}$ are not independent and we need a more sophisticated calculation procedure for the joint probability of these events.

It is obvious that

- the set of common elements is

$$E_j^{r+i-j} = \{ e_j, e_{j+1}, \dots, e_{i+r-1} \}, \tag{4a}$$

and the number of the elements involved is $r + i - j$,

- the set of no common elements in $A_i^{k,r}$ is

$$E_i^{j-i} = \{ e_i, e_{i+1}, \dots, e_{j-1} \}, \tag{4b}$$

and the number of the elements involved is $j - i$,

- the set of no common elements in $A_j^{k,r}$ is

$$E_{i+r}^{j-i} = \{ e_{i+r}, e_{i+r+1}, \dots, e_{j+r-1} \}, \tag{4c}$$

and the number of the elements involved is $j - i$.

So we have

$$E_i^r = E_i^{j-i} + E_j^{r+i-j}, E_j^r = E_j^{r+i-j} + E_{i+r}^{j-i}.$$

As the event $A_i^{k,r} A_j^{k,r}$ occurs if and only if at least k elements in both sets E_i^r and E_j^r are failing, in the set of common elements E_j^{r+i-j} at least max $(0,k - (j - i))$ elements must fail (as in any of the sets E_i^{j-i}, E_{i+r}^{j-i} at most $j - i$ elements can fail). The maximal number of the failing elements in the set E_i^{r+i-j} can be of course $r + i - j$ i.e. all of the elements can fail in it. So if k_1 denotes the exact number of the failing elements in E_j^{r+i-j} then by the theorem of total probability we get

$$\Pr(A_i^{k,r} A_j^{k,r}) = \sum_{k_1=l_1}^{u_1} \Pr\left(A_i^{k-k_1,j-i}\right) \Pr\left(B_j^{k_1,r+i-j}\right) \Pr\left(A_{i+r}^{k-k_1,j-i}\right), \tag{4d}$$

where

$$\begin{aligned} l_1 &= \max(0, k - (j - i)) \\ u_1 &= r + i - j \end{aligned}$$

and all of the probabilities involved in (4d) can be calculated by the algorithm described in the previous section.

5. Lower and upper bounds and approximations by simulation

5.1. The calculation of the best Boole-Bonferroni bounds

We want to give lower and upper bounds on the probability of the sum of events $A_1^{k,r} + A_2^{k,r} + \ldots + A_N^{k,r}$ when the first few M terms of the inclusion-exclusion formula:

$$\begin{aligned} S_1 &= \sum_{1\le i\le N} P(A_i^{k,r}) \\ S_2 &= \sum_{1\le i<j\le N} P(A_i^{k,r} A_j^{k,r}) \\ &\vdots \\ S_M &= \sum_{1\le i_1<\ldots<i_M\le N} P(A_{i_1}^{k,r} \ldots A_{i_M}^{k,r}) \end{aligned} \tag{5a}$$

are calculated before. These terms are called also binomial moments as we have

$$\begin{array}{rcl} p_1 + \binom{2}{1} p_2 + \ldots + \binom{M}{1} p_M + \ldots + \binom{N}{1} p_N &=& S_1 \\ p_2 + \ldots + \binom{M}{2} p_M + \ldots + \binom{N}{2} p_N &=& S_2 \\ \ddots \qquad\qquad \ddots && \vdots \\ p_M + \ldots + \binom{N}{M} p_N &=& S_M \end{array} \tag{5b}$$

where $p_l = \Pr(\mu = l)$, $l = 1, 2, \ldots, N$ and μ is a random variable expressing the number of those $A_1^{k,r}, A_2^{k,r}, \ldots, A_N^{k,r}$ events which occur. It is obvious that

$$\Pr(A_1^{k,r} + A_2^{k,r} + \ldots + A_N^{k,r}) = p_1 + p_2 + \ldots + p_N. \tag{5c}$$

So if we solve the following linear programming problems

$$\begin{array}{l} \min \{p_1 + p_2 + \ldots + p_M + \ldots + p_N\} \\ \text{subject to} \\ \begin{array}{rcl} p_1 + \binom{2}{1} p_2 + \ldots + \binom{M}{1} p_M + \ldots + \binom{N}{1} p_N &=& S_1 \\ p_2 + \ldots + \binom{M}{2} p_M + \ldots + \binom{N}{2} p_N &=& S_2 \\ \ddots \qquad\qquad \ddots && \vdots \\ p_M + \ldots + \binom{N}{M} p_N &=& S_M \end{array} \end{array} \tag{5d}$$

$$p_1 \geq 0,\ p_2 \geq 0\ ,\dots,\ p_M \geq 0\ ,\dots,\ p_N \geq 0$$

$$\begin{array}{llllllll}
\max \{p_1 + & p_2 + & \dots + & p_M + & \dots + & p_N\} & & \\
\text{subject to} & & & & & & & \\
p_1 + & \binom{2}{1} p_2 + & \dots + & \binom{M}{1} p_M + & \dots + & \binom{N}{1} p_N & = & S_1 \\
 & p_2 + & \dots + & \binom{M}{2} p_M + & \dots + & \binom{N}{2} p_N & = & S_2 \\
 & & \ddots & & \ddots & & & \vdots \\
 & & & p_M + & \dots + & \binom{N}{M} p_N & = & S_M
\end{array} \tag{5e}$$

$$p_1 \geq 0,\ p_2 \geq 0\ ,\dots,\ p_M \geq 0\ ,\dots,\ p_N \geq 0$$

we get lower and upper bounds on the system failure probability

$$P = \Pr(A_1^{k,r} + A_2^{k,r} + \dots + A_N^{k,r}).$$

We will use only the case $M = 2$. In this case we have closed formulae on the exact lower and upper bounds. These are

$$\frac{2}{k^* + 1} S_1 - \frac{2}{k^*(k^* + 1)} S_2 \leq P \leq S_1 - \frac{2}{N} S_2 \tag{5f}$$

where $k^* = 1 + \left[\frac{2S_2}{S_1}\right]$, as it was proved by many authors before.

5.2. Upper bound by Hunter and Worsley

Instead of the linear programming problems (5d) and (5e) we could regard their analogous versions involving $2^N - 1$ variables and having on the right hand side all of the individual probabilities compressed in the binomial moments $S_1, \dots, S_M$. The first result in this direction was achieved by Hailperin [3]. Unfortunately for not too large N values the size of the linear programming problems to be solved becomes huge. Hunter [5] and later on independently of him Worsley [14] gave a very effective upper bound using S_1 and some of the individual probabilities involved in S_2. This upper bound can be computed very quickly and easy to prove that it is always sharper than the upper bound (5f).

The Hunter-Worsley upper bounds is given by

$$P \leq S_1 - \sum_{(i,j) \in T^*} \Pr(A_i^{k,r} A_j^{k,r}), \tag{5g}$$

where T^* is a maximal spanning tree in a non-oriented complete graph with n nodes constructed in the following manner. Let us assign to node i the event $A_i^{k,r}$ (or the probability $\Pr(A_i^{k,r})$) and to arc (i,j) the weight $\Pr(A_i^{k,r} A_j^{k,r})$. We remark that the maximum weight spanning tree can be found with effective algorithms, one example is the well known algorithm by Kruskal [7]. We remark that lower resp. upper bounds on the system failure probabilities and upper resp. lower bounds on the reliability of the system easily can be transformed to each other. Consider the fact that the lower and upper bounds change each other in this transformation

5.3. The approximation by simulation

When applying the variance reduction technique of Szántai [13] we calculate four different approximations in one simulation procedure. These are based on the following random variables having expected value

$$\Pr(A_1^{k,r} + A_2^{k,r} + \ldots + A_N^{k,r}).$$

5.3.1. *Crude Monte-Carlo Simulation*

If μ denotes the number of those $A_1^{k,r}, A_2^{k,r}, \ldots, A_N^{k,r}$ events in a random trial which occur then

$$\nu_0 = \begin{cases} 0, & if\ \mu = 0 \\ 1, & if\ \mu \geq 1 \end{cases}$$

is a random variable with expected value $\Pr(A_1^{k,r} + A_2^{k,r} + \ldots + A_N^{k,r})$.

5.3.2. *Monte Carlo simulation of the difference from the best Boole-Bonferroni bounds involving S_1 and S_2*

Let us denote

$$L_1 = \frac{2}{k^*+1}S_1 - \frac{2}{k^*(k^*+1)}S_2 \leq P \leq S_1 - \frac{2}{N}S_2 = U_2,$$

then

$$P - L_1 = S_1 - S_2 + \ldots + (-1)^{N-1}S_N - \frac{2}{k^*+1}S_1 + \frac{2}{k^*(k^*+1)}S_2,$$

and

$$P - U_2 = -S_2 + \ldots + (-1)^{N-1}S_N + \frac{2}{N}S_2.$$

If μ denotes the number of those $A_1^{k,r}, A_2^{k,r}, \ldots, A_N^{k,r}$ events in a random trial which occur then

$$\nu_1 = \begin{cases} 0, & if\ \mu \leq 1 \\ \sum\limits_{i=1}^{N}(-1)^{i-1}\binom{\mu}{i} - \frac{2}{k^*+1}\binom{\mu}{1}1 + \frac{2}{k^*(k^*+1)}\binom{\mu}{2} & \\ = \frac{1}{k^*(k^*+1)}(\mu - k^*)(\mu - k^* - 1), & if\ \mu \geq 1 \end{cases}$$

and

$$\nu_2 = \begin{cases} 0, & if\ \mu \leq 1 \\ \sum_{i=2}^{N} (-1)^{i-1}\binom{\mu}{i} - \frac{2}{N}\binom{\mu}{2} = \frac{1}{N}(\mu-1)(\mu-N), & if\ \mu \geq 2 \end{cases}$$

are random variables with expected values $P - L_1$ and $P - U_2$, respectively.

5.3.3. *Monte Carlo simulation of the difference from the Hunter-Worsley bound*

Let us denote

$$P \leq S_1 - \sum_{(i,j)\in T^*} \Pr(A_i A_j) = U_3,$$

then

$$P - U_3 = -S_2 + S_3 - \ldots + (-1)^{N-1} S_N + \sum_{(i,j)\in T^*} \Pr(A_i^{k,r} A_j^{k,r}).$$

If μ denotes the number of those $\mathrm{A}_1^{k,r}, A_2^{k,r}, \ldots, A_N^{k,r}$ events in a random trial which occur and λ denotes the number of those $A_i^{k,r} A_j^{k,r}$ events with $(i,j) \in T^*$ in a random trial which occur then

$$\nu_3 = \begin{cases} 0, & if\ \mu \leq 1 \\ \sum_{i=2}^{N} (-1)^{i-1}\binom{\mu}{i} + \lambda = 1 - \mu + \lambda, & if\ \mu \geq 2 \end{cases}$$

is a random variable with expected value $P - U_3$.

5.3.4. *Calculation of the final estimator of P with minimal variance*

Let us denote by $\hat{P}_0, \hat{P}_1, \hat{P}_2, \hat{P}_3$ the four different estimators and let us suppose that in the Monte Carlo evaluation procedure their covariance matrix

$$\hat{\mathbf{C}} = \begin{pmatrix} \hat{c}_{00} & \hat{c}_{01} & \hat{c}_{02} & \hat{c}_{03} \\ \hat{c}_{01} & \hat{c}_{11} & \hat{c}_{12} & \hat{c}_{13} \\ \hat{c}_{02} & \hat{c}_{12} & \hat{c}_{22} & \hat{c}_{23} \\ \hat{c}_{03} & \hat{c}_{13} & \hat{c}_{23} & \hat{c}_{33} \end{pmatrix} \tag{5h}$$

also has been estimated.

Now if we define

$$\hat{P} = w_0 \hat{P}_0 + w_1 \hat{P}_1 + w_2 \hat{P}_2 + w_3 \hat{P}_3 \tag{5i}$$

with $w_0 + w_1 + w_2 + w_3 = 1$ then $\hat{P}$ also will be an unbiased estimator of the unknown system failure probability P. Then the variance of $\hat{P}$ is $\mathbf{W}^T\hat{\mathbf{C}}\mathbf{W}$ where $\mathbf{W}^T = (w_0, w_1, w_2, w_3)$ and the Lagrangian problem

$$\begin{aligned} &\min \mathbf{W}^T\hat{\mathbf{C}}\mathbf{W} \\ \text{subject to} & \\ &w_0 + w_1 + w_2 + w_3 = 1 \end{aligned} \tag{5j}$$

can easily be solved.

In fact, the gradient of $\mathbf{W}^T\hat{\mathbf{C}}\mathbf{W}$ equal $2\mathbf{W}^T\hat{\mathbf{C}}$ hence one has to solve the system of linear equations

$$\begin{aligned} \hat{c}_{00}w_0 + \hat{c}_{01}w_1 + \hat{c}_{02}w_2 + \hat{c}_{03}w_3 - \lambda &= 0 \\ \hat{c}_{01}w_0 + \hat{c}_{11}w_1 + \hat{c}_{12}w_2 + \hat{c}_{13}w_3 - \lambda &= 0 \\ \hat{c}_{02}w_0 + \hat{c}_{12}w_1 + \hat{c}_{22}w_2 + \hat{c}_{23}w_3 - \lambda &= 0 \\ \hat{c}_{03}w_0 + \hat{c}_{13}w_1 + \hat{c}_{23}w_2 + \hat{c}_{33}w_3 - \lambda &= 0 \\ w_0 + w_1 + w_2 + w_3 &= 1 \end{aligned} \tag{5k}$$

for the unknowns $w_0, w_1, w_2, w_3, \lambda$.

6. Numerical results

In the paper by Papastavridis and Koutras [8] the authors claim "... there is no direct way to compare our bounds with the bounds in [12], which rely on Bonferroni type inequalities. Numerical calculations indicate that the bounds in [12] are superior for components of low reliability whereas our bounds work better with components of high reliability." The results of this paper give the possibility of the above mentioned comparison. We remark however that our Bonferroni type bounds are based on the first two binomial moments only, whereas those in the paper [12] are based on the first three binomial moments. This is because of the computational burden of the third binomial moments when the element reliabilities are unequal. In the same time we apply the Hunter–Worsley type bounding technique, too. This technique produces an excellent lower bound on the system reliability. In Table I we give numerical results for test examples taken from the paper by Papastavridis and Koutras [8]. We give results only for the cases when $n = 20$, as the $n = 50$ cases give the same system reliability values. This is not surprising because the $p_i = 1 - 2^{-i}$ element probabilities

Table I. $p_i = 1 - 2^{-i}, i = 1, 2, \ldots, 20$

n	r	k	P-K lower	B2 lower	H-W lower	B2 upper	P-K upper
20	2	2	0.8388	0.8354	**0.8512**	**0.8526**	0.8543
20	5	2	0.7608	0.6910	**0.7630**	0.7894	**0.7646**
20	10	2	0.7530	0.6795	**0.7531**	0.7879	**0.7531**
20	20	2	0.7528	0.7528	–	0.7528	0.7528
20	3	3	0.9822	0.9823	**0.9832**	0.9832	0.9832
20	5	3	0.9666	0.9632	**0.0671**	0.9676	**0.9672**
20	10	3	0.9615	0.9566	0.9615	0.9624	**0.9615**
20	20	3	0.9613	0.9613	–	0.9613	0.9613
20	5	5	1.0000	1.0000	1.0000	1.0000	1.0000
20	10	5	0.9999	0.9999	0.9999	0.9999	0.9999
20	20	5	0.9999	0.9999	–	0.9999	0.9999
20	10	10	1.0000	1.0000	1.0000	1.0000	1.0000
20	20	10	1.0000	1.0000	–	1.0000	1.0000

Table II. $p_i = 1 - 2^{-i}, i = 1, 2, \ldots, 20; n = 20, r = 2, k = 2;$
No. of trials $= 10^5$

	Crude M-C	Variance reduction
Estimation	0.8515	0.8515
Computing time	2.97	3.35
Standard deviation	0.001125	0.000045
$\pm 3\sigma$ interval	(0.8481 – 0.8549)	(0.8514 – 0.8516)
Efficiency	–	513.88

are equal to 1 with a very high accuracy if $i > 20$. The original bounds by Papastavridis and Koutras are denoted by P–K. H–W denotes the Hunter–Worsley bounds while B2 denotes the Bonferroni type bounds of this paper. Bold faced figures in the table show the best lower and upper bounds achieved. The computation time was less than one second in every case on a PC 586 computer.

In the paper by Papastavridis and Koutras [8] the exact reliability was computed through a simulation algorithm performed on 10^5 systems for every value of n, m, k. In the next tables we compare the results of the crude Monte Carlo simulation and the variance reduction technique proposed in Section 5.3. As the simulation has no meaning in the cases when the calculated lower and upper bounds are equal we restrict ourselves on the cases $n = 20, r = 2, k = 2$; $n = 20, r = 5, k = 2$; $n = 20, r = 5, k = 3$, only. These are the first, second and sixth cases in Table I. In each cases the estimated value, the computer time, the σ standard deviation and the $\pm 3\sigma$ interval covering the exact system

Table III. $p_i = 1 - 2^{-i}, i = 1, 2, \ldots, 20; n = 20, r = 5, k = 2$;
No. of trials $= 10^5$

	Crude M-C	Variance reduction
Estimation	0.7633	0.7638
Computing time	2.97	3.63
Standard deviation	0.001344	0.000084
$\pm 3\sigma$ interval	(0.7593 – 0.7673)	(0.7635 – 0.7641)
Efficiency	–	209.02

Table IV. $p_i = 1 - 2^{-i}, i = 1, 2, \ldots, 20; n = 20, r = 5, k = 3$;
No. of trials $= 10^5$

	Crude M-C	Variance reduction
Estimation	0.9669	0.9671
Computing time	3.07	3.68
Standard deviation	0.000566	0.000014
$\pm 3\sigma$ interval	(0.9652 – 0.9686)	(0.9671 – 0.9671)
Efficiency	–	1,382.50

reliability with probability about 0.997 are given. In addition the efficiency index of the variance reduction is also given. This is calculated after Hammersley and Handscomb [4] as the ratio $t_0\sigma_0^2/t\sigma^2$, where t_0, σ_0 are the computing time and the standard deviation according to the crude Monte Carlo estimation and t, σ are the same according to the estimation by the variance reduction technique of Section 5.3.

Tables II–IV show that the calculated lower and upper bounds of the paper [8] produced sharper intervals for the system reliability than the $\pm 3\sigma$ intervals of the crude Monte Carlo estimations covering the exact system reliability with probability about 0.997. This means that the crude Monte Carlo simulation algorithm performed on 10^5 systems improves the deterministic bounds only with probability less than 0.997. In the same time the Monte Carlo simulation algorithm proposed in Section 5.3 reduces significantly the variance of the estimated system reliability whereas the computation time is not increased significantly according to the computation time of the crude Monte Carlo simulation algorithm. This results in high efficiency indices and proves the importance of the variance reduction technique of section 5.3 in the framework of system reliability estimations. We remark that there exist many other variance reduction techniques for the system reliability estimations. One of them by Psillakis [11] is just for the consecutive $k - out - of - r - from - n : F$ systems and it uses the variance reduction technique of the "use of antithetic variables". The comparison

Table V. $\mathbf{n = 100, r = 5, k = 4}$

n	J.C.lower	B2 lower	H-W lower	B2 upper	J.C. upper
0.70	**0.086538**	0.000000	0.000000	0.468529	**0.322445**
0.87	0.905488	0.878496	**0.916573**	0.939403	**0.929619**
0.96	0.999044	0.998817	**0.999090**	0.999133	**0.999126**

Table VI. $p_i = 1 - 2^{-i}, i = 1, 2, \ldots, 10, p_i = 2^{-(i-10)}, i = 11, \ldots, 20$

n	r	k	J.C. lower	B2 lower	H-W lower	B2 upper	J.C. upper
20	15	9	0.156958	0.017570	**0.240491**	**0.297331**	0.713023
20	15	10	0.683208	0.684486	**0.696942**	**0.699086**	0.905208
20	18	11	0.891983	0.921077	**0.924677**	**0.937106**	0.950708

of variance reduction techniques is not easy as the efficiency depends not only on the variance reduction technique itself but also on the computer, the uniform random number generator used and so on. However these two variance reduction techniques are not in conflict with each other, they easily can be combined in a unique algorithm.

Now we compare the Bonferroni type bounds of this paper with the bounds by Jun Cai [6]. Although Jun Cai's bounds are according to consecutive $k - out - of - r - from - n : F$ reliability systems with unequal element reliabilities, in the paper [6] there was given only three examples with equal element reliabilities. These are given in the next table. The notations of the bounds are the same as in Table I, except J.C. which means the bounds by Jun Cai [6].

From Table V one could believe the Jun Cai's bounds to be superior, mainly in the case of upper bounds. To show that this is not so, Table VI gives further examples with really unequal element reliabilities (those in the examples by Papastavridis and Koutras [8] cannot be regarded unequal in a strict sense as the element reliabilities tend to one very quickly).

Finally the efficiency of the variance reduction technique of Section 5.3 will be illustrated on the examples of Tables VII–IX.

Table VII. $p_i = 1 - 2^{-i}, i = 1, 2, \ldots, 20; n = 20, r = 5, k = 3;$
No. of trials $= 10^5$

	Crude M-C	Variance reduction
Estimation	0.9669	0.9671
Computing time	3.07	3.68
Standard deviation	0.000566	0.000014
$\pm 3\sigma$ interval	(0.9652 – 0.9686)	(0.9671 – 0.9671)
Efficiency	–	1,382.50

Table VIII. $p_i = 1 - 2^{-i}, i = 1, 2, \ldots, 20; n = 20, r = 5, k = 3;$
No. of trials $= 10^5$

	Crude M-C	Variance reduction
Estimation	0.9669	0.9671
Computing time	3.07	3.68
Standard deviation	0.000566	0.000014
$\pm 3\sigma$ interval	(0.9652 – 0.9686)	(0.9671 – 0.9671)
Efficiency	–	1,382.50

Table IX. $p_i = 1 - 2^{-i}, i = 1, 2, \ldots, 20; n = 20, r = 5, k = 3;$
No. of trials $= 10^5$

	Crude M-C	Variance reduction
Estimation	0.9669	0.9671
Computing time	3.07	3.68
Standard deviation	0.000566	0.000014
$\pm 3\sigma$ interval	(0.9652 – 0.9686)	(0.9671 – 0.9671)
Efficiency	–	1,382.50

References

1. C. C. Fong, J. A. Buzacott, "Improved bounds for system-failure probability", *IEEE Trans. Reliability*, vol R-36, NO.4 1987 Oct, pp 454–458.
2. A. Habib, T. Szántai, "New bounds on the reliability of the consecutive *k-out-of-r-from-n:F* system", submitted to *Journal of Reliability Engineering and System Safety.*
3. Th. Hailperin, "Best possible inequalities for the probability of a logical function of events", *The Amer. Math. Monthly*, vol 72, 1965, pp 343–359.
4. J. M. Hammersley and D. C. Handscomb, *Monte Carlo Methods*, Methuen, London, 1964.
5. D. Hunter, "An upper bound for the probability of a union", *J. Appl. Prob.*, vol 13, 1976, pp 597–603.
6. Jun Cai, "Reliability of a large consecutive *k-out-of-r-from-n:F* system with unequal component-reliability", *IEEE Trans. Reliability*, vol. R-43, 1994 Mar, pp 107–111.
7. J. B. Kruskal, "On the shortest spanning subtree of a graph and the travelling salesman problem", *Proc. of the Amer. Math. Soc.*, vol 7, 1956, pp 48–50.
8. S. Papastavridis, M. V. Koutras, "Bounds for reliability of consecutive-*k-within-m-out-of-n* systems", *IEEE Trans. Reliability*, vol R-42, 1993 Mar, pp 156–160.
9. A. Prékopa "Boole-Bonferroni inequalities and linear programming", *Operations Research*, vol 36, 1988, pp 145–162.
10. A. Prékopa, "Sharp bounds on probabilities using linear programming", *Operations Research*, vol 38, 1990, pp 227–239.
11. Z. M. Psillakis, "A simulation algorithm for computing failure probability of a consecutive *k-out-of-r-from-n:F* system", *IEEE Trans. Reliability*, vol R-44, 1995 Sep, pp 523–531.
12. M. Sfakianakis, S. Kounias, A. Hillaris, "Reliability of a consecutive-*k-out-of-r-from-n:F* system", *IEEE Trans. Reliability*, vol R-41, 1992 Sep, pp 442–447.
13. T. Szántai, "Evaluation of a special multivariate gamma distribution", *Math. Prog. Study*, vol 27, 1986, pp 1–16.
14. K. J. Worsley, "An improved Bonferroni inequality and applications", *Biometrika*, vol 69, 1982, pp 297–302.

A common root of three minimization problems

B. Uhrin
Computer and Automation Institute
Hungarian Academy of Sciences
1518, Budapest, Pf. 63, Hungary
e-mail: uhrin@sztaki.hu

Abstract. It is showed that three minimization problems from three different fields of mathematics (approximation theory, convex geometry, matrix theory) can be transformed easily to a simple problem of a linear algebraic type. The latter one is then solved in an elementary way using another field: the theory of inequalities (the Hölder inequality and its simple extensions).

1. Introduction

In the abstract of his opening plenary lecture at the 2nd European Congress of Mathematics, J.-P. Serre, [1], writes: "To construct dictionaries (or correspondences, or translations) between two different areas of mathematics has always been an important source of progress." This sentence might be the motto of the present paper.

The first problem is from *(discrete) approximation theory.*

PROBLEM 1. Given a real valued function f defined on a finite set $Q \subset R^1$ of cardinality m and a real number $1 \leq p \leq \infty$, find a polynomial y^* of degree at most $n-1 < m-1$ which is on Q "nearest" in the l_p-norm to f.

More exactly: find y^* (depending on p) such that

$$\left(\sum_{t\in Q} |f(t) - y^*(t)|^p\right)^{1/p} \leq \left(\sum_{t\in Q} |f(t) - y(t)|^p\right)^{1/p}, \qquad 1 \leq p < \infty,$$

$$\max_{t\in Q} |f(t) - y^*(t)| \leq \max_{t\in Q} |f(t) - y(t)|, \qquad p = \infty, \tag{1}$$

holds for all polynomials y of degree at most $n-1$. ∘

Two particular cases of this problem (for $p = 2$ and $p = \infty$) are discrete versions of two classical problems (the second one goes back to Čebyšev and was in fact one of the starting points of approximation theory, see, e.g., [2,3,4]).

F. Giannessi et al. (eds.), New Trends in Mathematical Programming, 305–314.

Introduce for y in a Euclidean space, say, $y \in R^d$, the notation

$$\begin{aligned} \|y\|_p &:= (\sum_{i=1}^{d} |y_i|^p)^{\frac{1}{p}}, \qquad 0 < p < +\infty\,; \\ \|y\|_{+\infty} &:= \max_{1\le i\le d} |y_i|, \qquad p = +\infty\,. \end{aligned} \tag{2}$$

(This notation will be used in R^d of different dimensions, the dimension will be clear from the context.)

The second problem is of a *(discrete and convex) geometric flavour.*

PROBLEM 2. Let $K_p \subset R^m$ be the unit ball of the norm $\|\cdot\|_p$, i.e., $K_p := \{y \in R^m : \|y\|_p \le 1\}$, $1 \le p \le \infty$. Given a linear subspace $L \subset R^m$ of dimension $n < m$ and $c \in R^m$, $c \notin L$, find $\tau_p > 0$ such that the boundary of the set $\tau_p K_p$ touches $(L + c)$, i.e.,

$$\tau_p = \inf\{\tau > 0 :\ \tau K_p \cap (L + c) \neq \emptyset\} \tag{3}$$

and describe (or characterize) the set $\tau_p K_p \cap (L + c)$. $\circ$

Finally, the third problem is from *matrix theory.*

PROBLEM 3. Let A be a real $m \times n$-matrix, $m > n$, $b \in R^m$ be a given vector and $1 \le p, r \le \infty$. Find an "l_p-approximative" solution $x^{(p)} \in R^n$ of the system of linear equations $Ax = b$ and among all such solutions find one having minimal l_r-norm.

More exactly, denoting

$$a(x) := Ax - b \in R^m, \quad x \in R^n\,,$$

find $x^{(p)} \in R^n$ such that

$$\|a(x^{(p)})\|_p \ \le\ \|a(x)\|_p, \quad \forall\, x \in R^n\,, \tag{4}$$

and find $x^{(p)}_{(r)} \in R^n$ such that $Ax^{(p)}_{(r)} = b + a(x^{(p)})$ and

$$\|x^{(p)}_{(r)}\|_r \ = \ \min\{\|x\|_r \ : Ax = b + a(x^{(p)})\}\,. \quad \circ \tag{5}$$

The aim of this paper is to demonstrate that the three above problems can be easily "translated" to a fourth one (of a linear algebraic type) which's solution is an easy consequence of some results from another area of mathematics: the *theory of inequalities.*

2. The common solution of the problems

Denote by $\langle a, b\rangle$ the usual scalar product of two vectors and by θ the zero vector in an Euclidean space. Let $\mathrm{sgn}(\alpha)$ be the usual signum function, i.e., it is $+1$ if $\alpha > 0$, -1 if $\alpha < 0$ and 0 if $\alpha = 0$.

Further, denote for $0 < p \leq \infty$ and for $\theta \neq z \in R^d$

$$Y_p(z) := \{y \in R^d : \langle y, z\rangle = \|y\|_p \cdot \|z\|_q\}\,, \tag{6}$$

where $q := p/(p-1)$ if $1 < p < \infty$, $q := \infty$ if $0 < p \leq 1$ and $q := 1$ if $p = \infty$.

Now we have

THEOREM 1. ([7], Theorem 2,6). *The sets (6) are exactly as follows.*

For $1 < p < \infty$: $Y_p(z) = \{ty(z) : t > 0\}$, *where*

$$y(z)_i := |z_i|^{q-1} \cdot \mathrm{sgn}(z_i), \qquad i = 1, 2, \ldots, d\,. \tag{7}$$

For $p = \infty$: $Y_p(z) = \{ty(z, v) : t > 0\}$, *where*

$$y(z, v)_i := \begin{cases} v_i & \text{if } z_i = 0, \\ \mathrm{sgn}(z_i) & \text{if } z_i \neq 0, \end{cases} \qquad i = 1, 2, \ldots, d\,, \tag{8}$$

and v_i *are any numbers such that* $|v_i| \leq 1$.

For $p = 1$: $Y_p(z) = \{ty(z, w) : t > 0\}$, *where*

$$y(z, w)_i := \begin{cases} 0 & \text{if } |z_i| < \|z\|_\infty\,, \\ w_i \cdot \mathrm{sgn}(z_i) & \text{if } |z_i| = \|z\|_\infty\,, \end{cases} \qquad i = 1, 2, \ldots, d, \tag{9}$$

and w_i *are any non-negative numbers such that* $\sum_i w_i = 1$.

For $0 < p < 1$: $Y_p(z) = \{ty(z, r) : t > 0\}$, *where*

$$(y(z, r))_i := \begin{cases} 0 & \text{if } i \neq r\,, \\ \mathrm{sgn}(z_i) & \text{if } i = r\,, \end{cases} \qquad i = 1, 2, \ldots, d\,, \tag{10}$$

and r *is any index such that* $|z_r| = \|z\|_\infty$. ∘

Let us remind here the proof from [7], which is very simple.

Proof. ([7], Proof of Theorem 2.6.) Firstly, observe that for $y, z \in R^d$ we have

$$\langle y, z\rangle \leq \sum_{i=1}^{d} |y_i||z_i|\,, \tag{11}$$

and equality is in (11) if and only if

$$y_i \cdot z_i \geq 0 \qquad \forall\ 1 \leq i \leq d\,. \tag{12}$$

Secondly, for $1 < p < \infty$ we have

$$\sum_{i=1}^{d} |y_i||z_i| \leq \|y\|_p \cdot \|z\|_q \ , \tag{13}$$

and equality in (13) is if and only if for some $\delta > 0$ we have

$$|y_i|^p = \delta |z_i|^q, \quad i = 1, 2, \ldots, d \ , \tag{14}$$

(the Hölder inequality, see, e.g., [9])

Thirdly, for $p = 1$ we have

$$\sum_{i=1}^{d} |y_i||z_i| \leq \|y\|_1 \cdot \|z\|_\infty \ , \tag{15}$$

and equality is in (15) if and only if

$$y_i = 0 \qquad \forall \ i \ \text{ s.t. } \ |z_i| < \|z\|_\infty \ . \tag{16}$$

For $p = \infty$ the situation is analogous, but now the roles of y, z in (15) and (16) are reversed.

Finally, for $0 < p < 1$, first using the fact

$$\|y\|_1 \leq \|y\|_p \ , \tag{17}$$

write a weakening of (15) and after that observe that equality is in (17) if and only if

$$y_i \neq 0 \qquad \text{for at most one } \ i \ . \tag{18}$$

The condition (18) is a consequence of the elementary fact that for $0 < p < 1$ and $0 < t < 1$ we have

$$1 + t < (1 + t^p)^{1/p} \ . \quad \bullet \tag{19}$$

Let us note that in the proof above only the Hölder inequality (13) and the conditions (14) of equality in it are not elementary.

Now, let $L \subset R^d$ be a linear subspace of dimension $1 \leq k < d$, let $c \in R^d$, $c \notin L$ and for $0 < p \leq \infty$ denote

$$\delta_p := \min_{y \in L+c} \|y\|_p \tag{20}$$

and

$$H_p := \{y \in L + c : \|y\|_p = \delta_p\}. \tag{21}$$

Let $L^{\perp}$ be the orthogonal complement subspace to L, i.e., $L^{\perp} := \{x \in R^d : \langle x, y\rangle = 0,\ y \in L\}$ and put $L_c^{\perp}$ for the elements z of $L^{\perp}$ such that $\langle z, c\rangle > 0$.

The set $L_c^{\perp} \cap (L+c)$ contains exactly one element, denote it by $z(c)$.

Now, let us formulate

PROBLEM 4. Given $0 < p \le \infty$, L and c as above, find δ_p and H_p . ∘

The following easy corollary of Theorem gives the solution of this problem.

PROPOSITION 1. ([7], Theorem 2.7.) *Let $0 < p \le \infty$ and assume that there is $\hat{z} \in L_c^{\perp}$ (depending on p) such that*

$$Y_p(\hat{z}) \cap (L+c) \neq \emptyset\,. \tag{22}$$

Then

$$\bigcup_{z \in L_c^{\perp}} (Y_p(z) \cap (L+c)) = H_p \tag{23}$$

and for any $z \in L_c^{\perp}$ such that $Y_p(z) \cap (L+c) \neq \emptyset$ we have

$$\delta_p = \frac{\langle c, z\rangle}{\|z\|_q}\,. \qquad \circ \tag{24}$$

Proof. For any $y \in L + c$ and $z \in L_c^{\perp}$ we have $\langle y, z\rangle = \langle c, z\rangle$, hence using the inequality $\langle y, z\rangle \le \|y\|_p \cdot \|z\|_q$ (see (13),(15),(17)), we get

$$\frac{\langle c, z\rangle}{\|z\|_q} \le \|y\|_p \qquad \forall\, y \in L+c, \quad \forall\, z \in L_c^{\perp}\,. \tag{25}$$

But for $g \in Y_p(z) \cap (L+c) \neq \emptyset$ we have $\langle g, z\rangle = \|g\|_p \cdot \|z\|_q$ and $\langle g, z\rangle = \langle c, z\rangle$, hence

$$\bigcup_{z \in L_c^{\perp}} (Y_p(z) \cap (L+c)) \subseteq H_p \tag{26}$$

and also (24) is true.

Let now $z \in L_c^{\perp}$ be such that $Y_p(z) \cap (L+c) \neq \emptyset$. Then (24) implies that for any $y \in H_p$ we have

$$\|y\|_p = \frac{\langle c, z\rangle}{\|z\|_q} = \frac{\langle y, z\rangle}{\|z\|_q}\,, \tag{27}$$

hence $\langle y, z\rangle = \|y\|_p \cdot \|z\|_q$, and this shows by the definition (6) of $Y_p(z)$ that $y \in Y_p(z)$. ●

The exact forms (7),(8),(9) and (10) of elements of $Y_p(z)$ show at once that any $g(z) \in Y_p(z) \cap (L+c)) \neq \emptyset$ must be of the form

$$g(z) = \frac{\langle c, z \rangle}{\|z\|_q^q} y(z) \ , \qquad 1 < p < \infty \ , \tag{28}$$

$$g(z) = \frac{\langle c, z \rangle}{\|z\|_1} y(z, v) \ , \qquad p = \infty \ , \tag{29}$$

$$g(z) = \frac{\langle c, z \rangle}{\|z\|_\infty} y(z, w) \ , \qquad p = 1 \ , \tag{30}$$

$$g(z) = \frac{\langle c, z \rangle}{\|z\|_\infty} y(z, r) \ , \qquad 0 < p < 1 \ . \tag{31}$$

The converse question, whether above $g(z)$ belong to $L + c$ is settled by

PROPOSITION 2. ([7], Theorem 2.8.)

(i) If the dimension of L *is* $k = d - 1$*, then for any* $z \in L_c^\perp$ *and any* $0 < p \leq \infty$ *all vectors (28)-(31) belong to* $L + c$*.*

(ii) If $1 \leq k < d-1$ *and* $1 \leq p \leq \infty$*, then there is* $z \in L_c^\perp$ *(depending on* p*) such that the vectors (28), (29) and (30), respectively, belong to* $L + c$*.*

(iii) If $1 \leq k < d - 1, \ \ 0 < p < 1$*, and* $\delta_1 < \delta_p$*, then in general no* $z \in L_c^\perp$ *exists such that (31) belongs to* $L + c$*.*

(iv) If $1 \leq k < d - 1$ *and* $0 < p < 1$*, then* $\delta_1 = \delta_p$ *holds if and only if there is* $z \in L_c^\perp$ *such that (31) belongs to* $L + c$*.* ∘

Proof. (i): If $k = d - 1$, then $L_c^\perp$ is a halfline going from the origin, hence $g \in L+c$ is equivalent to $\langle g-c, z \rangle = 0$ for $\theta \neq z \in L_c^\perp$. But using (7)-(10) one can check easily that $\langle g, z \rangle = \langle c, z \rangle$ holds for all (28)-(31).

(ii): In this case the existence of a wanted $z \in L_c^\perp$ is guaranteed by the following well known identity (the duality theorem, see, e.g., [2] or [3]):

$$\min_{y \in L+c} \|y\|_p = \max_{L^\perp} \frac{\langle c, z \rangle}{\|z\|_q} \ . \tag{32}$$

(iii), (iv): these are simple consequences of (ii) case $p = 1$ and (17) (for details see [7], p. 680). •

We see that Propositions 1 and 2 (that are easy consequences of the Theorem) completely solve the Problem 4 for all $0 < p \leq \infty$ and $k = d - 1$ (in the sense that (28)-(31) give exactly δ_p and H_p).

The solution of the Problem 4 via Propositions 1 and 2 is not so perfect already in the case $1 \leq k < d - 1$, because still remains to find all vectors of the forms (28)-(31) which belong to $L+c$ (a procedure for this is given in [7], pp. 680-681). Such vectors surely exist if $1 \leq p \leq \infty$ and if $\delta_1 = \delta_p$ holds for $0 < p < 1$ (by Proposition 2, (ii), (iv)), while for $0 < p < 1$, $\delta_1 < \delta_p$ it is not sure (Proposition 2, (iii)).

[Let us note that in [7], pp. 682-685, a special geometrical construction giving the set H_p for all cases $0 < p \leq 1, \quad 1 \leq k \leq d-1$, is showed.]

Remark 1. Theorem 1 (equation (7)) shows that for $1 < p < \infty$ and for all $\theta \neq z \in R^d$ the set $Y_p(z)$ consists of a single halfline going from the origin θ. This halfline is especially interesting for $p = 2$: in this case we have simply

$$Y_2(z) = \{tz : t > 0\} . \tag{33}$$

But $z(c)$ is the unique element of the set $L_c^{\perp} \cap (L+c)$, hence the condition (22), $p = 2$, is automatically satisfied with $\hat{z} := z(c)$ and by (23), (24) we have

$$H_2 = \{z(c)\}, \quad \delta_2 = \frac{\langle c, z(c)\rangle}{\|z(c)\|_2} . \tag{34}$$

We see that in this case the condition (22) is guaranteed by (7) itself for all $1 \leq k \leq d-1$, i.e., the duality theorem (32) is not needed at all for $p = 2$. ○

Remark 2. One can check easily that for the vectors (28)-(31) we have

$$g(\lambda z) = g(z), \quad \lambda > 0, \quad z \in L_c^{\perp}. \tag{35}$$

This means that for $k = d-1, 0 < p \leq \infty$, the elements of H_p are exactly the vectors (28)-(31) written with $z := z(c)$, implying among others that H_p is unique for $0 < p < \infty$.

The situation for $1 \leq k < d-1$ and $p \neq 2$, is not so nice already: it is not clear whether $Y_p(z(c)) \cap (L+c))$ is empty or not, i.e., whether (28)-(31) written with $z := z(c)$ belong to $L+c$ or not. ○

3. Solutions of Problems 1, 2 and 3

All three problems can be easily transformed to the Problem 4 and the solutions (28)-(31) of the latter one easily re-transform to their solutions. Well, the transformations in question are perhaps a "folklore", but some recent publications (e.g., concerning the Problem 3, see remarks below) indicate that some authors were not aware of a great benefit of such transformations.

Problem 1.: Let $Q := (t_1 < t_2 < \ldots < t_m)$ and denote by $f, b_i \in R^m$ the vectors $f := (f(t_1), f(t_2), \ldots, f(t_m))$ and $b_i := (t_1^i, t_2^i, \ldots, t_m^i)$, $i = 0, 1, \ldots, n-1$. Put $P \subset R^m$ for the linear subspace of dimension n spanned by the vectors b_i. Now (1) boils down to the problem of finding all $y^* \in P - f$ such that $\|y^*\|_p = \min_{y \in P-f} \|y\|_p$.

Problem 2.: It is clear that $\tau K_p \cap (L + c) = \{y \in L + c : \|y\|_p \leq \tau\}$. This shows that $\tau_p = \delta_p$ and $\tau_p K_p \cap (L + c) = H_p$.

Problem 3.: The set $\{Ax : x \in R^n\}$ is a linear subspace, say L in R^m (the subspace spanned by the columns of A) and denoting $c = -b$ the problem (4) boils down to two problems. The first is the Problem 4 with $d := m$ and the latter L and c. Solving it we get δ_p and H_p. The second problem is: for all $y \in H_p$, solving the system of linear equations $Ax = y + b$ (which is surely solvable), we get all $x^{(p)}$.

After that we solve the problem (5) as follows.

Let L_1 be the linear subspace of R^n spanned by the rows of A. Put $L := L_1^{\perp} \subset R^n$ (i.e., $L_1 = L^{\perp}$) and taking now $c := x^{(p)}$, we solve the Problem 4 with $p := r, d := n$ and the latter L and c to get $x_{(r)}^{(p)}$.

In the above way we get all results concerning the Problem 3, especially in the case when $n = m-1$ and the rank of the matrix A is $m-1$ (as it is assumed, e.g., in [5]). For example, the elements (28) written with $z := z(c)$ give [5], Theorem 5.1 at once. Similarly, (29) and (30) with $z := z(c)$ give [5], Theorem 5.2 and [5], Theorem 5.3, respectively. We can say that the results of [5] (and partly those of [6]) are simple consequences of the Assertions 1, 2, i.e., of the simple Theorem. This fact is a good argument in favour of the basic approach of this paper: having a problem (say in matrix theory) it is worthy of looking at it through more "glasses" (e.g., linear algebraic, geometric, analytic) and it can happen that in the course of transformations of the problem we come upon a "face" of the problem which's solution is very simple.

4. Concluding remarks

(r1). The Theorem above provides a tool for writing down all solutions of the Problem 4 (consequently, of the Problems 1, 2, 3).

If the dimension k of L is equal to $d-1$, then this tool is perfect in the sense that *all* information on the Problem 4 is contained in the formulae (28)-(31) and using the exact forms of $g(z)$ all results can be get by simple calculations.

A few examples to demonstrate this:

(a) The duality theorem (32) is a consequence of formulas.

(b) For $1 < p < \infty$ the set H_p consists of a unique point.

(c) H_∞ consists of a unique point if and only if L is a Haar-subspace (recall that L is Haar if $\theta \neq z \in L^\perp$ implies $z_i \neq 0$ for all i). If L is a Čebyšev-subspace then the unique element g of H_∞ is of the form: $g_i = e(-1)^i \delta_\infty$, $i = 1, \ldots, d$, $|e| = 1$ (the alternation property of best Čebyšev approximation; recall that L is Čebyšev if $\theta \neq z \in L^\perp$ implies $\mathrm{sgn}(z_i) = e(-1)^i$, $i = 1, \ldots, d$, for some $|e| = 1$).

(d) Interesting conclusions can be drawn at once also for $0 < p \leq 1$ (see [7]).

(r2). We know that H_p consists of a unique point in all cases $1 \leq k < d$, $1 < p < \infty$. This is a consequence of the Minkowski inequality for the l_p-norm and the conditions of equality in it, see, [9] (in other words, of the strict convexity of the l_p-norm). As we have seen, this fact is in the case $k = d - 1$ a consequence of the formula (28). An interesting task would be to prove the unicity of H_p form (28) also in the cases $1 \leq k < d - 1$.

(r3). Let us note that for $0 < p < 1$ the Problem 4 is not a "convex" but rather a "concave" minimization, which is a part of a new area called "global optimization" (for the latter field see, e.g., [10]).

(r4). All results here can be transformed to give solutions to the Problem 4 when instead of $\|y\|_p$ one take any weighted "l_p-norm" with positive weights (see [7] for more details). Moreover, all the technique extends to treat Problem 2 in a much more general setting, when instead of K_p we take any convex body K containing the origin (see [8]).

References

1. J.-P. Serre, "Correspondences and dictionaries in geometry and number theory", In: Abstracts of the 2nd European Congress of Mathematics, Budapest, July 22-26, 1996, J.Bolyai Math.Soc., 1996, p.5.
2. E. W. Cheney, Introduction to Approximation Theory, McGraw-Hill, New York, 1966.
3. H. S. Shapiro, Topics in Approximation Theory, Springer, Berlin- Heidelberg-New York, 1971.
4. I. Singer, Best Approximation in Normed Linear Spaces by Elements of Linear Subspaces, Springer, Berlin-Heidelberg-New York, 1970.
5. J. Miao, A. Ben-Israel, On l_p-approximate solutions of linear equations, Linear Alg. Appl., 199(1994), 305–327.
6. J. Miao, A. Ben-Israel, The Geometry of Basic, Approximate, and Minimum Norm Solutions of Linear Equations, Linear Alg. Appl., 216(1995), 25–41.
7. B. Uhrin, An elementary constructive approach to discrete linear l_p-approximation, $0 < p \leq +\infty$, In: Szabados J., Tandori K., Eds., Proc. Appr. Theory Conference held in Kecskemét (Hungary), 1990, Coll. Math. Soc. J. Bolyai, Vol. 58, 1991, pp.673–688.
8. B. Uhrin, Touching points of a star-shaped set with an affine subspace, Linear Alg. Appl., 171(1992), 149–159.

9. G. H. Hardy, J. E. Littlewood, G. Pólya, Inequalities, Cambridge Univ. Press, Cambridge, 1951.
10. R. Horst, Hoang Tuy, "Global Optimization, Deterministic Approaches", Springer, Berlin - Heidelberg - New York, 1990.

Applied Optimization

1. D.-Z. Du and D.F. Hsu (eds.): *Combinatorial Network Theory.* 1996
ISBN 0-7923-3777-8
2. M.J. Panik: *Linear Programming: Mathematics, Theory and Algorithms.* 1996
ISBN 0-7923-3782-4
3. R.B. Kearfott and V. Kreinovich (eds.): *Applications of Interval Computations.* 1996 ISBN 0-7923-3847-2
4. N. Hritonenko and Y. Yatsenko: *Modeling and Optimimization of the Lifetime of Technology.* 1996 ISBN 0-7923-4014-0
5. T. Terlaky (ed.): *Interior Point Methods of Mathematical Programming.* 1996
ISBN 0-7923-4201-1
6. B. Jansen: *Interior Point Techniques in Optimization.* Complementarity, Sensitivity and Algorithms. 1997 ISBN 0-7923-4430-8
7. A. Migdalas, P.M. Pardalos and S. Storøy (eds.): *Parallel Computing in Optimization.* 1997 ISBN 0-7923-4583-5
8. F.A. Lootsma: *Fuzzy Logic for Planning and Decision Making.* 1997
ISBN 0-7923-4681-5
9. J.A. dos Santos Gromicho: *Quasiconvex Optimization and Location Theory.* 1998
ISBN 0-7923-4694-7
10. V. Kreinovich, A. Lakeyev, J. Rohn and P. Kahl: *Computational Complexity and Feasibility of Data Processing and Interval Computations.* 1998
ISBN 0-7923-4865-6
11. J. Gil-Aluja: *The Interactive Management of Human Resources in Uncertainty.* 1998
ISBN 0-7923-4886-9
12. C. Zopounidis and A.I. Dimitras: *Multicriteria Decision Aid Methods for the Prediction of Business Failure.* 1998 ISBN 0-7923-4900-8
13. F. Giannessi, S. Komlósi and T. Rapcsák (eds.): *New Trends in Mathematical Programming.* Homage to Steven Vajda. 1998 ISBN 0-7923-5036-7

KLUWER ACADEMIC PUBLISHERS – DORDRECHT / BOSTON / LONDON

MIX
Papier aus verantwortungsvollen Quellen
Paper from responsible sources
FSC® C105338

If you have any concerns about our products,
you can contact us on
ProductSafety@springernature.com

In case Publisher is established outside the EU,
the EU authorized representative is:
Springer Nature Customer Service Center GmbH
Europaplatz 3, 69115 Heidelberg, Germany

Printed by Libri Plureos GmbH
in Hamburg, Germany